Basic Thermodynamics

Basic Thermodynamics

P.B. Nagaraj
D. Venkatesh

PUBLISHING FOR ONE WORLD

NEW AGE INTERNATIONAL (P) LIMITED, PUBLISHERS
New Delhi • Bangalore • Chennai • Cochin • Guwahati • Hyderabad
Jalandhar • Kolkata • Lucknow • Mumbai • Ranchi
Visit us at **www.newagepublishers.com**

Published by New Age International (P) Ltd., Publishers
First Edition: 2005
Reprint: 2011

Branches:

- No. 37/10, 8th Cross (Near Hanuman Temple), Azad Nagar, Chamrajpet, **Bangalore** - 560 018. Tel.: (080) 26756823, Telefax: 26756820, E-mail: bangalore@newagepublishers.com
- 26, Damodaran Street, T. Nagar, **Chennai** - 600 017. Tel.: (044) 24353401, Telefax: 24351463 E-mail: chennai@newagepublishers.com
- CC-39/1016, Carrier Station Road, Ernakulam South, **Cochin** - 682 016. Tel.: (0484) 2377004, Telefax: 4051303 E-mail: cochin@newagepublishers.com
- Hemsen Complex, Mohd. Shah Road, Paltan Bazar, Near Starline Hotel, **Guwahati** - 781 008.Tel.: (0361) 2513881 Telefax: 2543669, E-mail: guwahati@newagepublishers.com
- No. 105, 1st Floor, Madhiray Kaveri Tower, 3-2-19, Azam Jahi Road, Nimboliadda, **Hyderabad** - 500 027. Tel.: (040) 24652456, Telefax: 24652457, E-mail:hyderabad@newagepublishers.com
- RDB Chambers (Formerly Lotus Cinema)106A, 1st Floor, S.N. Banerjee Road, **Kolkata** - 700 014. Tel.: (033) 22273773, Telefax: 22275247, E-mail:kolkata@newagepublishers.com
- 16-A, Jopling Road, **Lucknow** - 226 001. Tel.: (0522) 2209578, 4045297, Telefax: 2204098 E-mail: lucknow@newagepublishers.com
- 142C, Victor House, Ground Floor, N.M. Joshi Marg, Lower Parel, **Mumbai** - 400 013.Tel.: (022) 24927869 Telefax: 24915415, E-mail: mumbai@newagepublishers.com
- 22, Golden House, Daryaganj, **New Delhi** - 110 002. Tel.: (011) 23262370, 23262368, Telefax: 43551305 E-mail: sales@newagepublishers.com

ISBN (10) : 81-224-1613-6
ISBN (13) : 978-81-224-1613-8

₹ 150.00

C-11-07-5755

Printed in India at Printman (India), Delhi.
Typeset at The Composer, Delhi.

PUBLISHING FOR ONE WORLD
NEW AGE INTERNATIONAL (P) LIMITED, PUBLISHERS
4835/24, Ansari Road, Daryaganj, New Delhi-110002
Visit us at **www.newagepublishers.com**

Dedicated to
Our Beloved Parents

Preface

This book titled "Basic Thermodynamics" makes an attempt to cover the portions keeping in view of the syllabus for IIIrd semester B.E., Mechanical, prescribed by Visveswaraiah Technological University. This book can also be useful for students of other engineering disciplines like B.E. in Industrial production, Industrial Engineering management, Automobile, Diploma in Mechanical and IP, IEM and Automobile engineering, AMIE etc. The whole book is written with precise explanations, neat sketches and good number of numericals. The numerical problems from VTU question papers have also been updated.

We express sincere thanks to our founder chairman Late Sri M.S. Ramaiah, Gokula Education Trust, Sri M.R. Jayaram, Chairman, Gokula Education Trust, Sri M.R. Seetharam, Director, MSRIT, for providing conducive environment and best facilities in the institution.

We express our sincere thanks to Dr. K.S. Ramanatha, Principal, MSRIT, Dr. R. Chandrashekar, Professor and Head, Department of Mechanical Engineering, MSRIT for their co-operation and encouragement.

We also express our sincere thanks to Dr. P. Dinesh, Dr. Kishore, T. Kashyap, and all teaching faculty of our department for their co-operation and encouragement.

The authors owe greatly to M/s New Age International for encouraging us and bringing out this book in this presentable form.

We take this opportunity to thank our family members Smt. M. R. Savitha Nagaraj, Mahima P.N., B.M. Basavaraj, M Rajappa, Suma Rajappa, Sindhu M.R., D. Nagaveni, D.V. Anirudh, who are the constant source of inspiration to us.

The authors will be happy to receive constructive criticisms from readers for improving this text book.

Authors

Contents

CHAPTER 1

Basic Concepts and Definitions

1.1 INTRODUCTION

Engineering is nothing but the application of knowledge of science and mathematics gained by study, experience and practice to develop ways to utilize, the materials and forces of nature economically for the benefit of mankind. The knowledge of engineering science gives solutions to various engineering problems, which are not necessarily beneficial to mankind. To decide whether our solution is good for mankind or not, the knowledge of social science and humanities is essential.

Thermodynamics is the study of energy and its transformation. It is one of the most fascinating branch of science. Thermodynamics discusses the relationship between heat, work and the physical properties of working substance. It also deals with equilibrium and feasibility of a process. The science of thermodynamics is based on observations of common experience which have been formulated into laws which govern the principle of energy conversion. Application of thermodynamic principles in practical design tasks, may be that of a simple pressure cooker or of a complex chemical plant. The applications of the thermodynamic laws and principles are found in all fields of energy technology, say in steam and nuclear power plants, gas turbines, internal combustion engines, air conditioning, refrigeration, gas dynamics, jet propulsion, compressors, etc. It is really difficult to identify any area where there is no interaction in terms of energy and matter. It is a science having it's relevance in every walk of life. Thermodynamics can be classified as classical thermodynamics and statistical thermodynamics. The classical thermodynamics is applied in engineering problems.

The word thermodynamics derives from two Greek words "therme" which means "heat" and "dynamikos" which means "power".

Thus, study of heat related to matter in motion is called "Thermodynamics". The study of engineering thermodynamics is mainly concerned with work producing or utilizing machines such as engines, turbines and compressors together with working substances used in such machines.

Another definition of thermodynamics is that, it is the science that deals with the various phenomena of energy transfer and its effects on the physical properties of substances. Energy transfer means conversion of heat into mechanical work as in the case of internal combustion engines employed in automobiles.

According to Van Wylen, "Thermodynamics is the science of energy, equilibrium and entropy" (3 E'S). He treated the subject in such a way that, it deals with energy, matter and the laws governing their interactions.

Hatsopoulos and Keenan defined the thermodynamics as the "science of states and changes in state of physical systems and the interactions between systems which may accompany changes in state."

The thermodynamic principles are embodied in two laws commonly called 'the first law' and 'the second law' both of which deals with energy transformations. The first law is nothing but the restatement of the law of conservation of energy and the second law puts a restriction on certain possible energy transformations.

A device involves many substances like gases and vapours while transforming and utilizing the energy.

The study and analysis of system can be done by considering system in two approaches. One is called microscopic approach in which the matter i.e. gases and vapour is composed of several molecules and behaviour each individual molecule is studied and the analysis is applied to collective molecular action by statistical methods and hence this approach is known as statistical approach or microscopic study. In statistical approach, average behaviour of molecules based on statistical behaviour of a system is considered. In the macroscopic approach, a certain quantity of matter composed of large number of molecules is considered without the events occurring at the molecular level being taken into account. Generally, we consider the behaviour of finite quantity of matter. This approach is also known as classical approach.

In general, we can say that macroscopic approach analysis = Σ (microscopic approach analysis).

1.2 MICROSCOPIC AND MACROSCOPIC DESCRIPTION OF MATTER

There are two approaches in the study of thermodynamics from which the working or behaviour of a system can be studied.

(1) microscopic or statistical approach.

(2) macroscopic or classical approach.

In microscopic approach, the knowledge of structure of matter is considered and a large number of variables are needed to describe the state of matter. The matter is composed of several molecules and behaviour of each individual molecule is studied. Each molecule is having certain position, velocity and energy at a given instant. The velocity and energy change very frequently due to collision of molecules. The analysis is made on the behaviour of individual atoms and molecules, for example, some studies in nuclear physics such as the atomic structure of a fissionable material like uranium.

Example

The gas in a cylinder is assumed to contain a large number of molecules each having same mass and velocity independent of each other. In order to describe the thermodynamic system in view of microscopic approach, it is necessary to describe the position of each and every molecule which is very complicated. Hence, this approach is rarely employed but, has become more important in recent years. The behaviour of gas is to be described by summing up the behaviour of each molecule.

In macroscopic approach the structure of matter is not considered, in fact it is simple, and only few variables are used to describe the state of matter. In this approach, a certain quantity of matter composed of large number of molecules is considered without the events occurring at the molecular level being taken into account. In this case, the properties of a particular mass of substance, such as it's pressure, temperature and volume are analysed. Generally, in engineering, this analysis is used for study of heat engines and other devices. This method gives the fundamental knowledge for the analysis of a wide variety of engineering problems.

Example

(1) Consider a piston and cylinder arrangement of an internal combustion engine as a thermodynamic system. At any instant, the system has certain volume depending upon the position of piston. At this volume, different pro-perties such as pressure, temperature, chemical composition can be easily described.

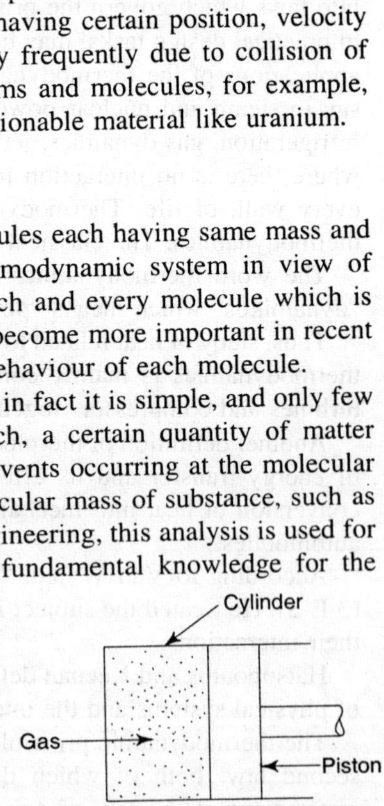

Fig. 1.1 Piston-cylinder machine

(2) Distance measurement in metre.

(3) Time measurement in seconds.

The results of macroscopic thermodynamics are obtained from microscopic study of matter.

For example, consider a cube of 25 mm side and containing a monatomic gas at atmospheric pressure and temperature.

Suppose, this volume contains 10^{20} atoms. In order to describe position and velocity of each atom, three co-ordinates and velocity components are required. Therefore in view of microscopic approach, at least 6×10^{20} equations are required to describe the behaviour of system. Even for a large digital computer, computation task would become difficult. However, to reduce complexity of the problem, two approaches have been adopted. In one approach, the average values of all the particles in the system are considered and the analysis is applied to collective molecular action by statistical methods. In the other approach, number of variables that can be handled are reduced to few only. Here, the gross or average effects of many molecules is considered.

1.3 CONCEPT OF CONTINUUM

According to macroscopic point of view, the substance is considered to be continuous and molecules are considered only in large volumes. The behaviour of individual molecule can be neglected. This concept is known as "continuum". The assumption of continuum is best suited for macroscopic approach where discontinuity at molecular level can be easily ignored as the scale of analysis is quite large.

1.4 THERMODYNAMIC SYSTEMS

In order to analyse the problem, it is necessary to specify objects which are under consideration. In thermodynamics, this is done by considering an imaginary envelope around the objects, thereby restricting the study to a specified region.

A thermodynamic system is defined as a quantity of matter of fixed mass or region in the space whose volume need not be constant and where energy changes are to be analysed. The attention is focussed on this region for study. The matter or region in the space is bounded by a closed surface or wall, which may be actual one (ex: tank containing fluid) or hypothetical one (boundary of some amount of fluid flowing in a pipe). It may change in shape and size.

Everything external to the system or real/hypothetical boundary is termed as the surroundings or the environment. The envelope which separates the system and surrounding is called boundary of the system. The boundary may be either fixed or moving. A system and its surroundings together is termed as universe.

$$\therefore \quad \text{Universe} = \text{System} + \text{Surroundings}$$

Sometimes, a system can be defined as the 'control system' and boundary of which may be treated as 'control boundary'. The 'control volume' is the volume enclosed within this boundary and the space within the boundary is called 'control space'.

Consider a piston and cylinder arrangement as a thermodynamic system as shown in figure 1.2. The gas temperature in the cylinder can be raised by external heating and this causes the piston to move and also changes the system boundary size. It means, both heat and work crosses boundary of the system during this process, but the matter that comprises the system can always be identified. It is essential that position of boundary be specified very carefully. For example, a system in which gas contained in a cylinder, boundary is to be specified within the cylinder to restrict the system under consideration to the gas itself.

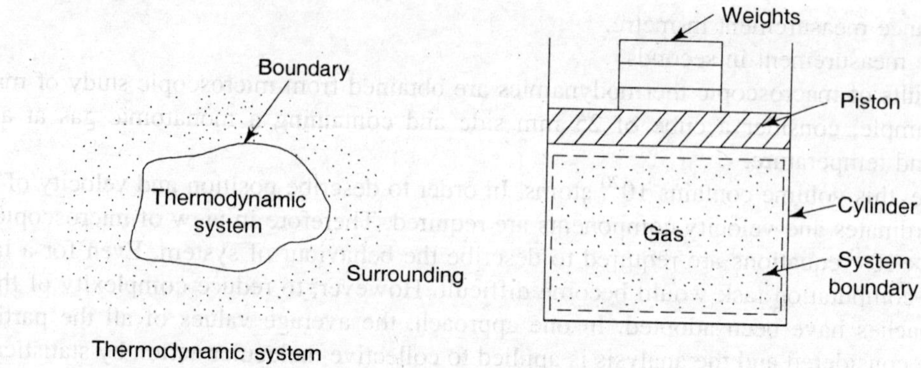

Fig. 1.2 Thermodynamic system

If the boundary is located outside the cylinder, the system includes both the gas and the cylinder. In case of a steam turbine, steam will cross the boundary as it enters and leaves the turbine and it is desirable to place the boundary outside the turbine.

The thermodynamic systems are classified based on energy and mass interactions of the system with surroundings or other systems into

(1) closed system (non-flow system)
(2) open system (flow system)
(3) isolated system

1.4.1 Closed System

The closed system is a system of fixed mass. In this system, energy may cross the boundary, and the total mass within the boundary is fixed. The system and it's boundary may contract or expand in volume.

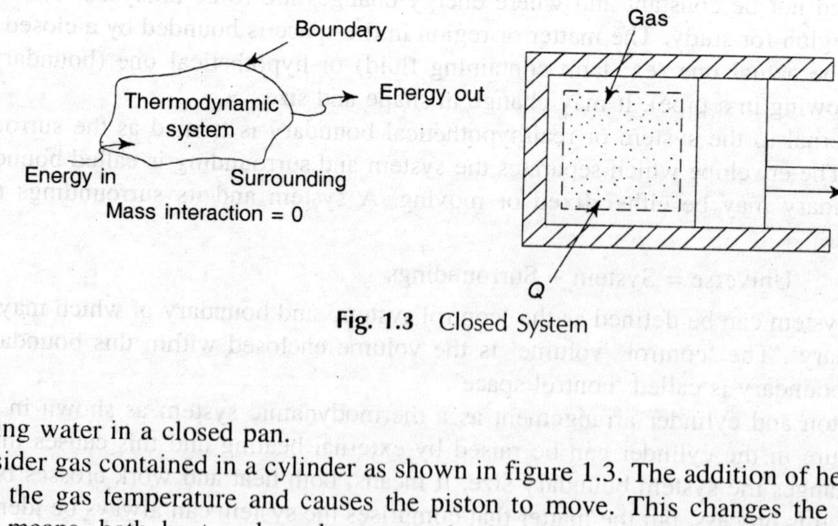

Fig. 1.3 Closed System

Example

(1) Boiling water in a closed pan.
(2) Consider gas contained in a cylinder as shown in figure 1.3. The addition of heat to cylinder will raise the gas temperature and causes the piston to move. This changes the system boundary. This means, both heat and work crosses the boundary of system. But the original mass of working substance (gas) remain unchanged.

1.4.2 Open System

The open system is one in which matter crosses the boundary of the system. There may be energy transfer also. i.e. both energy and mass crosses the boundary of the system. Most of the engineering devices belongs to this type.

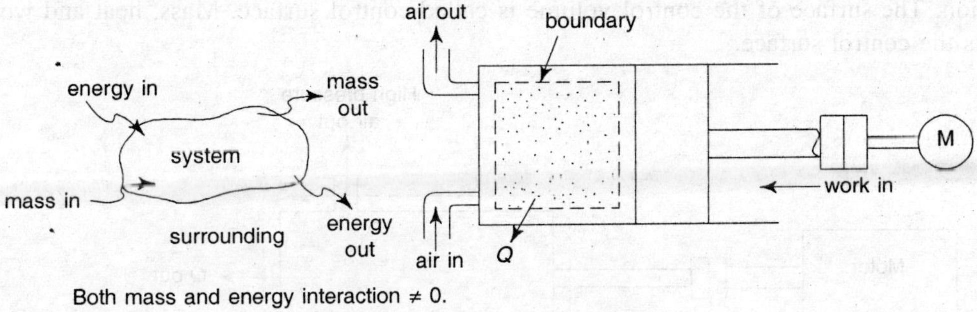

Fig. 1.4 Open system

If the inflow of mass is equal to out flow of mass, then the mass in the system is constant and the system is known as steady flow system.

Example

(1) Air compressor.
In an air compressor, air enters at low pressure and leaves at high pressure. The working substance (gas) crosses the boundary of the system. In addition to this mass transfer, heat and work interactions take place across the system boundary.

(2) Automobile engine.

1.4.3 Isolated System

In an isolated system, neither mass nor energy crosses the system boundary. It is of fixed mass and energy. The system is not affected by the surrounding i.e. there is no interaction between the system and surroundings. *Ex:* flow through pipe.

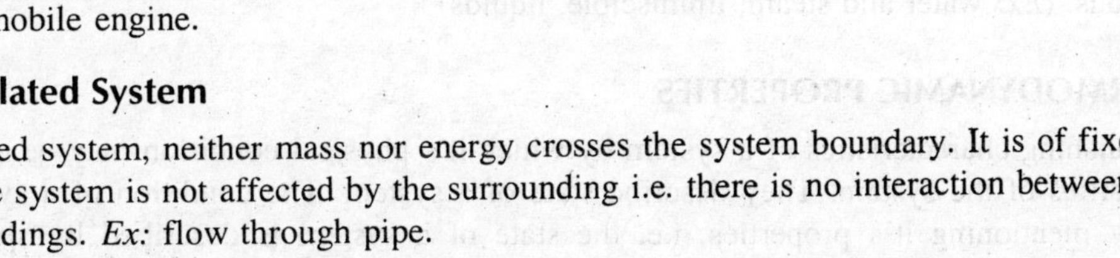

Fig. 1.5 Isolated system

1.5 CONTROL VOLUME AND CONTROL SURFACE

In a closed system, our attention is focussed on a fixed mass of matter for the analysis, whereas in the open system, the analysis is concentrated on the region in the space through which matter flows.

The space volume through which matter, momentum and energy flows is termed as 'control volume' and the surface or envelope of this control volume is known as 'control surface'. Mass, energy and work (momentum) can flow across the control surface.

The fluid flows through pipes can be analysed by using the concept of control volume. The control volume may be stationary or may be contracted/expanded to change in size and position as in the case of open systems. In closed systems, no mass transfer take place across the control surface.

Consider an air compressor, that involves flow of mass into/out of the device as shown in figure 1.6. For the analysis, it is required to specify a control volume that surrounds the device under consideration. The surface of the control volume is called control surface. Mass, heat and work can flow across the control surface.

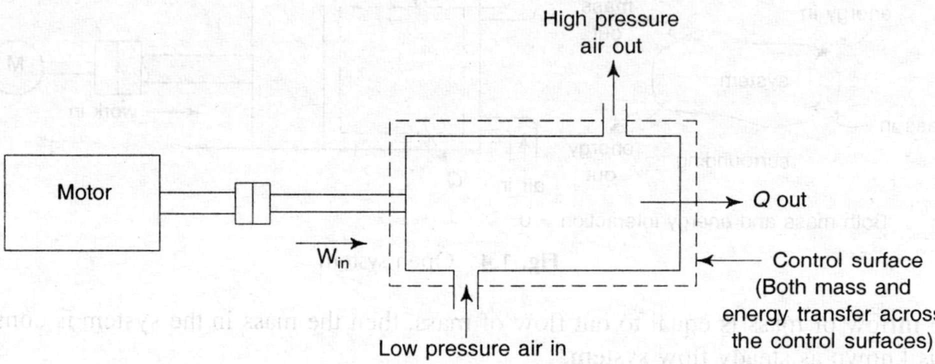

Fig. 1.6 Air Compressor

1.6 HOMOGENEOUS AND HETEROGENEOUS SYSTEMS

If the substance within the system exists in a single phase like air, steam, liquids then the system is called Homogeneous system. In these systems, the substance exists in only one phase.

If the substance within the system exists in more than one phase, then the system is called Heterogeneous. (*Ex:* water and steam, immiscible, liquids)

1.7 THERMODYNAMIC PROPERTIES

The distinguishing characteristics of a system by which it's physical condition may be described are called properties of the system. They describe state of a system. The condition of a system can be specified by mentioning it's properties, i.e. the state of a system is described by specifying it's thermodynamic coordinates. *Ex:* temperature, volume, pressure, chemical energy content etc. These co-ordinates are usually denoted as properties which are macroscopic in nature. The property must have a definite value when the system is at a particular state and the value of which should not depend upon the past history of the system.

A property can also be defined as any quantity that depends on state of the system and is independent of the path by which the system has reached the given state. The change in value of a property is dependent only on the end states of the system. It's differential must be exact.

The property which is dependent upon the physical and chemical structure of the substance is called an internal or thermostatic property.

In classical thermodynamics there are two types of properties:

(i) Intensive property
(ii) Extensive property

Properties which are independent of mass such as pressure and temperature, are known as intensive properties. Some of the other examples are density, velocity, specific volume.

Properties which are dependent upon mass, such as volume and energy in its various forms are called extensive properties. Some of the other examples are internal energy etc.

If mass is increased, the values of extensive properties also increases.

Specific extensive properties i.e. extensive properties per unit mass are intensive properties. *Ex:* specific volume, specific energy, density etc.

If a property can be varied at will, quite independently of other properties, then the property is termed as an "independent property". *Ex:* Temperature or pressure of a gas can be varied quite independently of each other.

Some of the properties cannot be varied independently, those properties are termed as dependent properties. *Ex:* While discussing vapour formation, temperature at which liquid vapourises depends on the pressure. Here pressure is an independent property, but temperature is a dependent property.

1.8 THERMODYNAMIC STATE OF A SYSTEM

The state of a system at any instant is it's condition or configuration of existence at that instant. The various properties of a system defines the state of the system. At any equilibrium condition, the state of a system can be described by few properties like pressure, temperature, specific volume, internal energy etc. The state of a system can be represented by a point on the diagram whose co-ordinates are thermodynamic properties. When a system changes from one equilibrium state (state 1) to another (state 2), due to energy interaction, the system attains a new state which is shown by point '2' on the property diagram.

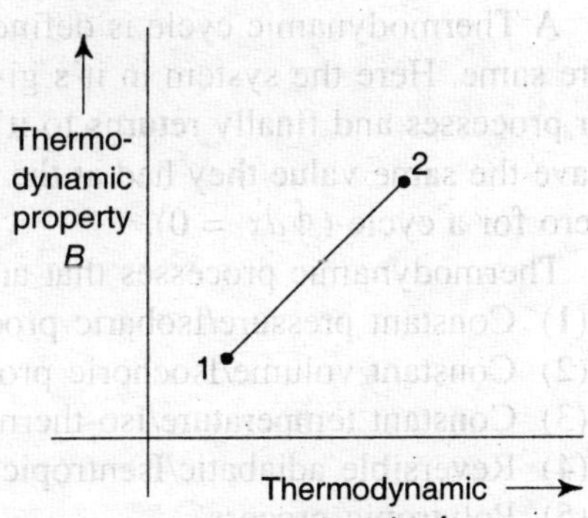

Fig. 1.7 Property Diagram

Example

Consider a given mass of water, which may become vapour or solid (ice) by heating or cooling. Each phase of water may exist at different pressures and temperatures or we can say water may exist in different states.

1.9 THERMODYNAMIC PROCESS

Whenever a system undergoes a change of state, then it is said to execute a process.

When the state of a system is changed by a number of operations (or one operation) having been carried out on the system, then the system is said to be undergone a process.

For example, consider a piston and cylinder arrangement in which some weights are placed on the piston. If one of the weight is removed, the piston rises and that changes the state of the system. Let P_1, V_1 be the initial values of pressure and specific volume, before removing the weight and P_2, V_2 are the corresponding values after the system has attained a new state.

From the diagram, it is clear that pressure decreases and specific volume increases during the change of state from 1 to 2 or else the system is said to have undergone a process '1-2'.

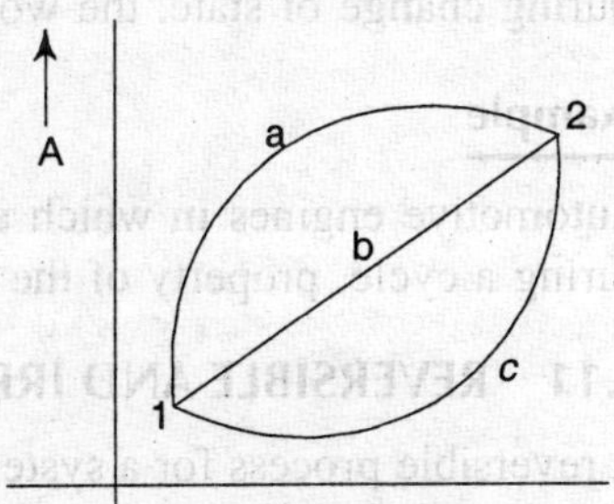

Fig. 1.8(a)

If the process goes on slowly that the equilibrium state exists at every moments, then such a process is called as "Equilibrium Process" otherwise it is referred as "Non Equilibrium Process". A process can change the system from one non-equilibrium state to another by following a path of non-equilibrium states.

If the properties do not depend on the path followed in reaching the state, but only on the equilibrium state itself, then the properties are called "Point functions". As in the figure 1.8(a), the change in property '*A*' between states '1' and '2' is same irrespective of the path '*a*', '*b*' or '*c*'.

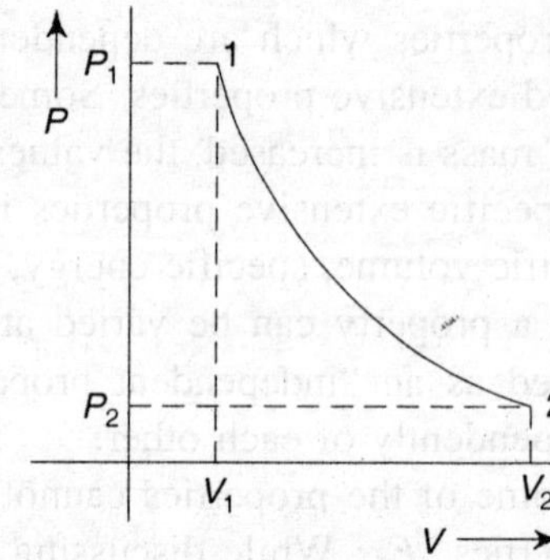

Fig. 1.8(b)

If the properties depends on the path followed in reaching the state, then the properties are called "path functions".

A Thermodynamic cycle is defined as a series of state changes such that the final and initial states are same. Here the system in it's given initial state goes through a number of different state changes or processes and finally returns to it's original state. Therefore, at the end of a cycle all the properties have the same value they had at the original state, i.e. the net change in any property of the system is zero for a cycle ($\oint dx = 0$).

Thermodynamic processes that are commonly experienced in engineering practice are:
(1) Constant pressure/Isobaric process
(2) Constant volume/Isochoric process
(3) Constant temperature/Iso-thermal process
(4) Reversible adiabatic/Isentropic process
(5) Polytropic process
(6) Throttling process/Iso-enthalpic process

The cycles are classified into

1.10 THERMODYNAMIC CYCLE

During change of state, the working substance does not change its chemical composition.

Example

(1) Water circulation in steam power plant
(2) Refrigerant in refrigeration process

Mechanical Cycle

During change of state, the working substance changes its chemical composition.

Example

Automotive engines in which air and fuel mixture is supplied and burnt gases leaves the engine, i.e. during a cycle, property of the working substance changes or their end states are not same.

1.11 REVERSIBLE AND IRREVERSIBLE PROCESS

A reversible process for a system is an ideal process which once having taken place can be reversed in such a way that the initial state and all energies transformed during the process can be completely regained in both systems and surroundings. This process does not leave any net change in the system or in the surroundings. A reversible process is always, quasi-static.

Some of the examples for reversible processes are : Motion without friction, expansion or compression with no pressure difference, heat transfer without temperature gradient, reversible adiabatic process, etc.

If the process is not reversible, i.e. if the initial state and energies transformed cannot be restored without net change in the system after the process has taken place, it is called irreversible. This process leaves traces of changes in the system and environment.

Some of the examples for irreversible processes are: Motion with friction, free expansion, compression or expansion due to finite pressure difference, heat transfer with finite temperature gradient, mixing of nonidentical gases, and all processes which involve dissipative effects.

Consider a process 1-2, i.e. expansion of gas in a cylinder. Let w_{12} be the amount of work done and Q_{12} be the quantity of heat transferred between system and surrounding.

If it is possible to change the system from state 2 to state 1 by supplying back w_{12} and Q_{12}, then process 1-2 is called reversible process. If there is any change in the requirement of work and heat to bring back the system from 2 to 1, then the process 1-2 is called "Irreversible process".

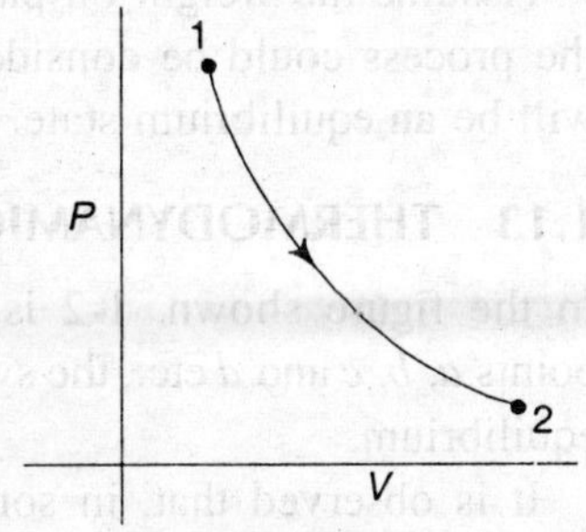

Fig. 1.9

A process is said to occur, when the system undergoes a change of state. The intermediate equilibrium conditions of the process cannot be defined if it occurs at a faster rate and is difficult to calculate heat and work transfer for such process.

1.12 QUASI-STATIC PROCESS

If a process take place at a faster rate, then the intermediate conditions cannot be defined. Therefore, an assumption is made such that the process is taking place at such a rate that the intermediate conditions can be defined and hence must be represented on a thermodynamic property diagram.

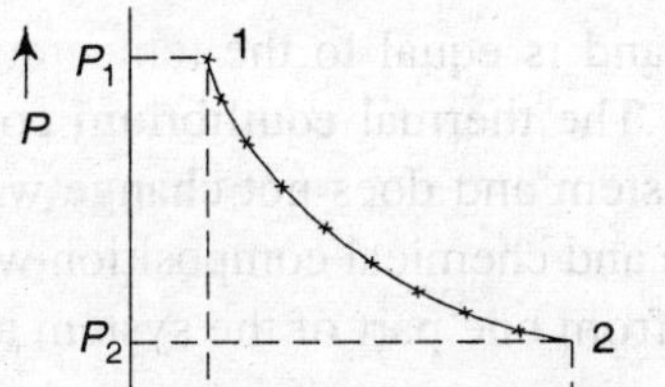

Fig. 1.10(a) Property diagram (transition between two states)

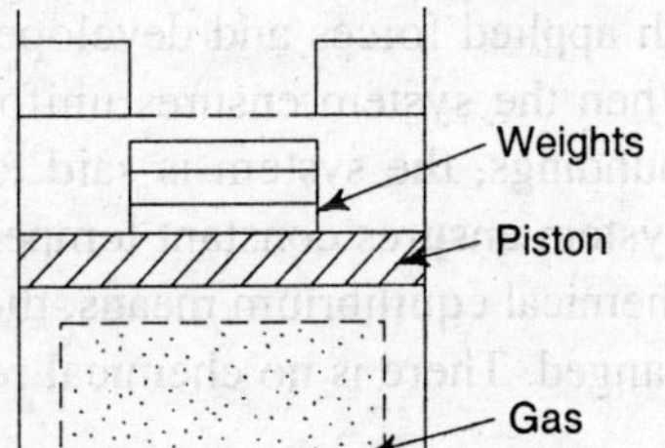

Fig. 1.10(b) System which may undergo quasi-static process

A quasi-static process is also known as quasi-equilibrium process in which the process is carried out in such a manner that, at every instant the system departs only infinitesimally from an equilibrium state. It is an ideal process in which the system changes very slowly it's state, under the influence of infinite simal pressure or temperature difference. Quasi means "almost". Infinite slowness is the characteristic feature of this process. It is also a reversible process.

Consider a system of gas contained in a cylinder. Initially at state 1 the system is in equilibrium and state of the system is represented by the properties P_1, V_1 and T_1. The upward force exerted by the gas is balanced by weight on the piston. The unbalanced force will set up between system and surrounding and under gas pressure by removing weights on the piston. As a result, piston will move up and as it hits the stops, the system regains equilibrium condition (state 2) and properties of the system are represented by P_2, V_2 and T_2. State 1 and state 2 are the initial and final equilibrium states. Let us consider intermediate points between 1 and 2. These points represent the intermediate states passed through by the system, and are called non-equilibrium states. These non-equilibrium states are not definable on thermodynamic co-ordinates.

Assume the weight on piston consists of many small pieces and are removed slowly one by one, the process could be considered as quasi-equilibrium. So every state passed through by the system will be an equilibrium state.

1.13 THERMODYNAMIC EQUILIBRIUM

In the figure shown, 1-2 is a quasi-static process in which at points *a*, *b*, *c* and *d* etc., the system is very close to thermodynamic equilibrium.

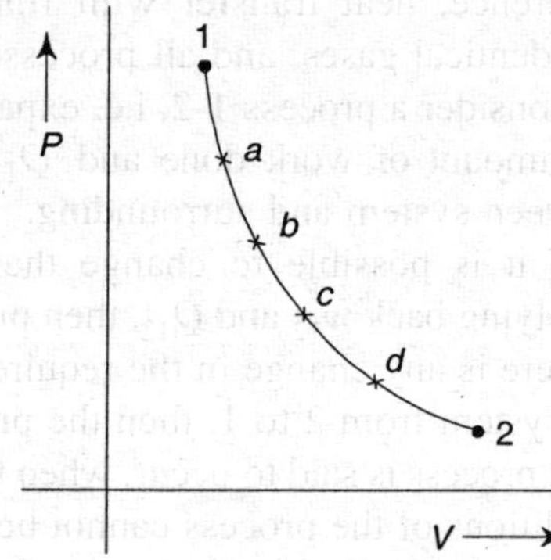

Fig. 1.11

It is observed that, in some situations a collection of matter experiences negligible changes. For example, temperature and pressure of gas in a tank exposed to atmospheric temperature will be essentially constant as the instruments measure these properties are insensitive to fluctuations. So thermodynamic equilibrium means, the collection of matter (system) experiences no changes in all it's properties. In other words, the system is said to be in thermodynamic equilibrium, when it satisfies mechanical, thermal and chemical equilibrium conditions.

When a system has no unbalanced force within it and when the force it exerts on it's boundary is balanced by external force, then the system is said to be in mechanical equilibrium. It ensures in the system that pressure is same at all points and does not change with time. It is the state of a system at which applied forces and developed stresses are fully balanced.

When the system ensures uniform temperature throughout and is equal to the temperature of the surroundings, the system is said to be in thermal equilibrium. The thermal equilibrium condition in the system ensures constant temperature at all points in that system and does not change with time.

Chemical equilibrium means, the system is chemically stable and chemical composition will remain unchanged. There is no chemical reaction or transfer of matter from one part of the system to another.

1.14 TEMPERATURE

Temperature is man's perceptions of 'hotness' or 'coldness' of a body. The hot body transfers energy to the cold body as molecules in it vibrate at a faster rate than that of cold body. The feel of a hot body is due to the impact of such vibration and energy transfer take place from hot body to the cold one or fingers of the hand. The ability to transfer energy is taken as the measure of hotness of the body.

When hot and cold bodies are brought into contact, the hot body becomes cooler and cold body becomes warmer. After some time, they appear to have same hotness or coldness. It is also seen that, different materials at the same temperature are appeared to be at different temperature. So it is very difficult to give the exact definition of temperature. Temperature is a measure of hotness of a body and may be defined as the ability of the body to transfer energy.

1.15 EQUALITY OF TEMPERATURE

Because of difficulty in defining temperature, we define equality of temperature. Consider two systems A and B which are at different temperatures T_1 and T_2 and are perfectly insulated from surroundings. When these two systems are in physical contact with each other, there will be heat exchange between these two systems due to temperature difference and this alters the physical properties like length, electrical resistance etc. of both the systems. After some period, when both the systems attain thermal equilibrium condition, no change in physical properties will be observed. It means, both the systems are at the same temperatures and the concept is known as "Equality of temperature".

1.16 ZEROTH LAW OF THERMODYNAMICS STATEMENT

When two systems are each in thermal equilibrium with a third system, they are also in thermal equilibrium with each other.

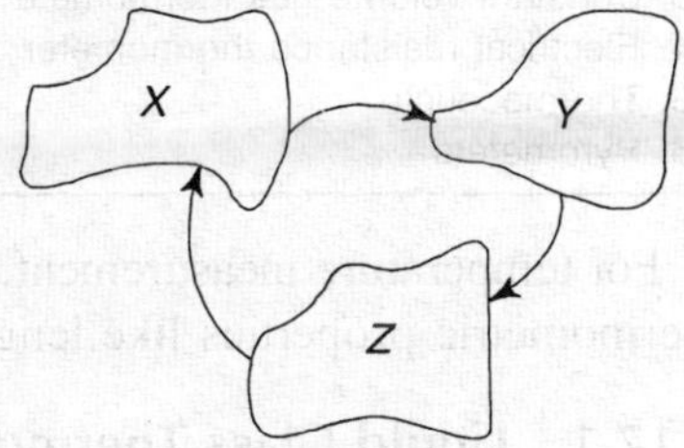

Fig. 1.12

Consider three systems X, Y and Z. Let 'X' is in thermal equilibrium with 'Y' and 'Y' is in thermal equilibrium with 'Z', then we can say that, X is in thermal equilibrium with 'Z'. Thermal equilibrium means the systems X, Y and Z will be at same temperature.

This law provides basis for temperature measurement. The temperature of a system may be determined by bringing it into thermal equilibrium with a thermometer by adopting zeroth law.

This law is useful in comparing the temperature of two systems 'X' and 'Z' with the help of 'Y' (thermometer). This is done without actually bringing 'X' and 'Z' in contact with each other.

The temperature and it's conversion factors are as follows:

$$°R = °F + 459.67$$

$$K = °C + 273.15$$

$$K = 1.8°R$$

where,

°R = degree Rankine (absolute degree Fahrenheit)

°F = degree Fahrenheit

K = degree Kelvin (absolute degree Celsius)

The concept of temperature can also be explained by considering two systems 'A' and 'B' having any two gases with thermodynamic properties P_1, V_1 and T_1 and P_2, V_2 and T_2 etc. When these two systems are in communication with each other through an adiabatic wall, thermodynamic properties of both the systems will remain unchanged even after a long period of time.

When a diathermic wall is placed between two systems, the thermodynamic properties of both the systems rapidly change till the systems regains equilibrium conditions.

When both the systems are in equilibrium, one property acquires a common numerical value for both of the systems. This property is called temperature. i.e. If T_1 and T_2 are the temperatures of systems A and B, under equilibrium condition T_1 become equal to T_2.

1.17 TEMPERATURE MEASUREMENTS

A reference system known as "thermometer" is used to measure temperature and the physical property that changes with temperature is called "thermodynamic property". Temperature cannot be measured directly. The change in temperature, causes the thermometric property to vary and this effect is used as a measure of temperature. Some of the thermometric properties are:

(1) Length of liquid column in a capillary connected to a bulb.

(2) The pressure of a fixed mass of gas kept at constant volume.

(3) Electric resistance of a metallic wire.
(4) Emf of a thermocouple.

Some different types of thermometers with their thermometric properties are given below.

Table 1.1 Types of Thermometers

Thermometers	*Thermometric property*	*Symbol*
1. Mercury-glass thermometer	Length	*L*
2. Constant pressure gas thermometer	Volume	*V*
3. Constant volume gas thermometer	Pressure	*P*
4. Electrical resistance thermometer	Resistance	*R*
5. Thermocouple	Thermal emf	*E*
6. Pyrometers	Intensity of radiation	*J*

For temperature measurement, a number of thermometers are available and all of them use different thermometric properties like length, volume, pressure, resistance, etc.

1.17.1 Liquid Glass Thermometer

These thermometers uses liquids as the thermometric substance and change in the length of liquid column in the capillary with heat interactions is the characteristics used for temperature measurement. Usually mercury and alcohol are used in these type of thermometers. The figure shows mercury in glass thermometer. It consists of a vertical tube with graduations marked on it to show the temperature, one end of which is connected to a thermometric bulb. A small quantity of mercury is filled in a capillary tube. Mercury has lower specific heat and hence absorbs little heat from the body or source. When the bulb is brought in contact with a hot system, there is change in volume of mercury which results in rise or fall of mercury level in the capillary tube. The length of liquid column is used as a thermometric property and is a measure of temperature.

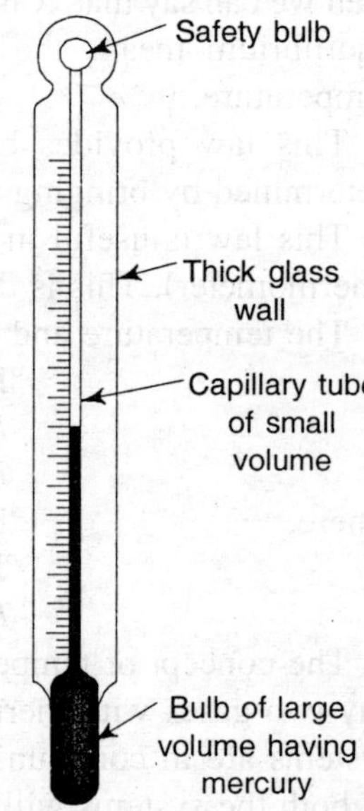

Fig. 1.13 Mercury glass thermometer

Advantages of mercury over other thermometric liquids are:

(1) Lower specific heat, hence absorbs little heat from the source.
(2) It can be conveniently seen in the capillary tube.
(3) It is a good conductor of heat, does not adhere to the wall of tube.
(4) It has uniform coefficient of expansion over a wide range of temperature.

In liquid-glass thermometer, the variation in temperature may not cause uniform change of properties. Hence the various thermometers will not indicate the same temperature between ice and steam points and in some cases they cannot be ignored. In such cases, gas thermometers are used.

1.17.2 Gas Thermometers

These thermometers are more sensitive and uses gaseous thermometric substance like Oxygen, Nitrogen, Hydrogen, Helium, etc. These gases have high coefficient of expansion and even a small change in temperature can also be recognised accurately.

(i) Constant Volume Gas Thermometers

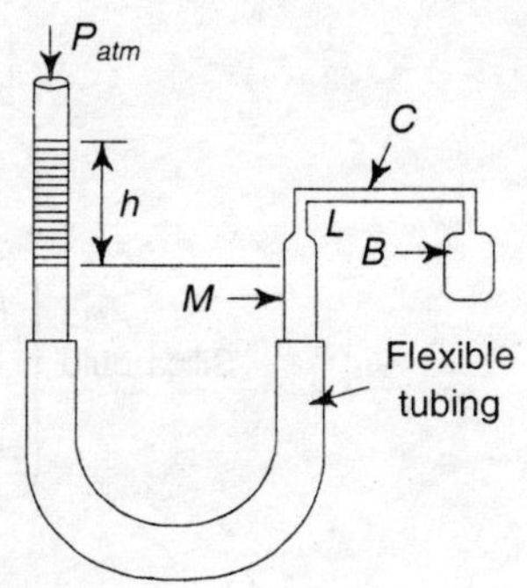

Fig. 1.14 Constant volume gas thermometer

This thermometer consists of a capillary tube (C), which connects thermometer bulb with a *U*-tube manometer. A small amount of helium gas is contained in the bulb '*B*'. The left limb of manometer is kept open to atmosphere and can be moved vertically and mercury level on the right limb can be adjusted so that it just touches lip '*L*' of the capillary. The pressure of the gas in the bulb is used as a thermometric property and is given by

$$P = P_{atm} + \rho_m h$$

where, P_{atm} = Atmospheric pressure

ρ_m = density of mercury

When the bulb is brought in contact with the system whose temperature is to be measured, it comes in thermal equilibrium with the system. The gas in the bulb will be heated and expanded, pushes the mercury column downward on the right limb. This rises mercury column on the left limb. The flexible limb is then adjusted so that the mercury again touches the lip '*L*'. The difference in mercury level '*h*' is recorded and the pressure '*P*' of the gas in the bulb is estimated. Since the gas volume in the bulb is constant, from ideal gas equation we can write,

$$\Delta T = \frac{V}{R} \Delta P \qquad (\because PV = mRT)$$

($\because$ *V* is constant, *R* is constant)

$$\Delta T \propto \Delta P$$

i.e. the temperature rise is proportional to pressure rise.

Since for an ideal gas at constant volume, $T \propto P$

$$\frac{T}{Ttp} = \frac{P}{Ptp}$$

$$T = 273.16 \frac{P}{Ptp}$$

Ttp: Triple point temperature of water

or

$$T = 273.16 \lim_{Ptp \to 0} \left[\frac{P}{Ptp}\right]$$

(ii) Constant Pressure Gas Thermometer

This thermometer is very similar to constant volume gas thermometer except change in thermometric property. In this type, pressure of the gas is kept con-stant and volume is directly proportional to it's absolute temperature. It consists of a reservoir '*R*' which is filled with mercury and is connected to a silica bulb '*B*' through a connecting tube. The bulb '*C*' is called compensating bulb and is connected with compensating tube and the volume of which is equal to that of connecting tube. The manometer is usually filled with sulphuric acid.

Initially, the bulbs '*B*', '*R*' and '*C*' are immersed in melting ice. The mercury level in the reservoir must be zero and the stop valve must be closed. The level of sulphuric acid in the limbs of manometer

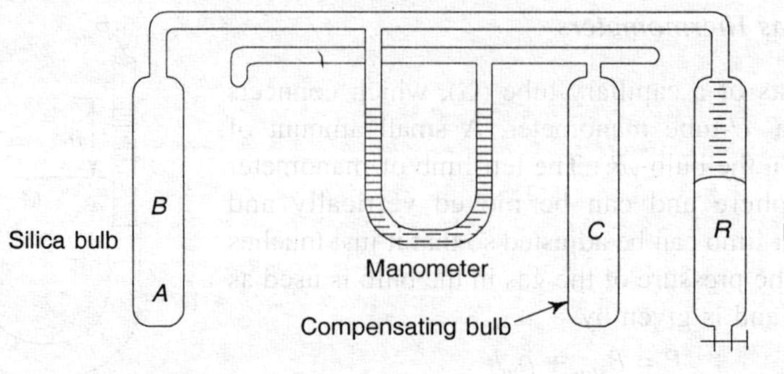

Fig. 1.15 Constant pressure gas thermometer

will be same which indicates that the pressure in the bulb '*B*' and '*C*' are same. Hence the gas and air are maintained at same pressure.

Now consider bulb '*B*' which has definite number of air molecules and bulb '*C*' and compensating tube contain same number of molecules of air. If the bulb '*B*' is placed in a bath whose temperature is to be measured, then both connecting tube and compensating tubes are maintained at room temperature. The air in bulb '*B*' attain temperature equal to the temperature to be measured.

This thermometer is similar to constant volume gas thermometer, but change in volume of gas due to temperature variation is used as a thermometric property. The height of mercury column, *h* is kept constant and the volume of gas is used as a thermometric property.

At the limiting condition of temperature

$$T = 273.16 \cdot \frac{V}{Vtp}$$

$$T = 273.16 \lim_{Vtp \to 0} \left(\frac{V}{Vtp} \right)$$

1.17.3 Thermocouple

When two dissimilar metal wires are joined at their ends and the junctions are maintained at different temperatures, an emf is generated. By knowing temperature at one junction, other junction temperature can be measured in terms of emf.

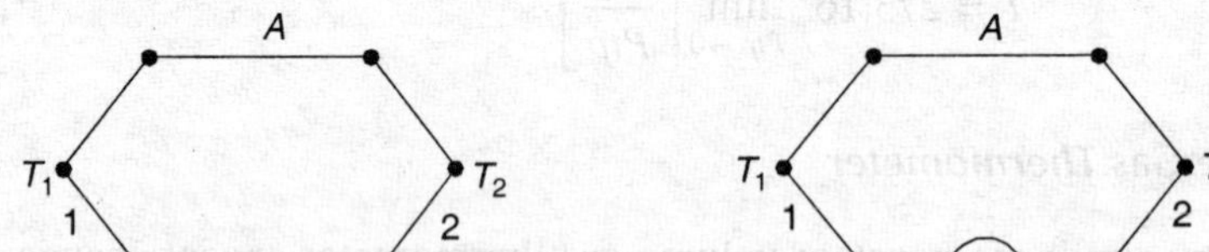

Fig. 1.16

When two dissimilar metals *A* and *B* are joined at the ends 1 and 2 with their temperatures 'T_1' and 'T_2', produces an emf E.

Then $$T(E) = 273.16 \left[\frac{E}{Etp} \right]$$

where *E* and *Etp* are thermometric properties

1.17.4 Electric Resistance Thermometer

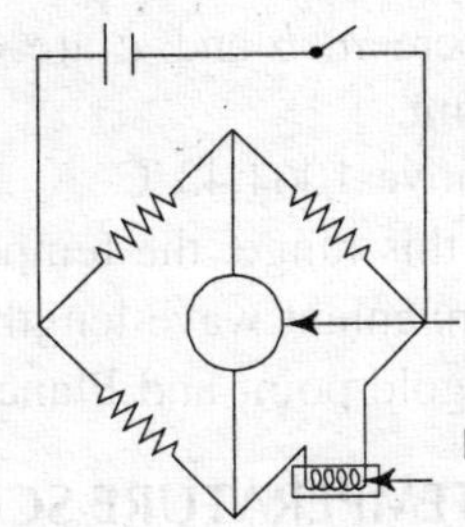

Fig. 1.17 Wheat-stone bridge

It was first developed by Siemen in 1871 and is also known as "Platinum resistance thermo-meter". It works on the principle of Wheatstone bridge.

In this thermometer, temperature change causes change in resistance of a metal wire which is the thermometric property. It may also be used as a standard for calibrating other thermometers, as it measures temperature to a high degree of accuracy and is more sensitive.

We can write

$$R = R_0\,[1 + At + Bt^2]$$

where R_0 = Platinum wire resistance when it is immersed in melting ice.

A and B = constants

1.18 INTERNATIONAL TEMPERATURE SCALE

It was adopted at the seventh general conference on weights and measures held in 1927. In 1968, slight modifications were introduced into the scale and now it is well accepted standard scale of practice. It is based on a number of fixed and easily reproducible, points that are assigned definite values of temperatures.

Table 1.2 Temperatures of Fixed Points
Primary Points for the International Practical Temperature Scale of 1968

	Temperature °C
Triple point of oxygen	−218.78
Normal boiling point of oxygen	−182.97
Triple point of water (standard)	+0.01
Normal boiling point of water	100.00
Normal boiling point of sulphur	444.6
Normal freezing point of zinc	419.58
Normal melting point of antimony	630.56
Normal melting point of silver	960.80
Normal melting point of gold	1063.00

The means available for measurement and interpolation are as follows:

1. The range from −259.34°C – 0°C

$$R = R_0\,[1 + At + Bt^2 + C(t - 100) + t^3]$$

where R_0, A, B and C are the constants obtained by finding resistance at oxygen, ice, steam and sulphur points respectively.

2. The range from 0°C to 630.74°C

It is also based on platinum resistance thermometer

$$R = R_0\,(1 + At + Bt^2)$$

where R_0, A and B are computed by measuring resistance at ice point, steam point and sulphur point.

3. The range from 630.74°C – 1064.43°C. It is based on measurement of temperature on a standard platinum against rhodium-platinum thermocouple, in terms of emf.

$$E = a + bt + ct^2$$

where a, b and c are computed from measurements at antimony point, silver point and gold point.

4. Above 1064.43°C

In this range, the temperature measurement is done by comparing intensity of radiation of any convenient wave length with intensity of radiation of same wavelength emitted by a black body at gold point and Planck's equation is used to measure temperature.

1.19 TEMPERATURE SCALE

To perform the measurement of temperature it is required to set up standards which may be used for calibration of different thermometers. The boiling and freezing points of water are two such "standards", but they do not cover the whole range of temperatures.

The two temperatures scale normally used for temperature measurements are Fahrenheit and Celsius scales. Until 1954, the Celsius scale was based on two fixed points, ice point and steam point. The ice point is defined as the temperature of mixture of ice and water which is in equilibrium with saturated air at 1 atm pressure and is assigned a value of 0°C on Celsius scale and 32°F on Fahrenheit scale. The steam point is the temperature of water and steam which are in equilibrium at 1 atmospheric pressure. It is assigned a value of 100°C on Celsius scale and 212°F on Fahrenheit scale.

In 1954, single fixed point method was adopted and Celsius scale was defined in terms of the ideal gas temperature scale. The triple point of water is used as a single fixed point (triple point means, it is the state at which solid, liquid and vapour phases of water exist together in equilibrium). The magnitude of degree is defined in terms of ideal gas temperature scale. A value of 0.01°C is assigned to triple point of water and steam point is found to be 100.00°C by experimental methods.

The Celsius scale has 100 units between ice and steam points, whereas Fahrenheit scale has 180 units. The absolute temperature scale has only positive values. The absolute Celsius scale is termed as Kelvin scale and absolute Fahrenheit scale is called as the Rankine scale. The same physical state represents zero points on both of these absolute scales and gives same values for the ratio of two temperature values, irrespective of the scale used, *i.e.*

$$\left(\frac{T_2}{T_1}\right) \text{Rankine} = \left(\frac{T_2}{T_1}\right) \text{Kelvin}$$

The relationship between these two scales is given by

(1) $°\text{F} = 32.0 + \frac{9}{5}\,°\text{C}$

$°\text{F} = 32.0 + 1.8°\text{C}$

(2) $°\text{R} = \frac{9}{5} K$

$°\text{R} = 1.8\ \text{K}$

$°\text{R} = °\text{F} + 459.67$

$\text{K} = °\text{C} + 273.15$

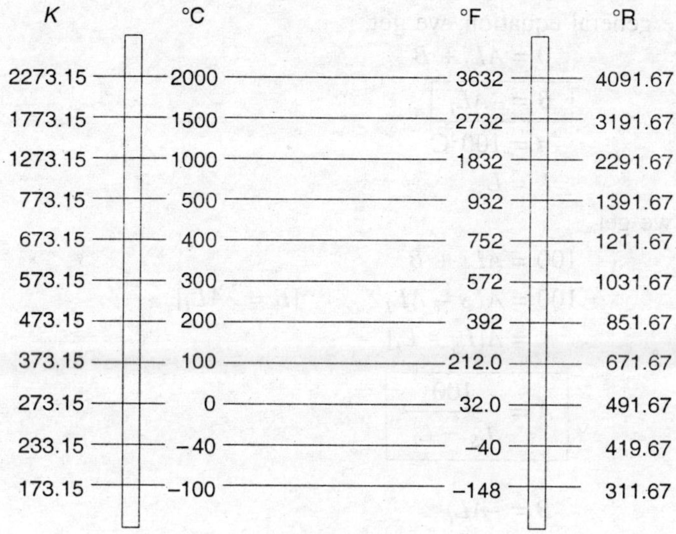

Fig. 1.18

Kelvin Scale

This is an absolute scale. The ice point is assigned with a value of 273.15 *K* and steam point is assigned with a value of 373.15 *K*. The triple point of water is 273.16 *K*.

Rankine Scale

This is also an absolute scale and the corresponding values are:

Ice point – 491.67 *R*
Steam point – 671.67 *R*
Triple point of water – 491.69 *R*

Celsius Scale

On this scale, the freezing point (ice point) corresponds to 0°C, and boiling point is referred as 100°C (steam point). The scale has 100 divisions between these two, each representing 1°C. The corresponding triple point of water is 0.01°C.

Fahrenheit Scale

Ice point – 32°F
Steam point – 212°F
Triple point of water – 32.02°F

The scale has 180 divisions each representing 1°F.

1.20 COMPARISON OF TEMPERATURE SCALES

Let us consider a thermometric property '*L*', such that '*t*' is in °C and '*t*' is a linear function of '*L*'. Then the general equation is

$t = AL + B$, where *A* and *B* are constants for Celsius scale.

At ice point, $t = 0°C$

$L = L_I$ [L_I = Thermometric property at ice point]

Substitute these, in the general equation, we get,

$$0 = AL_I + B$$

$$\boxed{B = -AL_I}$$

At steam point, $t = 100°C$

$$L = L_S$$

∴ After substitution, we get

$$100 = AL_S + B$$

$$100 = AL_S - AL_I \qquad [B = -AL_I]$$

$$= A[L_S - L_I]$$

$$\boxed{A = \frac{100}{L_S - L_I}}$$

Now

$$B = -AL_I$$

$$\boxed{B = -\frac{100L_I}{L_S - L_I}}$$

Now $t = AL + B$

Substitute the values of A and B, we get,

$$t = \frac{100}{L_S - L_I} L + \left(\frac{-100L_I}{L_S - L_I}\right)$$

$$t = \frac{100}{L_S - L_I}(L - L_I)$$

∴

$$\boxed{t°C = \frac{100(L - L_I)}{(L_S - L_I)}}$$

Fahrenheit scale;

we know that $t = AL + B$

At ice point, $t = 32°F$

$$L = L_I$$

$$32 = AL_I + B \qquad (1)$$

At steam point, $t = 212°F$

$$L = L_S$$

∴

$$212 = AL_S + B \qquad (2)$$

Solving (1) and (2), we get,

$$212 = AL_S + B$$

$$(-)32 = AL_I + B$$

$$180 = A(L_S - L_I)$$

$$\boxed{A = \frac{180}{(L_S - L_I)}}$$

Substitute the value of 'A' in eqn. (1), we get,

$$32 = \frac{180}{(L_S - L_I)} L_I + B$$

$$B = 32 - \frac{180 L_I}{L_S - L_I}$$

Now $$t = AL + B$$

$$t = \left(\frac{180}{L_S - L_I}\right) L + 32 - \frac{180 L_I}{(L_S - L_I)}$$

$$t = \frac{180}{L_S - L_I} (L - L_I) + 32$$

$$\therefore \quad \boxed{t\,°\text{F} = 32 + 180 \left[\frac{L - L_I}{L_S - L_I}\right]}$$

Similarly for Rankine scale,

$$\boxed{T°\text{R} = \left[\frac{L - L_I}{L_S - L_I}\right] \cdot 180 + 491.67}$$

$$\boxed{T\text{K} = \left[\frac{L - L_I}{L_S - L_I}\right] \cdot 100 + 273.15}$$

Relation between Celsius and Fahrenheit Scale

We know that,

$$t°\text{C} = \left(\frac{L - L_I}{L_S - L_I}\right) \cdot 100$$

$$\frac{t\,°\text{C}}{100} = \left(\frac{L - L_I}{L_S - L_I}\right) \quad (1)$$

also, $$t°\text{F} = 32 + 180 \left[\frac{L - L_I}{L_S - L_I}\right] \quad (2)$$

from equations (1) and (2), we can write

$$t°\text{F} = 32 + 180 \left[\frac{t\,°\text{C}}{100}\right]$$

$$\boxed{t°\text{F} = 32 + \frac{9}{5}\, t°\text{C}}$$

and $$\boxed{t°\text{C} = (t°\text{F} - 32) \cdot \frac{5}{9}}$$

1.21 IDEAL GAS TEMPERATURE SCALE

We know that,

$$T = 273.16 \cdot \frac{P}{Ptp} \qquad (A)$$

Suppose a number of measurements were made with different amount of gas in the gas bulb of a constant volume gas thermometer, depending on the amount of gas in the bulb, the pressure at triple point (Ptp = 1000, 500, 250, 100 mm of Hg) and system temperature T will change. For different gases, the graph shown in the figure can be obtained by plotting T verses Ptp. From the graph, it is clear that, all gases indicate the same temperature as Ptp is decreased and approaches zero.

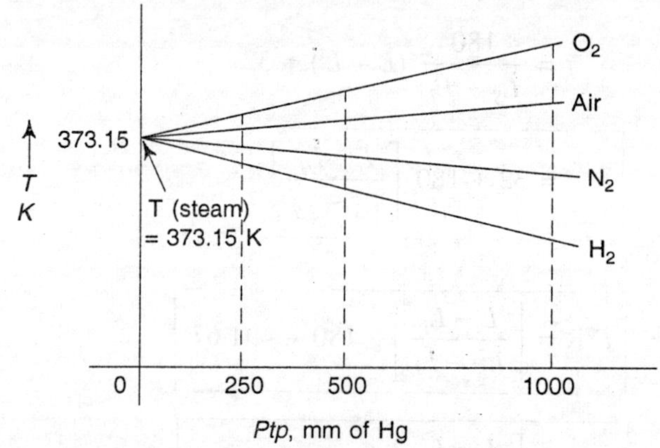

Fig. 1.19

A similar type of test may be made with a constant pressure gas thermo-meter. The values of 'P' are taken as 1000 mm of Hg, 500 mm of Hg etc., and in each trial, V and Vtp may be recorded when the bulb is surrounded by steam condensing at 1 atm and the triple point of water, respectively.

Then, $$T = 273.16 \frac{V}{Vtp}$$

Then, plot T v/s P, as shown in figure. It is clear from the experiments that all gases indicates same value of T as P approaches zero.

Since the real gas in the bulb, behaves like an ideal gas as $P \to 0$, the ideal gas temperature T can be defined by using any of these two equations

$$T = 273.16 \lim_{Ptp \to 0} \left(\frac{P}{Ptp}\right) \quad \text{or}$$

$$T = 273.16 \lim_{Ptp \to 0} \left(\frac{V}{Vtp}\right)$$

T = Ideal gas temperature scale expressed in K.

Formulae Used in Solving Problems

(1) $t = AL + B$

(2) °R = °F + 459.67
(3) K = °C + 273.15
(4) K = 1.8°R

1.22 NUMERICAL PROBLEMS

Problem 1 : *In 1701, Newton proposed a linear temperature scale in which the ice point temperature was taken as 0°N and the human body temperature was taken as 12°N. Find the conversion scale between Newton scale of temperature and centigrade scale of temperature, if the temperature of human body in centigrade scale is 37°C.*

Solution :
We know that,

$$t = AL + B \tag{1}$$

(a) For Newton scale
At Ice point,

$$t = 0°\text{N}$$

$$L = L_I \; [L_I = \text{Thermometric property at ice point}]$$

The equation (1) becomes

$$0 = A\,L_I + B$$

$$\boxed{B = -AL_I}$$

At human body temperature, t = 12°N

$$L = L_h$$

$[L_h$ = Thermometric property at human body temperature] Substitute these values in equation (1)

$$12 = A \cdot L_h + B$$

$$12 = A\,L_h - A\,L_I \qquad [\text{But } B = -AL_I]$$

$$= A\,[L_h - L_I]$$

$$\therefore \quad \boxed{A = \frac{12}{L_h - L_I}}$$

Now,

$$B = -A\,L_I$$

$$B = \left[-\frac{12}{L_h - L_I}\right] L_I$$

$$\therefore \quad \boxed{\text{B} = \frac{-12\,L_I}{L_h - L_I}}$$

We know that, t°N = $AL + B$
Substitute the values of 'A' and 'B'

$$t°\text{N} = \left(\frac{12}{L_h - L_I}\right) L + \left(\frac{-12\,L_I}{L_h - L_I}\right)$$

$$\therefore \quad \boxed{t°\text{N} = \frac{12}{L_h - L_I}\,(L - L_I)}$$

Repeat the same procedure for centigrade scale

At ice point, $t = 0°C$

$$L = L_I$$

$$0 = AL_I + B$$

$$\boxed{B = -AL_I}$$

At human body temperature, $t = 37°C$

$$L = L_h$$

$$\therefore \quad 37 = AL_h + B$$

$$= AL_h - AL_I \; [B = -AL_I]$$

$$37 = A[L_h - L_I]$$

$$\therefore \quad \boxed{A = \frac{37}{L_h - L_I}}$$

$$B = -AL_I$$

$$= \left(-\frac{37}{L_h - L_I}\right) L_I = \boxed{-\frac{37 L_I}{L_h - L_I}}$$

$$\therefore \quad t°C = AL + B$$

$$= \left(\frac{37}{L_h - L_I}\right) L + \left(-\frac{37 L_I}{L_h - L_I}\right)$$

$$\boxed{t°C = \left(\frac{37}{L_h - L_I}\right)(L - L_I)}$$

Now,

$$\frac{t\,°N}{t\,°C} = \frac{\frac{12}{L_h - L_I}(L - L_I)}{\left(\frac{37}{L_h - L_I}\right)(L - L_I)} = \frac{12}{37}$$

$$\therefore \quad t°C = \frac{37}{12}\, t°N$$

$$\therefore \quad \boxed{t°C = 3.083\, t°N}$$ *Ans.*

Problem 2: *A thermometer is calibrated with ice and steam points as fixed points referred to as 0°C and 100°C respectively. The equation used to establish the scale is* $t = a \log_e x + b$.

(a) Determine the constants 'a' and 'b' in terms of x_S *and* x_I.

(b) Prove that $t°C = 100 \dfrac{\log_e\left(\dfrac{x}{x_I}\right)}{\log_e\left(\dfrac{x_S}{x_I}\right)}$

Solution : The given equation is

$$t = a \log_e x + b$$

At ice point, $t = 0°C$, $x = x_I$

$$0 = a \log_e x_I + b$$

$$\boxed{b = -a \log_e x_I}$$

At steam point, $t = 100°C$, $x = x_S$

$$100 = a \log_e x_S + b$$

$$100 = a \log_e x_S - a \log_e x_I$$

$$100 = a \log_e \left(\frac{x_S}{x_I}\right)$$

$$\boxed{a = \frac{100}{\log_e \left(\frac{x_S}{x_I}\right)}} \qquad \textit{Ans.}$$

But

$$b = -a \log_e x_I = -\frac{100}{\log_e \left(\frac{x_S}{x_I}\right)} \cdot x_I$$

$$\boxed{b = \frac{-100\, x_I}{\log_e \left(\frac{x_S}{x_I}\right)}} \qquad \textit{Ans.}$$

Now,

$$t = a \log_e x + b$$

$$t = \frac{100}{\log_e \left(\frac{x_S}{x_I}\right)} \cdot x + \left(\frac{-100\, x_I}{\log_e \left(\frac{x_S}{x_I}\right)}\right)$$

$$t = \frac{100}{\log_e \left(\frac{x_S}{x_I}\right)} [\log_e x - \log_e x_I]$$

$$\therefore \qquad \boxed{t°C = \frac{100 \log_e (x / x_I)}{\log_e (x_S / x_I)}} \qquad \textit{Ans.}$$

Problem 3 : *Define a new temperature scale say degree N, in which the freezing and boiling points of water are 100°N and 300°N respectively. Correlate this temperature scale with centigrade scale.*

Solution : We know that $t = AL + B$ (1)

(i) *For Newton Scale*

At freezing and boiling points of water,

$$100 = AL_I + B \quad (2)$$

$$300 = AL_S + B \quad (3)$$

(3) – (2) gives $200 = A(L_s - L_I)$

$$\therefore \quad \boxed{A = \frac{200}{L_S - L_I}}$$

Substitute this value in equation 2, we get,

$$100 = \left(\frac{200}{L_S - L_I}\right) L_I + B$$

$$B = 100 - \frac{200\,L_I}{L_S - L_I}$$

∴ Substitute 'A' and 'B' in Eqn. (1), we get

$$t = \frac{200}{L_S - L_I} L + 100 - \frac{200\,L_I}{L_S - L_I}$$

$$\boxed{t°\text{N} = \left(\frac{L - L_I}{L_S - L_I}\right) 200 + 100} \qquad \textit{Ans.}$$

also we know that

$$\boxed{t°\text{C} = \left(\frac{L - L_I}{L_S - L_I}\right) 100 + 0}$$

$$\therefore \quad t°\text{N} = \left(\frac{L - L_I}{L_S - L_I}\right) 2 \times 100 + 100$$

$$\boxed{t°\text{N} = 2 \times t°\text{C} + 100} \qquad \textit{Ans.}$$

Problem 4 : *Define a new temperature scale of °B in which the boiling and freezing points of water are 500°B and 100°B respectively. Co-relate this temperature scale with centigrade scale of temperature.*

Solution : We know that, $t = AL + B$

for °B scale, at ice point, $t = 100°\text{B}$

$$L = L_I$$

$$100 = AL_I + B$$

$$\text{B} = 100 - AL_I$$

At steam point, $t = 500°\text{B}$

$$L = L_S$$

$$500 = AL_S + B$$

$$500 = AL_S + 100 - AL_I \qquad (\because B = 100 - AL_I)$$

$$400 = A\,(L_S - L_I)$$

$$\boxed{A = \frac{400}{L_S - L_I}}$$

$$\therefore \quad \boxed{B = 100 - \left(\frac{400}{L_S - L_I}\right) L_I}$$

$$\therefore \quad t°B = \left(\frac{400}{L_S - L_I}\right) L_I + 100 - \left(\frac{400\,L_I}{L_S - L_I}\right)$$

$$\boxed{t°B = \frac{400}{L_S - L_I}(L - L_I) + 100} \qquad (1)$$

for °C scale, at ice point, $t = 0°C$, $L = L_I$

$$0 = AL_I + B$$

$$B = -AL_I$$

At steam point, $t = 100°C$, $L = L_S$

$$100 = AL_S + B$$

$$100 = AL_S - AL_I$$

$$\therefore \quad \boxed{A = \frac{100}{L_S - L_I}}$$

$$\boxed{B = -\frac{100}{L_S - L_I} L_I}$$

$$t°C = \frac{100}{L_S - L_I} L - \frac{100}{L_S - L_I} L_I$$

$$\boxed{t°C = \frac{100}{L_S - L_I}(L - L_I)} \qquad (2)$$

from equation (1)

$$t°B = \frac{4 \times 100(L - L_I)}{(L_S - L_I)} + 100$$

$$\therefore \quad \boxed{t°B = 4 \times t°C + 100} \qquad \textit{Ans.}$$

$$\left[\text{where } t°C = \frac{100(L - L_I)}{(L_S - L_I)} \text{ from equation (2)}\right]$$

Problem 5 : *A centigrade and Fahrenheit thermometers are both immersed in a fluid, and the numerical value recorded on both thermometers is same. Determine the temperature of the fluid expressed as °K and °R and also find that identical value shown by thermometers.*

Solution : Given that, $t°C = t°F$

Writing °C and °F scale in terms of °K and °R scales, we get

$$T°K - 273.16 = T°R - 459.17$$

$$\therefore \quad T°R - T°K = 459.17 - 273.16$$

$$T°R - T°K = 186.54$$

But, $$T°R = 1.8\ T°K$$

$\therefore$ $1.8\ T°K - T°K = 186.54$

$$0.8\ T°K = 186.54$$

$$T°K = \frac{186.54}{0.8} = 233.17°K$$

$$\therefore \quad \boxed{T°K = 233.17°K} \qquad \textit{Ans.}$$

Now $$T°R = 1.8\ T°K$$

$$= 1.8 \times 233.17$$

$$\boxed{T°R = 419.70°R} \qquad \textit{Ans.}$$

We know that $$T°C = T°K - 273.16$$

$$= 233.17 - 273.16$$

$$\boxed{T°C = -39.99 \approx -40°C} \qquad \textit{Ans.}$$

and, $$T°F = T°R - 459.7$$

$$= 419.7 - 459.7$$

$$\boxed{T°F = -40°F} \qquad \textit{Ans.}$$

Problem 6 : *Fahrenheit and centigrade thermometers are both immersed in a fluid. If °F reading is twice that of °C reading, what is the temperature of fluid in terms of °R and °K.* *(VTU, Aug. 2000)*

Solution : For the given condition

$$T°F = 2\ T°C$$

$$(T°R - 459.7) = 2\ (T°K - 273.16)$$

$$1.8\ T°K - 459.7 = 2\ (T°K - 273.16) \qquad [\because T°R = 1.8\ T°K]$$

$$2\ T°K - 1.8\ T°K = 546.32 - 459.7$$

$$\therefore \quad T°K = \frac{86.62}{0.2} = 433.1°K$$

$$\boxed{T°K = 433.1°K}$$

$$T°R = 1.8\ T°K$$

$$= 1.8 \times 433.1$$

$$\boxed{T°R = 779.58°R}$$

Hence temperature of the fluid is

433.1°K or

779.58°R *Ans.*

Problem 7 : *A thermometer using pressure as a thermo metric property gives values of 1.86 and 6.81 at ice and steam point respectively. If ice point and steam point are assigned the values 10 and 120 respectively, determine the temperature corresponding to P = 2.3. The equation corresponding to temperature is t = a + b ln (P)* *(VTU, March, 2001)*

Solution : Given $t = a + b \ln (P)$

$$\text{At } t = 10,\ 10 = a + b \ln (P_I) \qquad (1)$$

$$\text{At } t = 120,\ 120 = a + b \ln (P_S) \qquad (2)$$

(2) – (1) gives

$$110 = b[\ln P_S - \ln P_I] = b \ln \frac{P_S}{P_I}$$

$$\therefore \qquad b = \frac{110}{\ln P_S/P_I}$$

from eqn (1)

$$a = 10 - b \ln (P_I)$$

$$= 10 - \frac{110}{\ln P_S/P_I} \cdot \ln (P_I)$$

$$\therefore \qquad t = 10 - \frac{110 \ln(P_I)}{\ln P_S/P_I} + \frac{110}{\ln \frac{P_S}{P_I}} P$$

$$t = 10 + \frac{110 \ln P/P_I}{\ln P_S/P_I}$$

given that, $P_I = 1.86,\ P_S = 6.81,\ P = 2.3$

$$\therefore \qquad t = 10 + \frac{110 \ln \frac{2.3}{1.86}}{\ln \frac{6.81}{1.86}}$$

$$\boxed{t = 28°C}$$ *Ans.*

Problem 8 : *Two Celsius thermometers 'A' and 'B' agree at the ice point (0°C) and steam point (100°C) and the related equation is $t_A = L + m\, t_B + n\, t_B^2$, where t_A and t_B are thermometer readings and L, m and n are constants. When both thermometers are immersed in an oil bath, thermometer A indicates 51°C and B registers 50°C. Determine the reading of A, when B reads 30°C.*

(VTU, Feb, 2002)

Solution :

At ice point, $t_A = t_B = 0°C$

The two thermometers are related by

$$t_A = L + m\,t_B + n\,t_B^2$$

$$0 = L + 0 + 0$$

$$\boxed{L = 0}$$

At steam point, $t_A = t_B = 100°C$

$\therefore$ $100 = L + m(100) + n(100)^2$

$\therefore$ $\boxed{100 = 0 + m(100) + 10{,}000\,n}$ (1) [$\because$ L = 0]

Given, when $t_A = 51°C$, while $t_B = 50°C$

$$51 = 0 + m(50) + n(50)^2$$

$$\boxed{51 = 50m + 2500n} \quad (2)$$

Solving equations (1) and (2) for constants m and n

$$100 = 100m + 10{,}000n$$

$$51 = 50m + 2{,}500n \times 2 \quad \text{(multiply by 2)}$$

$\therefore$

$$100 = 100m + 10{,}000n$$

$$(-)102 = 100m + 5{,}000n$$

$$-2 = 0 + 5000n$$

$\boxed{n = -4 \times 10^{-4},}$ substitute in any one of equation, we get

$$102 = 100m + 5{,}000\,(-4 \times 10^{-4})$$

$\therefore$ $\boxed{m = 1.04}$

When $t_B = 25°C$

$$t_A = L + m\,t_B + n\,t_B^2$$

$$= 0 + 1.04(30) + (-4 \times 10^{-4})\,(30^2)$$

$\therefore$ $\boxed{t_A = 30.84°C}$ *Ans.*

Problem 9 : *The relation between temperature 't' and property 'K' on a thermometric scale is given by t = a 1n k + b. The values of K are found to be 1.83 and 6.5 at the ice point and steam point. Determine the temperature, when K reads 2.42 on the thermometer. Take temperature values as 0° and 100°C at ice and steam point respectively*

Solution : The given equation is

$$t = a \ln k + b$$

$$0 = a \ln 1.83 + b \quad (1)$$

$$100 = a \ln 6.5 + b \quad (2)$$

At ice point
$t = 0°C$
$K = 1.83$
At steam point
$t = 100°C$
$K = 6.5$

(2) – (1) gives

$a \ln 6.5 - a \ln 1.83 = 100 - 0$

$$a = \frac{100}{1.267} = 78.92$$

from (1)

$$0 = 78.92\ 1\text{n}\ 1.83 + b$$

$\therefore$ $b = -47.69$

We know that

$$t = a \ln k + b$$

$$t = 78.92 \ln 2.42 + (-47.69)$$

$$\boxed{t = 22.056°C}$$ *Ans.*

Problem 10 : *The resistance in the windings of a motor is 78 ohms at room tem-perature (25°C). When operating at full load under steady conditions, the motor is switched off and the resistance of windings is found to be 95 ohms. The resistance of windings at temperature t°C is given by $R_t = R_0[1 + 0.00393\ t]$, where R_0 is the resistance at 0°C. Find the temperature of the coil at full load.*

Solution : We know that,

$$R_t = R_0\ [1 + 0.00393\ t]$$

At room temperature $t = 25°C$, resistance $R_t = 78$ ohms

$$78 = R_0[1 + 0.00393\ (25)]$$

$$R_0 = \frac{78}{1.0982} = 71.025$$

$$\boxed{R_0 = 71.025 \text{ ohm}}$$

At full load, $R_t = 95$ ohms and $t = ?$

$$R_t = R_0\ (1 + 0.00393\ t)$$

$$95 = 71.025\ (1 + 0.00393\ t)$$

$$\boxed{t = 85.87°C}$$ *Ans.*

Problem 11 : *The equation $R_t = R_0\ (1 + \alpha t)$ is used to a resistance thermometer, in which R_t and R_0 are the resistance values at t°C and 0°C respectively. The thermometer is calibrated by immersing in boiling water (100°C) and boiling sulphur (445°C) and the indicated resistance values are 14.7 ohm and 29.2 ohm respectively. Determine fluid temperature when resistance thermometer reads 25 ohm.*

Solution : When thermometer is immersed in boiling water,

$$t = 100°C,\ R_t = 14.7 \text{ ohm}$$

$$\therefore \quad 14.7 = R_0\ [1 + \alpha\ (100)] \quad (1)$$

When the same thermometer is immersed in boiling sulphur,

$$t = 445°C \text{ and } R_t = 29.2$$

$$\therefore \quad 29.2 = R_0\ [1 + \alpha\ (445)] \quad (2)$$

By solving equations (1) and (2),
(2)-(1) gives

$$14.5 = 345 \cdot \alpha \cdot R_0$$

$$R_0 \cdot \alpha = 0.042$$ [substitute this value in equation (1)]

$$14.7 = R_0 + R_0\alpha \times 100$$

$$= R_0 + 0.042 \times 100$$

$$\therefore \quad \boxed{R_0 = 10.5}$$

When $R_t = 25$ ohm, the temperature 't'

$$R_t = R_0(1 + \alpha t)$$

$$25 = R_0 + R_0 \alpha t$$
$$25 = 10.49 + 0.042 \times t$$
$$\boxed{t = 345.47°C}$$ **Ans.**

Problem 12 : *The emf of a thermocouple having one junction kept at the ice point and test junction is at the Celsius temperature t, is given by*

$\varepsilon = at + bt^2$, where $a = 0.2$ mv/deg, $b = -\ 5.0 \times 10^{-4}$ mv/deg.

(i)Sketch the graph of ε against t.

(ii)Suppose the emf ε is taken as a thermometric property and that a temperature scale t^1 is defined by the linear equation $t^1 = a^1\ \varepsilon + b^1$ such that $t^1 = 0$ at ice point and $t^1 = 100$ at steam point. Find the numerical values of a^1 and b^1 and plot a graph of ε against t^1.

Solution : The temperature t in terms of emf is given by

$$\varepsilon = at + bt^2 \qquad [\because\ a = 0.2 \text{ mV/deg},\ b = -5.0 \times 10^{-4} \text{ mV/deg}]$$

(i) $$\varepsilon = 0.2t + (-5.0 \times 10^{-4})\, t^2$$

At ice point $t = 0°C$

$$\varepsilon = 0 + 0 = 0$$
$$\boxed{\varepsilon = 0 \text{ mV}}$$

At steam point $t = 100°C$

$$\varepsilon = 0.2 \times 100 - 5 \times 10^{-4}\,(100)^2$$
$$\boxed{\varepsilon = 15 \text{ mV}}$$

The following values are obtained for different values of t.

t°C	0	10	20	30	40	50	60	70	80	90	100
e, mV	0	1.95	3.8	5.55	7.2	8.75	10.2	11.55	12.8	13.95	15

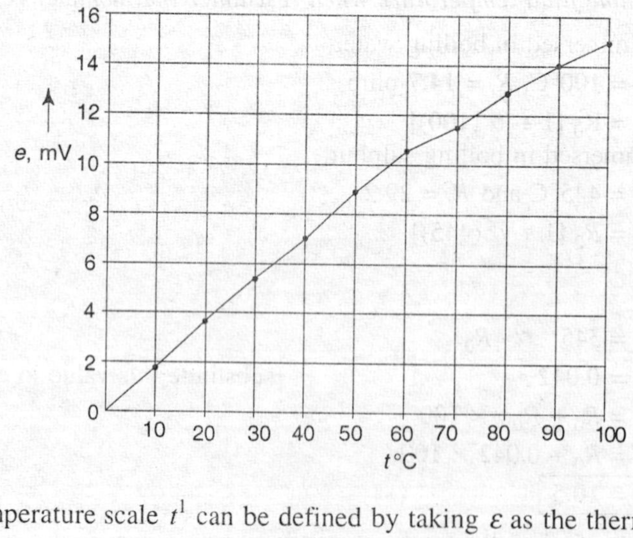

(ii) The temperature scale t^1 can be defined by taking ε as the thermometric property

$$t^1 = a^1\varepsilon + b^1$$

At ice point $t^1 = 0,\ \varepsilon = 0$

$$0 = 0 + b \qquad \boxed{\therefore\ b^1 = 0}$$

At steam point $\quad t^1 = 100°C,\ \varepsilon = 15 \quad$ [from graph]

$$100 = a^1 \times 15 + 0 \qquad [\because\ b^1 = 0]$$

$$\therefore \quad a^1 = \frac{100}{15} = 6.66$$

$\therefore$ The equation becomes

$$t^1 = 6.66\ \varepsilon + 0$$

$$\therefore \quad \boxed{t^1 = 6.66\ \varepsilon}$$

The values of ε for different values of t^1 are tabulated as follows:

t^1, °C	0	10	20	30	40	50	60	70	80	90	100
e, mV	0	1.5	3.00	4.5	6.00	7.50	9.00	10.5	12.01	13.51	15.00

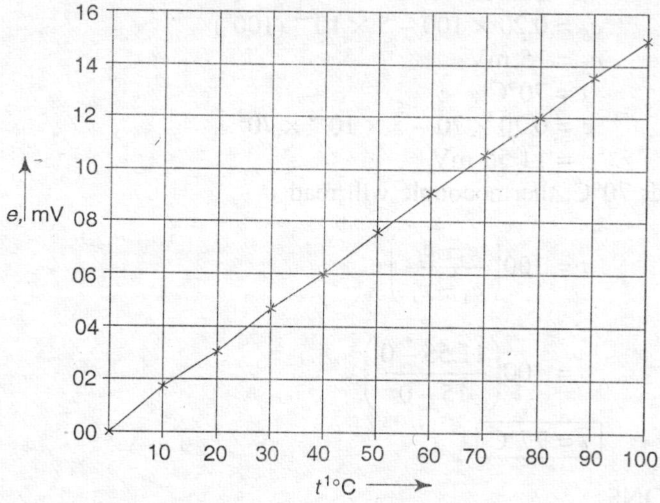

Problem 13 : *The emf in a thermo couple with test junction at ice point is given by*

$$\varepsilon = 0.2t - 5 \times 10^{-4}\ t^2\ mV.$$

The millivoltmeter is calibrated at ice point and steam points. What will this thermometer read in a place where gas thermometer reads 50°C *(VTU Feb. 2003)*

Solution : At freezing or ice point

$$\varepsilon_I = 0.2 \times 0 - 5 \times 10^{-4} \times 0$$
$$= 0 \text{ mV}$$

At boiling or steam point

$$\varepsilon_S = 0.2 \times 100 - 5 \times 10^{-4}\ (100^2)$$
$$= 15 \text{ mV}$$

At

$$t = 50°C$$
$$\varepsilon = 0.2 \times 50 - 5 \times 10^{-4}\ (50^2)$$
$$\varepsilon = 8.75 \text{ mV}.$$

$\therefore$ The temperature 't' can be calculated as

$$t = 100 \left[\frac{\varepsilon - \varepsilon_I}{\varepsilon_S - \varepsilon_I}\right]$$

$$= 100\left[\frac{8.75 - 0}{15 - 0}\right]$$

$$\boxed{t = 58.34°C}$$ *Ans.*

Problem 14 : *A thermocouple with test junction at t°C on a gas thermometer and cold junction at 0°C gives output emf as per the following relation.*

$$e = 0.20\, t - 5 \times 10^{-4} t^2,\ mV$$

where t is the temperature. The millivoltmeter is calibrated at ice and steam points. What temperature would this thermometer show when gas thermometer reads 70°C (VTU, Feb. 2004)

Solution : Given $e = 0.20\, t - 5 \times 10^{-4} t^2$, mV

At ice point $t = 0$

$e_I = 0.20 \times 0 - 5 \times 10^{-4} \times 0 = 0$ mV

At steam point, $t = 100°C$

$e_S = 0.20 \times 100 - 5 \times 10^{-4}\ (100^2)$

$e_S = 15$ mV

when $t = 70°C$

$e = 0.20 \times 70 - 5 \times 10^{-4} \times 70^2$

$= 11.55$ mV

when gas thermometer reads 70°C, thermocouple will read

$$t = 100\left(\frac{e - e_I}{e_S - e_I}\right)$$

$$= 100\left(\frac{11.55 - 0}{15 - 0}\right)$$

$$\boxed{t = 77°C}$$ *Ans.*

1.23 REVIEW QUESTIONS

1. Define thermodynamics and state its scope in the energy technology.
2. Distinguish between classical and statistical description of matter.
3. Explain the concept of macroscopic and microscopic view point as applied to stu dy of thermodynamics.
4. Explain the concept of continuum.
5. What are the different thermodynamic systems? Explain them with examples.
6. Differentiate between
   Homogeneous and Heterogeneous systems
   Intensive and Extensive properties
   Reversible and Irreversible processes
7. Define the following:
   Thermodynamic state
   Thermodynamic cycle and process
   Quasi-static process
8. What do you mean by thermodynamic equilibrium? Explain; how does it differ from thermal equilibrium.

9. State the concept of temperature and equality of temperature.
10. State and explain zeroth law of thermodynamics.
11. Name and define the law that forms basis for temperature measurement.
12. Define thermometric property and thermometric substance.
13. What are the different types of thermometers used for temperature measurement?
14. Explain a liquid glass type thermometer.
15. Explain the working of a
    (i) constant volume gas thermometer
    (ii) constant pressure gas thermometer
16. Explain the working of electric resistance thermometer.
17. Establish a correlation between Centigrade and Fahrenheit scales.
18. What do you understand by the ideal gas temperature scale?
19. What is the significance of international temperature scale?

CHAPTER 2

Work and Heat

2.1 INTRODUCTION

Energy is defined as the capacity to do work. When energy is added or removed from a system it's thermodynamic properties changes. The energy transferred to or from a system is referred as energy in transit and it may be in one of these three forms; work, electrical energy and heat. A potential or driving force is necessary to cause energy transfer to or from the system. The potentials associated may be mechanical, electrical and thermal respectively.

The energy of a closed system can be changed by two mechanisms (i) work interaction, (ii) heat interaction. It is necessary to define heat and work clearly, the analysis of thermodynamic problems depend upon distinguishing between them.

2.2 DEFINITION OF WORK

In mechanics, work is defined as a force '*F*' acting through a displacement '*S*' and the displacement being in the direction of force.

$$W = \int F ds$$

The above equation enables us to find the work required to raise a weight, to stretch a wire or to move a charged particle through a magnetic field.

If a system exists in which a force at the boundary of the system is moved through a distance, then work is done by or on the system.

In thermodynamics, work is defined as energy in transition between system and surroundings, without transfer of mass. Work is said to be done by a system on its surroundings if the sole effect external to the system could be the raising of a weight.

Example 1

Let us consider a system which includes a battery and a motor as shown in figure 2.1(a). The motor drives a fan i.e. the effect of work transfer from the system is driving the fan.

In order to define the work completely, suppose the fan is replaced by a pulley and weight arrangement as shown in figure 2.1(a). Instead of driving the fan, the motion will raise the weight and that the sole effect is raising of weight. Thus for the system shown in figure, the sole effect external to the system can be reduced to raise a weight and thus, work is transferred from the system.

Example 2

Consider expansion of gas, inside a cylinder. During expansion of gas in the cylinder, work will be done by the gas on piston and a weight can be lifted up through suitable linkages. So work will be done by the gas and sole effect external to the system could be raising of a weight.

As energy crosses system boundary, it disappears and becomes part of the energy content of the system or surroundings. Work is transient phenomena and the system never possesses certain quantity

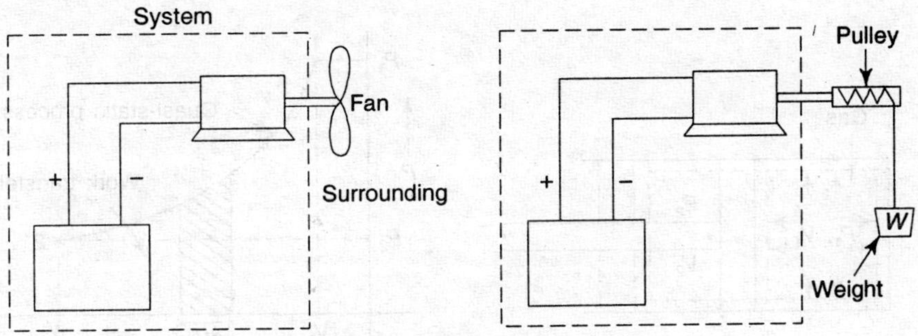

Fig. 2.1(a) Work Transfer

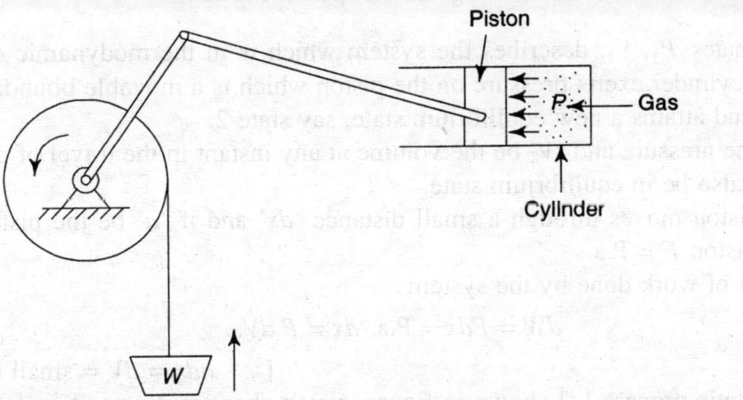

Fig. 2.1 (b) Lifting of Weight Due to Work done by the Gas During Expansion.

of work at a given state i.e. quantity of work is not a thermodynamic property. Work interaction is possible only when a change of state occurs.

2.3 SIGN CONVENTION FOR WORK

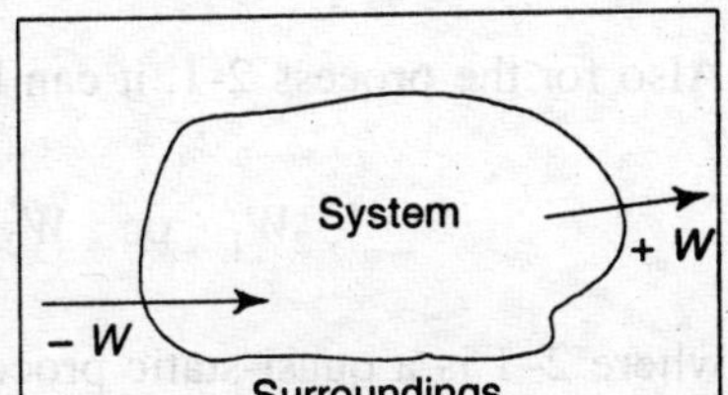

Fig. 2.2

Work done by the system on it's surroundings is considered as positive.

When work is done on the system it is considered as negative.

Thus

$+W$ — work done by the system.

$-W$ — work done on the system.

when the surroundings do positive work, the system does an equivalent of negative work and vice versa.

Thus for any process,

$$W \text{ system} + W \text{ surroundings} = 0.$$

Consider an I.C. engine in which work is considered negative for compression, as piston compresses the charge. Some amount of work is done on the system. Similarly work is considered as positive for expansion and engine delivers power. The work is done by the system to deliver power.

2.4 P.dV-DISPLACEMENT WORK IN A QUASI-STATIC PROCESS

Let us consider gas in the cylinder as a system.

Let P_1 and V_1 – Initial pressure and volume of the system.

P_2 and V_2 – Final pressure and volume of the system.

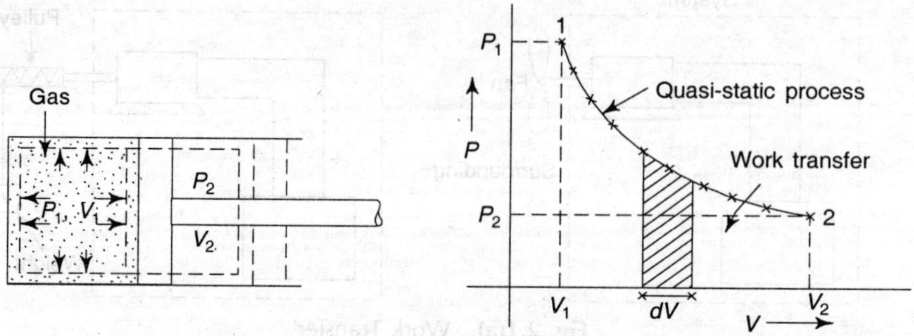

Fig. 2.3 Displacement work

The co-ordinates P_1, V_1, describes the system which is in thermodynamic equilibrium at state 1. The gas in the cylinder exerts pressure on the piston which is a movable boundary of the system. The piston moves and attains a new equilibrium state, say state 2.

Let '*P*' be the pressure and '*V*' be the volume at any instant in the travel of piston. This intermediate point must also be in equilibrium state.

When the piston moves through a small distance '*dx*' and if '*a*' be the piston area, the force '*F*' acting on the piston F = P.a.
and the amount of work done by the system

$$dW = Fdx = \text{P.a. } dx = P.dV.$$

$$[\because \quad adx = dV = \text{small displacement volume}]$$

For a quasistatic process 1-2 shown in figure, piston changes it's position from 1 to 2 and volume increases from V_1 to V_2. The amount of work done by the system will be

$$_1W_2 \quad \text{or} \quad W_{1-2} = \int_{V_1}^{V_2} P \cdot dV.$$

Also for the process 2-1, it can be written as

$$_2W_1 \quad \text{or} \quad W_{2-1} = \int_{V_2}^{V_1} PdV.$$

where 2–1 is a quasi-static process and viscous, gravitational and magnetic effects are neglected.

The area under path 1–2 on *P–V* diagram gives magnitude of work done by the system. The integral $\int PdV$ can be used only for a quasi-static process.

2.5 PATH FUNCTION AND POINT FUNCTION

Let us consider a system which undergoes a process 1-2 as shown in the *P-V* diagram. This process between the points 1 and 2 may take place along different quasi-static paths *A*, *B* and *C*.

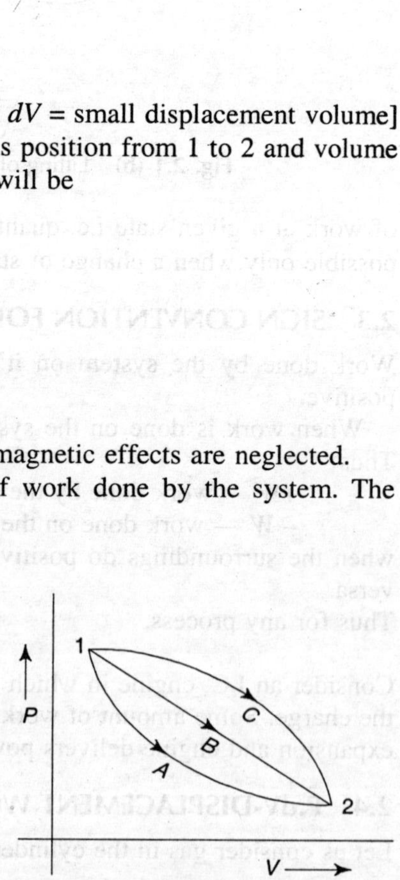

Fig. 2.4

Path Function

It is clear from the *P-V* diagram that, area under each curve, (say curve 1-*A*-2, curve 1-*B*-2, curve 1-*C*-2) represents work for each process. Since area under each curve is different, the amount of work obtained in each case will be different and is not a function of end states 1 and 2. As area under quasi-static path on a PV diagram gives magnitude of the work, and it is different for each paths i.e. the magnitude of work depends on the path, the system follows in changing from state 1 to state 2. Hence work is a path function and dw is an inexact or imperfect differential.

Point Function

During a process, if the values of variables depends on the end states, they are known as point functions. All thermodynamic properties say, pressure, temperature, volume, etc., are point functions. During a process, the variation in thermodynamic properties of a system are dependent only on the initial and final states, but not on the path followed by the system during change of state. For a given state, each property has a definite value. The differentials of point functions are exact differentials.

We have $$\int_{V_1}^{V_2} dV = V_2 - V_1$$

Thus, the change in volume depends only on the initial and final states. On the other hand, since work is a path function.

$$\int_1^2 dW \neq W_2 - W_1$$

$$\int_1^2 dW = W_{1-2} \quad \text{or} \quad {}_1W_2$$

For a cyclic process, the end conditions of the system are same and hence change in any property is zero, i.e.

$$\oint dV = 0, \quad \oint dP = 0, \quad \oint dT = 0.$$

The symbol '$\oint$' denotes cyclic integral of a closed path.

∴ The cyclic integral of any property is always zero.

2.6 OTHER TYPES OF WORK TRANSFER

Other than displacement work, (PdV), the additional types of work transfer between system and surroundings are:

2.6.1 Electrical work

Consider flow of current through a resistor as a system, in which work is transferred into the system. An electric force is necessary to drive the current through the resistor.

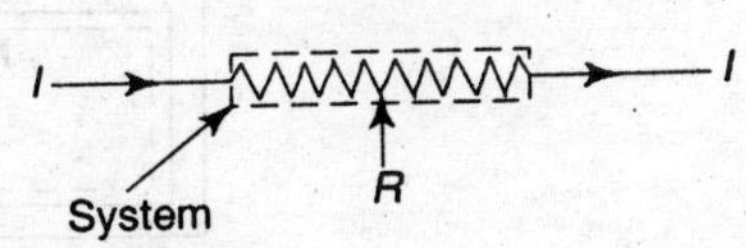

Fig. 2.5

Let I be the flow of current in pumps through the resistor and is given by $I = \dfrac{dc}{dt}$.

Where C = charge in coulombs

t = time in seconds

If E = voltage potential, then

dW = Electrical work done by the flow of current, then $dW = Edc$

$dW = E\,Idt$ $\quad$ [$\because$ $dc = I\,dt$]

After Integration

$$\int dW = W = \int_1^2 EIdt$$

$$W = \lim_{dt \to 0} \frac{dW}{dt} = EI$$

as $dt \to 0$, $W = EI$, is the rate at which work is transferred.

If a battery is used to supply current to the system, work is considered to be done by the cell when it is discharging current and work is done to it when it is being charged.

2.6.2 Shaft Work

Consider a system in which a shaft is driven by the motor as shown in figure 2.6. In this case, the work is transferred into the system. When torque T is applied to the shaft, it experiences an angular displacement $d\theta$.

∴ The shaft work is

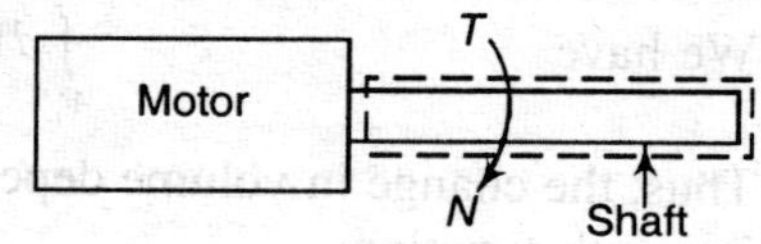

Fig. 2.6

$$dW = T\,d\theta.$$

for finite changes

$$\int dW = \int_1^2 Td\theta = \int_1^2 T\frac{d\theta}{dt}dt$$

$$W = TW, \quad \text{where} \quad W = \frac{d\theta}{dt} = \text{angular velocity.}$$

2.6.3 Paddle Wheel Work or Stirring Work

Consider a system in which a fluid is stirred in a random manner by a paddle wheel attached to the end of the shaft. The work is transferred into the fluid by lowering the weight which causes paddle wheel to turn. Due to work transfer, the fluid in the system gets stirred. Since the volume of the fluid in the container remains constant, $\int P \cdot dV = 0$.

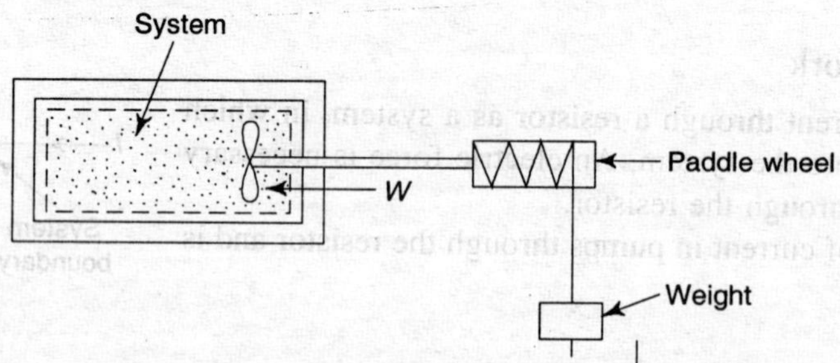

Fig. 2.7 Paddle work

Let m = mass of the weight lowered through the distance 'dz'.

T = Torque transmitted by the shaft, $d\theta$ = angle of rotation of the shaft.

Then work transfer into the fluid is given by $dW = mgdz = T \cdot d\theta$.
The total work transfer is

$$\int dW = W = \int_1^2 mgdz = \int_1^2 T \cdot d\theta.$$

2.6.4 Flow Work

The flow work is analogous to displacement work. The amount of work required to move the fluid in or out of the system is known as flow work. In an open system, it represents energy transfer across the boundary of the system. A blower, pump or compressor is used to send fluid across control volume.

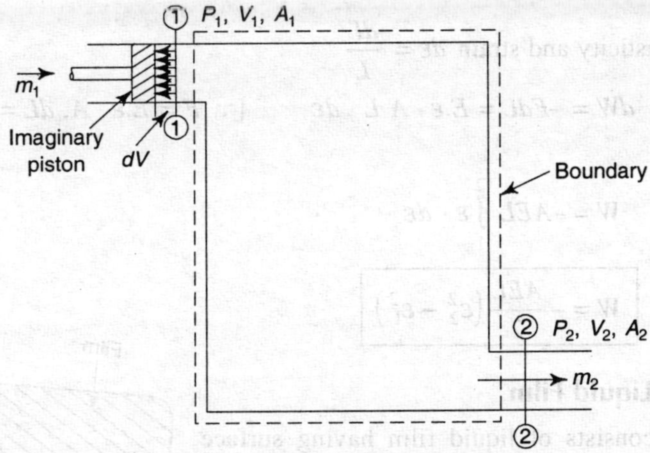

Fig. 2.8 Flow Work

Consider the work done by an imaginary piston on a fluid element in a fluid flow.
The work done on this imaginary piston by external pressure as the piston moves forward is

$$dW_{\text{flow}} = PdV.$$

where P–fluid pressure on the piston
dV–volume of fluid element which is ready to enter the system

$$\therefore \quad dW_{\text{flow}} = P \cdot Vdm \quad [dV = Vdm]$$

For Inlet section, section ①–①, we can write

$$(dW)_{\text{flow in}} = P_1 V_1\, dm_1.$$

Similarly for the fluid element leaving the system at section ②-②

$$(dw)_{\text{flow out}} = P_2 V_2\, dm_2.$$

$\therefore$ Per unit mass, flow work, $dW = P \cdot V$.
(for infinite final work)

$$\therefore \quad \boxed{dW = P \cdot V}$$

2.6.5 Work Done in Stretching a Wire

If we consider wire as a system, force of tension (F) and length 'L' are the two parameters used to describe the system. When external force of tension 'F' is applied to stretch the wire, its length changes from 'L' to '$L + dL$'. If the stretching process is carried out quasi-statically, the work involved is

$$dW = -F \cdot dL$$

The negative sign is used because of positive value of dL. i.e expansion of the wire. In this case work is to be done on the wire i.e. negative work for a finite change of length.

$$W = -\int_1^2 F \cdot dL$$

We know that, stress $\sigma = \dfrac{F}{A} = E \cdot \varepsilon \left[E = \dfrac{\sigma}{\varepsilon} \right]$

where E – modulus of elasticity and strain $d\varepsilon = \dfrac{dL}{L}$

$\therefore$ $$dW = -FdL = E.\varepsilon \cdot A\,L \cdot d\varepsilon \qquad [\because F = E.\varepsilon \cdot A,\ dL = L \cdot d\varepsilon]$$

After Integration

$$W = -AEL \int_1^2 \varepsilon \cdot d\varepsilon$$

$$\boxed{W = -\frac{AEL}{2}\left(\varepsilon_2^2 - \varepsilon_1^2\right)}$$

2.6.6 Stretching of a Liquid Film

Consider a system that consists of liquid film having surface tension 'τ' and is held between wire frame, one side of which can be moved as shown in figure. As the movable wire slides along the frame, the film area changes and the work is done on or by the film. If 'dA' be the change in area of the film, then the work done by the system is

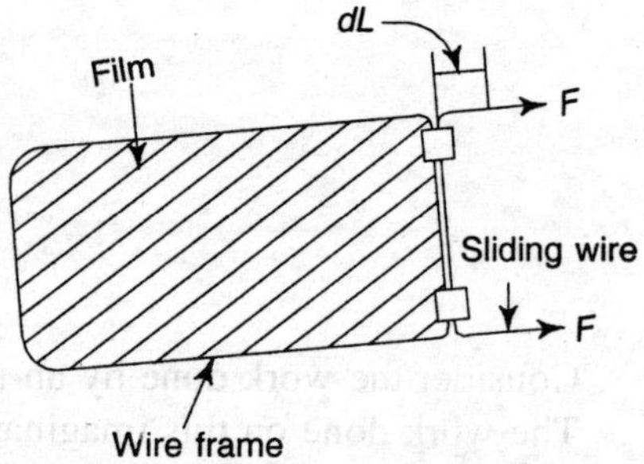

Fig. 2.9

$$dW = -\,\sigma \cdot dA$$

$\left[\sigma = \text{Surface tension } \dfrac{N}{m} \right]$

$$\boxed{W = -\int_1^2 \sigma \cdot dA}$$

2.6.7 Magnetic Work

For a magnetic substance, the system may be described by two parameters, external field 'H' and its magnetization 'μ'. If magnetization or demagnetization is a quasi-static process, then the magnetic work involved is

$$dW = -\,H \cdot d\mu$$

$$W = -\int_{l_1}^{l_2} H \cdot d\mu$$

Note: Work is equal to the integral of the product of an intensive property and change in it's related extensive property. The work done expressions are only valid for infinitesimally slow quasi-static processes.

2.7 HEAT TRANSFER

We have discussed several mode of work. But work is not the only mode of energy transfer. As a result of temperature difference, energy transfer takes place between one body and another or even within the body itself. This energy so transferred is called "Heat". Heat is a form of energy in transit and it occurs only at the boundary of the system (boundary phenomenon).

Heat is defined as the form of energy that is transferred across the boundary of the system without transfer of mass across the boundary at a given temperature to another system or the surroundings at a lower temperature by virtue of temperature difference between them. The temperature difference is the potential or force and heat transfer is the flux. It can also be written as the integral of product of intensive property '*T*' and differential change of an extensive property '*S*',

$$Q_{1-2} = \int_1^2 dQ = \int_1^2 Tds$$

Far example, consider a system say copper block at temperature '*Ts*'. If this block is kept in cold air and let '*Ta*' be the temperature of air. As temperature '*Ts*' is higher then '*Ta*', heat transfer take place from copper block to cold air by virtue of temperature difference, thus cools the block.

Heat always flows in the direction of decreasing temperature and for the heat transfer to occur, temperature gradient must exist.

The heat transfer not necessarily causes a temperature rise. When heat is added to an Ice and water mixture, the temperature of the mixture does not rise until all the ice has melted.

The temperature rise in a system need not be due to heat transfer alone, and may be caused by work transfer also. Heat is not a conserved quantity and is not a property of the system and hence it is a path function and '*dQ*' must be an inexact differential.

In an adiabatic process, there is no heat transfer between system and surroundings. The system interacts in terms of work only with its surroundings. The surface which is impermeable to heat flow is called adiabatic surface and the diathermic surface is one which allows flow of heat through it. The unit of heat is Joule in SI systems and the rate of heat transfer or work transfer is in Watts or Kilowatts.

2.8 MODES OF HEAT TRANSFER

Heat transfer take place by conduction, convection and thermal radiation or by the combined action of all these three modes.

1. *Conduction:* It is the mode of heat transfer in which matter providers passage for the heat flow. The energy exchange take place from high temperature region to low temperature region by the kinetic motion or direct impact of molecules (as in case of fluid at rest) and by the drift of electrons (as in the case of metals). When temperature gradient exists in a system comprising molecules of liquid, gas or solid, the higher energy level molecules imparts energy to adjacent molecules which are relatively at lower energy level. In metallic solids, the concentration of free electrons decides ability of the metallic alloys to conduct heat. The good conductors (pure metals like copper, silver, etc.) posses higher concentration of free electrons and heat is conducted through them due to drift of free electrons.

2. *Convection:* It is the process of energy transfer which is affected by circulation or mixing motion of fluid medium. It is mainly due to capacity of moving fluid to carry away the heat. If the fluid motion is setup by buoyancy effects resulting from the density variation caused by temperature gradient in the fluid, then flow of heat is said to be Natural convection. If the fluid motion is artificially created by means of an external agency like pump, blower or a fan, then flow of heat is said to be forced convection.

3. *Radiation:* If two bodies at different temperatures are placed in an evacuated adiabatic enclosure, so that they are not in contact through a solid or fluid medium, then the temperature of two bodies will tend to become equal. This mode of heat flow is said to be thermal radiation. The radiation heat transfer take place even without any medium through electromagnetic waves or quanta.

2.9 SIGN CONVENTION FOR HEAT TRANSFER

The symbol 'Q' is used to represent heat transfer in to or out of the system. Heat transfer in to the system is taken as 'Positive' and heat rejected from the system to the surroundings is taken as 'Negative'. That means when surroundings have higher temperature '$+Q$' is to be considered and when system is at higher temperature '$-Q$' is to be considered.

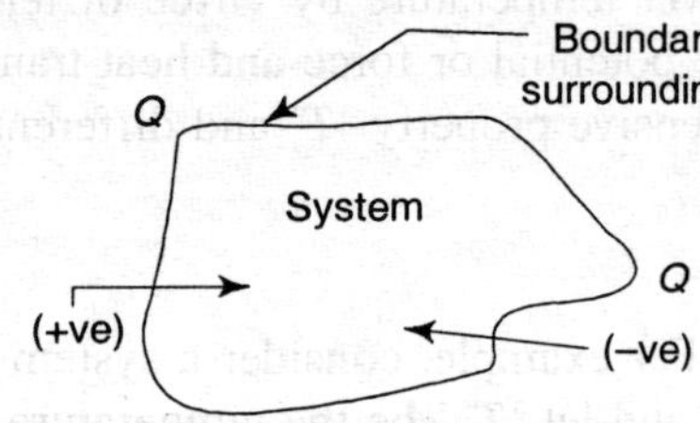

Fig. 2.10

In thermodynamics, the quantity of heat 'Q' is determined from the concept of entropy (s) which is an extensive property and can be plotted on X-co-ordinate on T-s diagram. This is similar to the term 'V' in the displacement work.

2.10 WORK AND HEAT-PATH FUNCTIONS

Consider a system which undergoes a process 1-2 along different quasi-static paths A, B and C as shown in figure 2.11.

The area under each path will give work involved in that process. The work involved in each process will be different (as area under each path is different) and it depends upon the path followed by the process, i.e. the magnitude of work does not depends on the end states 1 and 2 and is dependent on the path followed to reach the end states. Hence, work is a path function and 'dW' is an inexact or imperfect differential.

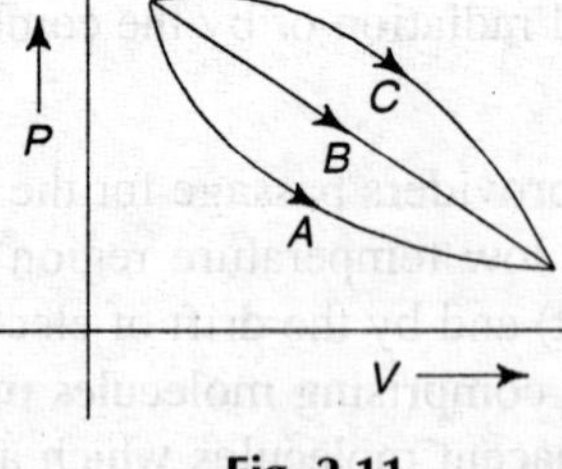

Fig. 2.11

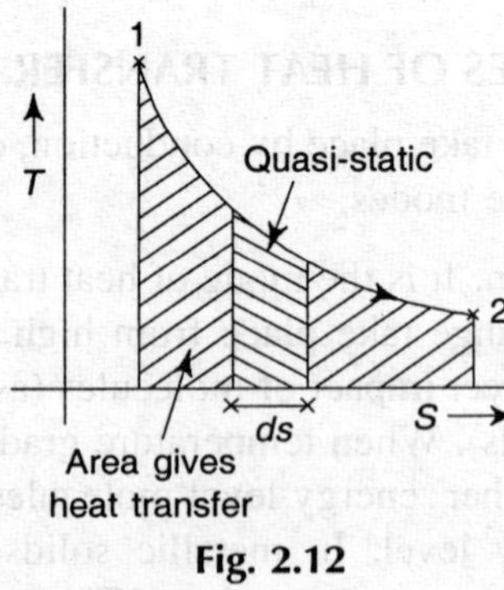

Fig. 2.12

Heat transfer is a path function. When the system changes from state '1' to state '2' the amount of heat transferred, depends on the intermediate states through which the system passes, i.e. it depends on the path which the system follows. Hence 'dQ' is an inexact differential.

$$\int_1^2 dQ = Q_{1-2} \quad \text{or} \quad {}_1Q_2.$$

2.11 COMPARISON OF HEAT AND WORK

1. Both heat and work are boundary phenomena. Both heat and work represents energy crossing system boundaries and are observed only at boundaries.
2. Both heat and work are transient phenomena. The systems never posses certain quantity of heat or work. When a process has taken place in a system, both heat and work crosses boundary of the system.
3. Both heat and work are path functions and inexact differentials. They can be written as integral of product of intensive property and differential change of an extensive property i.e. both heat and work are thermodynamic properties.

2.12 FORMULAE USED IN SOLVING PROBLEMS

(1) $W_{1-2} = {}_1W_2 = \int_{V_1}^{V_2} pdV$

(2) W = Electrical work = $\int_1^2 EI\, dt$, as $dt \to 0$, $W = EI$

(3) Shaft work = $W = T \cdot W$, where $W = \dfrac{d\theta}{dt}$ = angular velocity

(4) Paddle work = $W = \int_1^2 Td\theta$

(5) Flow work = $dW = PV$

(6) Work done in stretching a wire = $W = -\dfrac{AEL}{2}\left(\varepsilon_2^2 - \varepsilon_1^2\right)$

(7) Stretching of a liquid film = $W = -\int_1^2 \sigma dA$

(8) Magnetic work = $W = -\int_{l_1}^{l_2} H \cdot d\mu$

(9) $PV = mRT$

(10) $W = 2\pi NT$

(11) For sphere, $V = \dfrac{\pi D^3}{6}$

(12) $W = \dfrac{PmLAN}{60 \times 1000}$

2.13 NUMERICAL PROBLEMS

Problem 1: *Consider gas contained in a cylinder as a system, the initial pressure is being 210 kPa and the corresponding volume is 0.04 m³. Calculate the work done, when the volume of gas increases to 0.15 m³, (i) by heating at constant pressure, (ii) pressure varies inversely with volume, and (iii) the gas in the cylinder is considered to be ideal gas.*

Solution:

(i) For the constant pressure heat addition process 1–2,

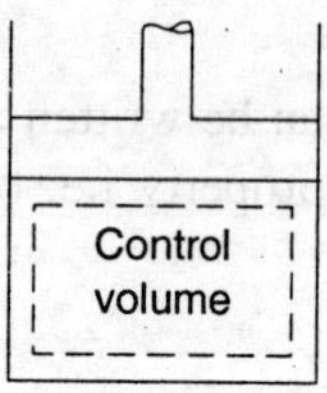

Fig. 2.13

$$W_{1-2} = \int_1^2 PdV$$

$$= P(V_2 - V_1)$$

$$W_{1-2} = 210\ (0.15 - 0.04)$$

$$\boxed{W_{1-2} = 23.1 \text{ kJ}} \quad \textit{Ans.}$$

$1Pa = 1\text{N/m}^2$

$\dfrac{kN}{m^2} \cdot \text{m}^3 = \text{KNm : kJ}$

(ii) Since $P = \dfrac{C}{V}$ i.e $P_1V_1 = P_2V_2 = C$

$$\therefore \quad P_2 = \frac{P_1V_1}{V_2} = \frac{210 \times 0.04}{0.15} = 56 \text{ kPa}$$

$$\therefore \quad W_{1\text{-}2} = \int_1^2 PdV = P_1V_1 \int_1^2 \frac{dV}{V} = P_1V_1 \ln \frac{V_2}{V_1}$$

$$\left[\because \quad PV = P_1V_1, \quad P = \frac{P_1V_1}{V}\right]$$

$$\therefore \quad W_{1\text{-}2} = 210 \times 0.04 \times \ln \frac{0.15}{0.04}$$

$$\boxed{W_{1\text{-}2} = 11.102} \quad \textit{Ans.}$$

(iii) For an ideal gas,

$$W_{1-2} = \int_1^2 PdV = \int_1^2 mRT\frac{dV}{V},\ P = .\frac{mRT}{V}$$

$$W_{1-2} = mRT \ln \frac{V_2}{V_1} = mRT \ln \frac{0.15}{0.04}$$

$$[PV : mRT,\ P_1V_1 = mRT_1]$$

$$w_{1-2} = \left(\frac{P_1V_1}{RT}\right) RT \ln 3.75$$

$$: P_1V_1 \ln 3.75$$

$$: 210 \times 0.04 \ln 3.75$$

$$\boxed{W_{1-2} : 11.10 \text{ kJ}} \quad \textit{Ans.}$$

Problem 2: *During one cycle the working fluid in an engine engages two work interactions 15 kg to the fluid and 44 kg from the fluid, and three heat interactions, two of which are known 75 kJ to the fluid and 40 kJ from the fluid. Evaluate the magnitude and direction of the third heat interaction.*

(VTU, Sept 2000)

Solution: wkt, from I[st] law of thermo dynquics

$$\oint \delta W = \oint \delta Q$$

$$W_1 + W_2 = Q_1 + Q_2 + Q_3$$

$$\oint \delta W = -15 + 44 = 29 \text{ kJ}$$

$$29 = 75 - 40 + Q_3$$

∴ Third heat interaction

$$\boxed{Q_3 = 6 \text{ kJ}}$$ *Ans.*

Problem 3: *A Vander walls gas is compressed reversibly at constant temperature from V_1 to V_2 (from initial to final specific volumes). The equation of state is given by the equation*

$$P = \frac{RT}{V-b} - \frac{a}{V^2}$$

Determine the work done per mol of the gas.

Solution: The given equation is

$$P = \frac{RT}{V-b} - \frac{a}{V^2}$$

We know that work done during the process is

$$W_{1-2} = \int_1^2 pdV = \int_1^2 \left(\frac{RT}{V-b} - \frac{a}{V^2}\right) dV$$

$$= \int_1^2 \frac{RT}{V-b} dV - \int_1^2 \frac{a}{V^2} dV$$

$$W_{1-2} = RT\,[\log_e (V-b)]_1^2 - a\left[\frac{1}{V}\right]_1^2$$

$$= RT\,\{\log_e (V_2 - b) - \log_e (V_1 - b)\} - a\left[\frac{1}{V_2} - \frac{1}{V_1}\right]$$

$$\boxed{W_{1-2} = RT \log_e \left(\frac{V_2 - b}{V_1 - b}\right) - a\left(\frac{1}{V_2} - \frac{1}{V_1}\right)}$$ *Ans.*

Problem 4: *For the system shown in figure 2.14, initial conditions of the gas are $V_1 = 0.15\ m^3$ and $P_1 = 205$ kPa. The atmospheric pressure is 100 kPa. The force exerted by the spring is proportional* *to the displacement from it's equilibrium position. The gas in the cylinder is heated until the volume is doubled and pressure of the gas is 600 kPa. Determine the work done by the gas.*

(VTU, Aug, 2001)

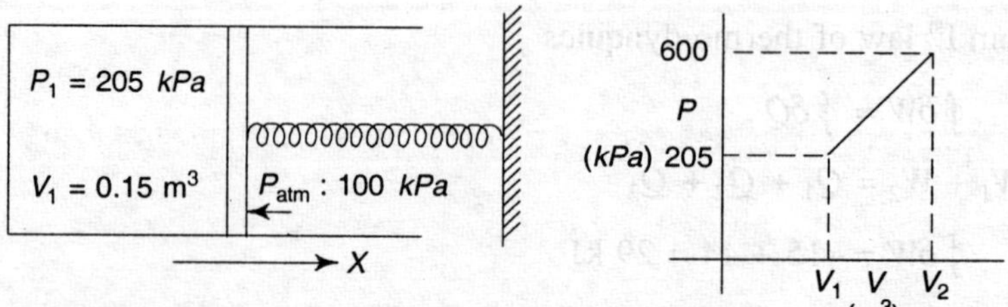

Fig. 2.14

Solution: Force balance at any instant gives

$$P \cdot A = P_{atm} A + KX = P_{atm} A + \frac{K}{A} (V - V_0)$$

$$\therefore \quad P = P_{atm} + \frac{K}{A^2} (V - V_0)$$

V_0 = gas volume when spring has natural length.

We know that, $$W = \int_1^2 PdV = \int_1^2 \left[P_{atm} + \frac{K}{A^2} (V - V_0) \right] dV$$

$$W_{1-2} = \int_1^2 P_{atm}\, dV + \int_1^2 \frac{K}{A^2} \cdot (V - V_0)\, dV.$$

$$= P_{atm} (V_2 - V_1) + \frac{K}{2A^2} [(V_2 - V_0)^2 - (V_1 - V_0)^2]$$

$$= P_{atm} (V_2 - V_1) + \frac{K}{2A^2} (V_2 - V_1) \{(V_2 - V_0) + (V_1 - V_0)\}$$

$$: (V_2 - V_1) \left[P_{atm} + \frac{(P_2 - P_{atm} + P_1 - P_{atm})}{2} \right] \quad \left[\because \frac{K}{A^2} (V - V_0) = P - P_{atm} \right]$$

$$= (V_2 - V_1) \frac{(P_1 + P_2)}{2}$$

$$= (2V_1 - V_1) \left(\frac{205 + 600}{2} \right) \quad : V_1 (402.5)$$

$$: 0.15 (402.5) \quad = \boxed{60.37 \text{ kJ}} \quad \textit{Ans.}$$

Problem 5: *A fluid is contained in a cylinder which has a stirring device turning at 8500 revolutions with an average torque of 1.75 Nm. The piston is frictionless and is held against fluid due to atmospheric pressure equal to 100 kPa. The piston diameter is 50 cms and if it moves outward by 65 cms, determine net work transfer for the system.*

Solution:

Consider fluid contained in the cylinder as a system.

Let W_1 be the work done by the stirrer on the system.

$$W_1 = 2\pi NT.$$
$$= 2\pi \times 8500 \times 1.75 \text{ Nm}$$
$$= 93.41 \text{ kJ}$$
$$\boxed{W_1 = -93.41}$$

(Negative sign is due to work done on the system)

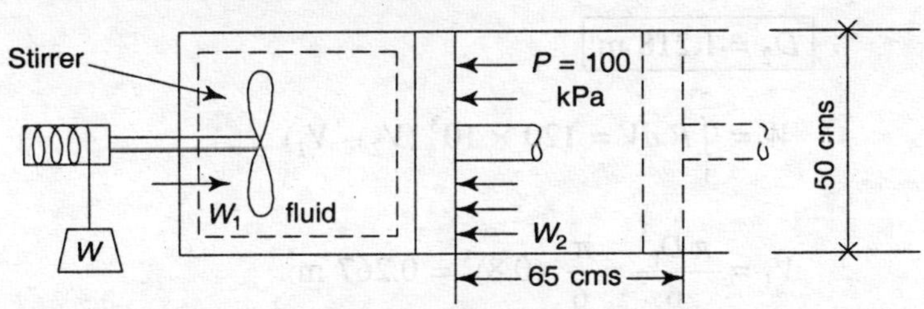

Fig. 2.15

Let W_2 be the displacement work done by the system due to piston movement.

$$W_2 = P.dV \qquad [1 \text{ Pascal} = 1\text{N/m}^2]$$
$$= PAL \qquad \left[\left(\frac{kN}{m^2}\right) \times m^2 \times m\right]$$
$$= 100 \times \frac{\pi}{4}(0.5)^2 \times 0.65 \qquad : kNm = \text{kJ}$$
$$\boxed{W_2 = +12.75 \text{ kJ}}$$

[Positive sign is due to work done by the system]

$\therefore$ Net work transfer for the system = W

$$W = W_1 + W_2 = -93.41 + 12.75$$
$$\boxed{W = -80.66 \text{ kJ}}$$

Problem 6: *A spherical ballon of 0.8 m diameter contains gas at 120 kPa. The gas in the ballon is heated until the pressure reaches 425 kPa, during which gas pressure is proportional to the cube of diameter of the ballon. Determine work done by the gas inside the ballon.*

Solution: When diameter of ballon $D_1 = 0.8$ m,

gas pressure $P_1 = 120$ kPa.

Also it is given that $P \alpha D^3$.

$$P = c\,D^3 \qquad [\therefore \; c = \text{constant}]$$

$\therefore$ $$P_1 = c.\,D_1^{\,3}.$$

$$120 \times 10^3 = c.(0.8)^3 \qquad \left[c : \frac{\text{Pascal}}{\text{m}^3}\right]$$

$$c = \frac{120 \times 10^3}{0.8^3}$$

$$c = 234.37 \times 10^3 = 234.37\text{kPa/m}^3$$

Now $$P_2 = c\,D_2^{\,3}$$

$$425 \times 10^3 = 234.37 \times 10^3.\, D_2^{\,3}$$

$$D_2 = \left(\frac{425 \times 10^3}{234.37 \times 10^3}\right)^{1/3}$$

$$= (1.81)^{0.33}$$

$$\boxed{D_2 = 1.218 \text{ m}}$$

We know that $$W = \int_1^2 P.dV = 120 \times 10^3\,(V_2 - V_1)$$

$$V_1 = \frac{\pi D_1^{\,3}}{6} = \frac{\pi}{6}\,(0.8)^3 = 0.267 \text{ m}^3$$

$$V_2 = \frac{\pi D_2^3}{6} = \frac{\pi}{6}\,(1.218)^3 : 0.946 \text{ m}^3$$

$\therefore$ $$W = 120 \times 10^3\,(0.946 - 0.267): 81.48 \text{ kJ}$$ ***Ans.***

Problem 7: *A spherical ballon 1 m diameter contain gas at 240 kPa and 290 K. When the gas inside the ballon is heated, its pressure raises to 485 kPa and pressure is being proportional to the diameter. Determine the work done by the gas inside the ballon.*

Solution: The given condition is

$$P \propto D$$

$$P = CD \qquad [C = \text{constant}]$$

$$240 = C(1)$$

$$C = 240 \quad \text{kPa/m}$$

Also $$P_2 = C \cdot D_2$$

$$485 = C \cdot D_2$$

$$485 = 240\, D_2$$

$$\boxed{D_2 = 2.02 \text{ m}}$$

For a spherical ballon

$$V = \frac{\pi}{6}\, D^3; \quad dV = \frac{\pi D^2}{2}\, dD$$

$\therefore$ Work done $$W_{1-2} = \int_1^2 PdV$$

$$= CD \cdot \int_1^2 dV = \int_1^2 CD \cdot \frac{\pi D^2}{2} dD$$

$$= \frac{C\pi}{8}\left(D_2^4 - D_1^4\right)$$

$$= \frac{240 \times \pi}{8} (2.02^4 - 1^4)$$

$$\boxed{W_{1-2} = 1474.198 \text{ kJ}}$$

Problem 8: *0.115 m^3 of gas has a pressure of 138 kN/m^2. It is compressed to 680 kN/m^2 according to the law $PV^{1.4} = C$. Determine the volume of gas after compression.*

Solution: Since the compression process follows the law $PV^{1.4} = C$.

We can write

$$P_1V_1^{1.4} = P_2V_2^{1.4} = C.$$

$$\left(\frac{P_1}{P_2}\right) = \left(\frac{V_2}{V_1}\right)^{1.4}$$

or

$$V_2 = V_1 \left(\frac{P_1}{P_2}\right)^{1/1.4}$$

$$= 0.115 \left(\frac{138}{680}\right)^{0.714}$$

$$\boxed{V_2 = 0.0368 \text{ m}^3}$$

Ans.

Problem 9: *A flexible ballon is to be filled with air from a storage bottle till the volume reaches 0.8 m^3. The atmospheric pressure is 1.013 × 10^5 N/m^2. Determine work done by the system comprising the air initially in the bottle.*

Solution: By considering ballon fabric light S_1–Initial system boundary S_2-final system boundary.

The displacement work for the bottle surface is zero, as it is rigid.

$$\therefore \quad W = \underset{\text{Ballon}}{\int P \cdot dV} + \underset{\text{bottle}}{\int pdV}$$

$$= P_{atm} \int dV$$

$$= 1.013 \times 10^5 \times (0.8\text{–}0)$$

$$= 81040 \text{ Nm}$$

Ans.

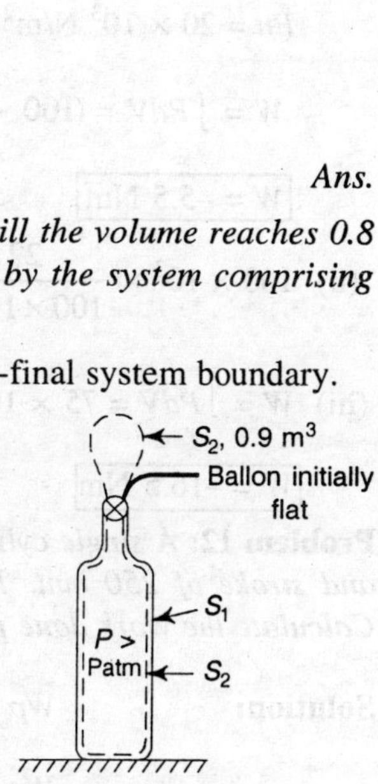

Fig. 2.16

Problem 10: *Determine the work done by the air which enters an evacuated bottle from the atmosphere. Consider the volume of air entered as 0.3 m^3 and atmospheric pressure is 1.013 × 10^5 $\frac{N}{m^2}$.*

Solution:

S_1— Initial system boundary
S_2—Final system boundary

$$W = \int_{\text{free air boundary}} PdV + \int_{\text{bottle}} PdV$$

$$= P_{atm} \int dV + 0$$

$$= 1.013 \times 10^5 (0 - 0.3)$$

$$= -30390 \text{ Nm} \qquad \textit{Ans.}$$

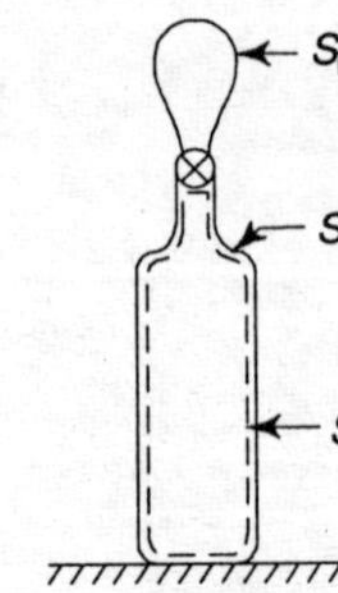

Fig. 2.17

$\int dV$ for free air boundary is negative, as it is contracting. The surroundings do positive work at the boundary and work done by the system is negative.

Problem 11: *A friction less leak proof piston of area 22 cm² closes a vertical cylinder which contains air at 75 × 10³ N/m². Atmospheric pressure of 100 × 10³ N/m² is acting on the lower side of the piston. A rigid vessel having nitrogen is connected to the cylinder and vessel is at a pressure of 20 × 10³ N/m² in the initial stage. Calculate the work done when the piston moves slowly through a distance of 10 cms from it's initial position for the following systems: (i) the piston, (ii) the atmosphere, (iii) the air plus Nitrogen.*

Solution:

(i) $A = 22 \text{ cm}^2$, $P = 75 \times 10^3 \text{ N/m}^2$

$l = 10$ cms, $P_{atm} = 100 \times 10^3 \text{ N/m}^2$

$Pn = 20 \times 10^3 \text{ N/m}^2$

$$W = \int PdV = (100 - 75) \times 10^3 \left[\frac{22}{100 \times 100} \times \frac{10}{100}\right]$$

$\boxed{W = -5.5 \text{ Nm}}$ *Ans.*

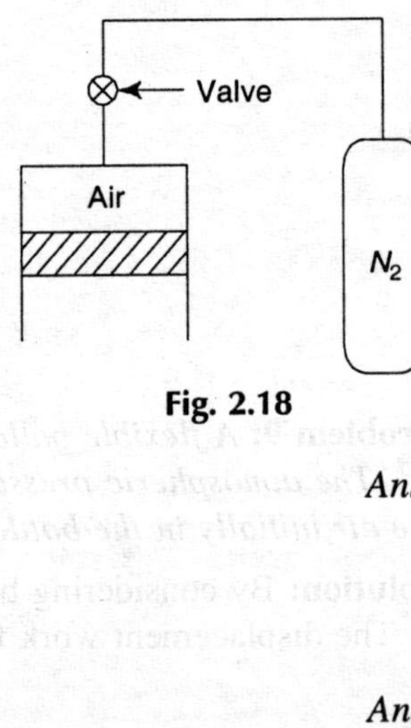

Fig. 2.18

(ii) $100 \times 10^3 \times \dfrac{22 \times 10}{100 \times 100 \times 100}$: $\boxed{22 \text{ Nm}}$ *Ans.*

(iii) $W = \int PdV = 75 \times 10^3 \times \dfrac{22 \times 10}{100 \times 100 \times 100}$

$\boxed{W = -16.5 \text{ Nm}}$ *Ans.*

Problem 12: *A single cylinder, double acting, 2 stroke oil engine has a cylinder diameter of 205 mm and stroke of 250 mm. The engine rpm is 300 and mean effective pressure is 900 × 10³ N/m². Calculate the work done per cycle and the indicated power.*

Solution: $Wp = \dfrac{PmLAN}{60} = 900 \times 10^3 \times 0.25 \times \dfrac{\pi}{4} (0.205)^2 \times \dfrac{300}{60}$

$Wp = 37132.15$ Nm/s.

$= 37.132$ kW

For double acting multiply by 2

$W = 37.137 \times 2 = 74.26$ kW.

Problem 13: *The pressure-volume relation for a non-flow reversible process is P = (8 – 4V) bar, where V is in m³. If 130 kJ of work is supplied to the system, Calculate final pressure and volume of the system. Take initial volume = 0.5 m³.*

Solution: Given $P = (8 - 4V) \text{ bar} = (8 - 4V) \times 10^5 \text{ N/m}^2$

$$W_{1-2} = + PdV = -\int_1^2 (8 - 4V) \times 10^5\, dV.$$

$$= 10^5 \left| 8V - \frac{4V^2}{2} \right|_{0.5}^{V_2}$$

$$= 10^5 [8 (V_2 - 0.5) - 2 (V_2^2 - 0.5^2)]$$

$$= 10^5 \{8V_2 - 4 - 2V_2^2 - 0.5\}$$

$$= -10^5 \{+2V_2^2 - 8 V_2 + 4.5\}$$

Since work is supplied to the system, it is negative

$$\therefore \quad -130 \times 10^3 = -10^5 [2V_2^2 - 8V_2 + 4.5]$$

$$2V_2^2 - 8V_2 + 3.2 = 0$$

$$V_2 = -\frac{b \pm \sqrt{b^2 - 4ac}}{2a}$$

$$= \frac{8 \pm \sqrt{8^2 - 4(2)(3.2)}}{2 \times 2} = \frac{+8 \pm 6.196}{4} = 3.549 \text{ or } 0.451$$

$$\boxed{V_2 = 0.451 \text{ m}^3}$$

∴ Final pressure $P_2 = 8 - 4V_2$

$$= 8 - 4 \times 0.451$$

$$\boxed{P_2 : 6.196 \text{ bar}}$$ ***Ans.***

Problem 14: *During a process in a system, the specific heat is given by the equation*

C = (0.2 + 0.002 T) kJ/kgK

where T = absolute temperature in kelvin.

Determine the quantity of heat required to raise temperature of system from 310 k to 415 k. Also determine the mean value of specific heat. Take mass of the system as 1.5 kg

Solution: We know that,

$$Q = m \int_1^2 cdT = 1.5 \int_{310}^{415} (0.2 + 0.002T)\, dT$$

$$= 1.5 \left| 0.2T + \frac{0.002T^2}{2} \right|_{310}^{415}$$

$$= 1.5 \{0.2 (415 - 300) + 0.001 (415^2 - 310^2)\}$$

$$= 148.68 \text{ kJ}$$

also $$Q = m\,Cm\;(T_2 - T_1);$$

$$148 = 1.5\,Cm\,(415 - 310) \quad \therefore \quad Cm = 0.939 \text{ kJ/kg k}$$

Problem 15: *State whether the heat and work are positive, negative or zero in the following cases. The system to be considered is shown underlined.*

(a) *1 kg of* **gas** *contained in an insulated cylinder expands moving the piston slowly outwards.*

Solution: As cylinder is insulated $Q = 0$.
As piston moves outwards, the gas does work on the system.

(b) *An electric current flows steadily through a* **resistor** *which is immersed in running water.*

Solution: Q is negative as heat is lost to water work is done on the resistor, hence work is negative.

(c) ***Liquid in a non conducting vessel*** *comes to rest from an initial state of turbulent motion*

Solution: Heat is contained in the vessel, $Q = 0$ system boundary remains constant $W = 0$.

(d) ***One kg of air*** *flows adiabatically from the atmosphere in to an evacuated bottle through a valve.*

Solution: As air flow is adiabatic Q is zero. Since the air boundary is contracting, W is negative.

Problem 16: *Classify the following as heat and work interactions. The system to be considered is shown underlined.*

(a) *A* **honeybee** *sucking nector from a flower*

Solution: Work interaction

(b) *Baking of* **bread** *in an oven*

Solution: Heat interaction

(c) *A person drawing money from his* ***ATM cash counter***

Solution: Work interaction

(d) *Rising of mercury column in a thermometer placed in the mouth of a* ***patient***

Solution: Heat interaction

2.14 REVIEW QUESTIONS

1. Give thermodynamic definition of work with examples.
2. Show that work is a path function and is not a property.
3. Define displacement work.
4. Differentiate between path function and point function.
5. Explain briefly how displacement work can be measured in the following cases:
   (i) Electrical work
   (ii) Stretching of a wire
   (iii) Flow work and shaft work
   (iv) Magnetic work.
6. Heat may not cause temperature rise, explain the statement with suitable examples.
7. Show that heat is a path function
8. Explain the different modes of heat transfer.
9. Differentiate between heat and work.

CHAPTER 3

First Law of Thermodynamics

3.1 INTRODUCTION

The first law of thermodynamics has been stated as the conservation of energy principle. The law of conservation of energy states that the total energy of an isolated system, in all it's form remains constant (energy can neither be created nor be destroyed). Heat and work are different forms of same entity called energy. The system may receive energy in the form of heat and energy may leave the system in the form of work or system may absorb work and reject the energy in the form of heat. For example, consider a bicycle wheel when it is paddled very fast, it starts rotating very fast. When the brakes are applied, the wheel gets stopped and the brake shoe temperature has been increased. That is work is done on the system in the form of paddle work to rotate the wheel. By applying the brakes, the wheel may be brought to rest and kinetic energy of wheel is converted in to heat energy which is being dissipated through the brake shoes and brake drum and thus heating up the brake shoe.

The first law is applied to a closed system first and then the concept is extended to an open system.

3.2 FIRST LAW FOR A CLOSED SYSTEM UNDERGOING A CYCLE

Statement: "Whenever a closed system executes a cyclic process or cyclic change, the algebraic sum of work transfer is proportional to the algebraic sum of heat transfer" *i.e.* algebraic summation of all energy transfer across system boundaries is zero.

The another form of statement is "whenever a thermodynamic system undergoes a cyclic change, both heat and work are mutually convertable one in to another".

At the end of a cycle, system experiences no net change, so algebraic sum of heat and work during a cycle is zero.

Suppose a system operate in a cycle consisting of processes 1-2, 2-3, 3-4 and 4-1 as shown in figure 3.1.

We can write

$$Q_{1\text{-}2} = E_2 - E_1 + W_{1\text{-}2}$$
$$Q_{2\text{-}3} = E_3 - E_2 + W_{2\text{-}3}$$
$$Q_{3\text{-}4} = E_4 - E_3 + W_{3\text{-}4}$$
$$Q_{4\text{-}1} = E_1 - E_4 + W_{4\text{-}1}$$

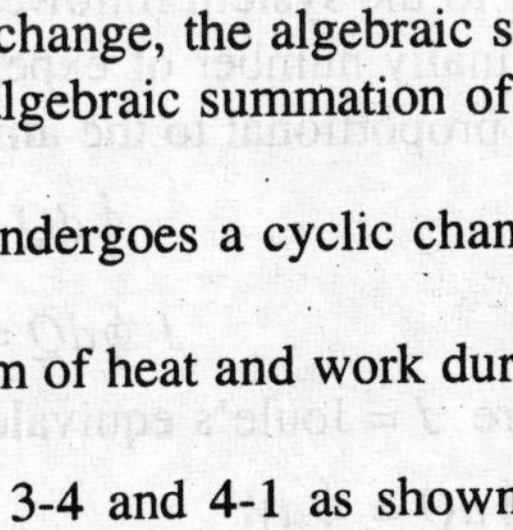

Fig. 3.1 A cycle

for the cycle,

$$Q_{12} + Q_{23} + Q_{34} + Q_{41} = W_{12} + W_{23} + W_{34} + W_{41}$$
$$\oint dQ = \oint dW$$

where $\oint$ means summation around the cycle.

i.e. In a work producing cycle the net heat addition is equal to net work output and in a work absorbing cycle the net heat rejection is equal to the net work input.

Illustration: Joule's Experiment

Consider a closed system which consists of known mass of water contained in an adiabatic vessel having thermometer and a paddle wheel as shown in figure 3.2. The water in the vessel is stirred by paddle wheel. The work of paddle wheel is obtained by a falling weight and let it be $W_{1\text{-}2}$.

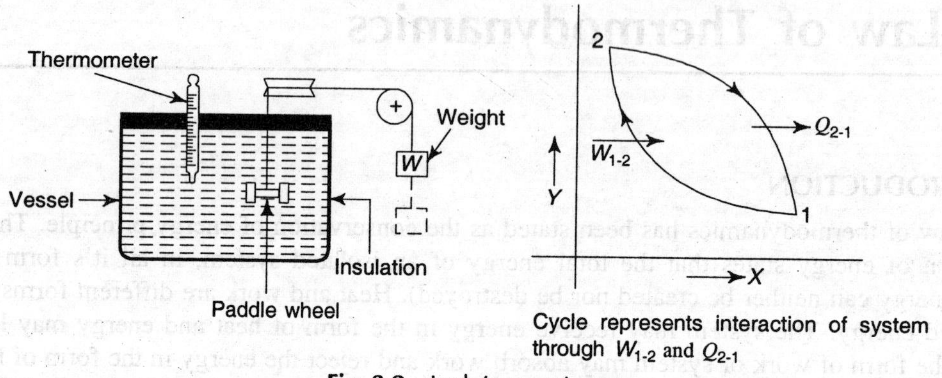

Fig. 3.2 Joule's experiment

Considering water in the vessel as a system, due to addition of paddle work, system temperature will rise. The amount of work done can be measured by the product of weight and the vertical distance through which the weight descends. Now the entire system is immersed in a water bath. The system and surroundings (water bath) interact by heat transfer, till the system returns to its original temperature *i.e.* it is in thermal equilibrium with the surroundings. The amount of heat transfer $Q_{2\text{-}1}$ can be measured. The system thus executes a cycle which consists of definite amount of work input $W_{1\text{-}2}$ to the system followed by the transfer of an amount of heat $Q_{2\text{-}1}$ from the system.

Finally number of experiments gave a conclusion that the amount of work supplied to the system was proportional to the amount of heat removed from the system in water bath.

i.e $\oint dQ \propto \oint dW$

$$J \oint dQ = \oint dW$$

where J = Joule's equivalent or mechanical equivalent of heat,

$$\oint dQ = \oint dW$$

when both work and heat transfer are in the same units.

3.3 FIRST LAW FOR A CLOSED SYSTEM UNDERGOING A CHANGE OF STATE

We know that $(\Sigma W)_{cycle} = (\Sigma Q)_{cycle}$ is applicable only to the systems which execute cycles. When we consider first law of thermodynamics for a system that undergoes a change of state which involves both heat and work interactions, the net energy transfer will be accumulated within the system. Consider a system shown in figure.

Let Q = Amount of heat transferred to the system
W = Work done on the system
$Q\text{-}W$ = Energy stored in the system

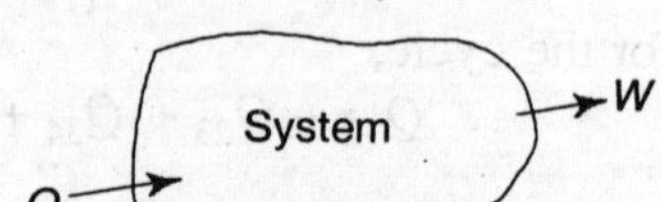

System with heat and work transfer during a process

Fig. 3.3

Energy in storage is neither heat nor work, but energy stored in the system is termed as "Internal energy".

$\therefore \quad Q - W = \Delta E \quad [\Delta E = \text{change in internal energy} = E_2 - E_1]$

$\therefore \quad Q = \Delta E + W$

3.4 ENERGY–A PROPERTY OF THE SYSTEM

Consider a system that executes a cycle and changes it's state from 1 to 2 by following path '*A*' and change of state from 2-1 follows the path 'B' to complete a cycle.

From first law of thermodynamics, we can write

$$\oint dQ = \oint dW$$

$$\int_{1A}^{2A} dQ + \int_{2B}^{1B} dQ = \int_{1A}^{2A} dW + \int_{2B}^{1B} dW \qquad (1)$$

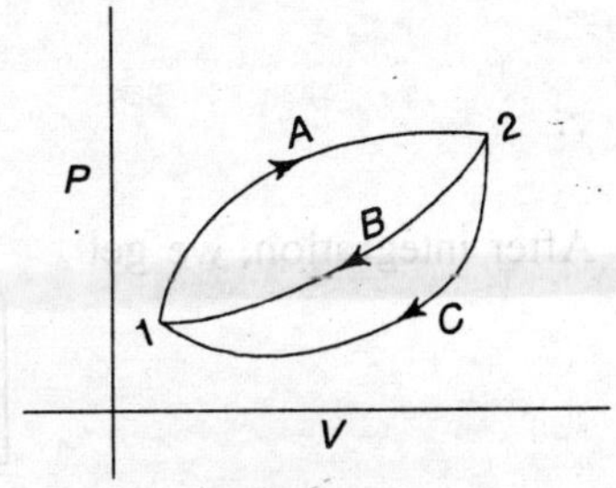

Fig. 3.4

Now consider another cycle in which system changes it's state from 1 to 2 by following path '*A*' and change of state 2-1 follows the path '*C*'.

Apply first law of thermodynamics, we can write

$$\oint dQ = \oint dW$$

$$\int_{1A}^{2A} dQ + \int_{2C}^{1C} dQ = \int_{1A}^{2A} dW + \int_{2C}^{1C} dW \qquad (2)$$

(2) – (1) gives,

$$\int_{2C}^{1C} dQ - \int_{2B}^{1B} dQ = \int_{2C}^{1C} dW - \int_{2B}^{1B} dW$$

$$\therefore \quad \boxed{\int_{2C}^{1C} dQ - dW = \int_{2B}^{1B} dQ - dW}$$

So we can conclude that the quantity ($dQ - dW$) is same for all the processes between the states 1 and 2. This value depends only on the initial and final states and not on the path followed between the two states. It is a point function and hence property of the system.

i.e $\quad dQ - dW = dE \quad$ (where dE = differential of property of the system *i.e.* internal energy)

$$\therefore \quad dQ = dW + dE \qquad (3)$$

After integration

$$\int_P^2 dQ = \int_P^2 dW + \int_P^2 dE$$

$$\boxed{Q_{1\text{-}2} = W_{1\text{-}2} + E_2 - E_1}$$

The energy '*E*' has a definite value and may be present in various forms, potential, kinetic etc.,

i.e. $\quad E$ = Internal energy + Kinetic energy + Potential energy

$$E = U + KE + PE$$

$$dE = dU + d(KE) + d(PE) \qquad (4)$$

Substitute (4) in Eqn. (3), we get

$$dQ = dW + dU + d(KE) + d(PE)$$

Expressions for *KE* and *PE*

$$KE = \frac{1}{2}\ mV^2/g_c$$

$$PE = \frac{mg}{g_c} \cdot 2$$

$$\therefore \quad dQ = dW + dU + d\left(\frac{mv^2}{2g_c}\right) + d\left(\frac{mg}{g_c} \cdot 2\right)$$

After integration, we get

$$\boxed{Q_{1\text{-}2} = W_{1\text{-}2} + (U_2 - U_1) + \frac{m\left(v_2^2 - v_1^2\right)}{2g_c} + \frac{mg}{g_c}\ (z_2 - z_1)}$$

3.5 INTERNAL ENERGY-A THERMODYNAMIC PROPERTY

Internal energy is associated with mass of the system and hence is an extensive property. The change in the value of internal energy is the algebraic sum of heat supply and work done.

$$\Delta U = Q - W$$

$$U_2 - U_1 = \int_1^2 \delta Q - \delta W$$

where U–Internal energy of a given mass of a substance
u–Internal energy per unit mass (specific internal energy)

3.6 EXPRESSIONS FOR KINETIC ENERGY AND POTENTIAL ENERGY

Both Potential and Kinetic energy of a system are extensive properties.

Consider a system which is initially at rest. Let '*F*' be the horizontal external force acted upon the system and moves it through a distance dx in the direction of force. Thus, there is no change in potential energy and heat transfer and change in internal energy can be neglected.

Thus from the first law,

$$\delta W = -Fdx = -dKE$$

But, $$F = \frac{ma}{g_c}$$

$$= m \cdot \frac{dV}{dt} = \frac{m}{g_c} \cdot \frac{dx}{dt} \cdot \frac{dV}{dx}$$

$$= \frac{m}{g_c} \cdot V \cdot \frac{dV}{dx}$$

$$\therefore \quad dKE = Fdx = \frac{mV}{g_c}\ dV$$

On integration, we get

$$\int_0^{KE} dKE = \int_0^{V} \frac{mV}{g_c}\, dV$$

$$\therefore \qquad \boxed{KE = \frac{mV^2}{2g_c}}$$

Now consider a system which is initially at rest and at some elevation from a reference. Let '*F*' be the vertical force acted upon the system and raises it by an amount '*dz*'. Let 'g' be the acceleration due to gravity at this point. Then

$$\delta W = -Fdz = -dPE$$

$$F = \frac{ma}{g_c} = \frac{mg}{g_c}$$

$$\therefore \qquad dPE = Fdz = \frac{mg}{g_c} \cdot dz$$

On integration, we get,

$$\int_{PE_1}^{PE_2} dPE = \frac{m}{g_c}\, g \int_{z_1}^{z_2} dz$$

$$\therefore \qquad \boxed{PE_2 - PE_1 = \frac{mg}{g_c}\,(z_2 - z_1)}$$

(Assuming 'g' constant)

$$\therefore \qquad dE = dU + d(KE) + d(PE)$$

$$\boxed{dE = dU + d\left(\frac{mVdV}{g_c}\right) + d[\mathrm{mg}\,(z_2 - z_1)]}$$

3.7 ENTHALPY (H)

It is an extensive property and denoted by 'H'. The sum of internal energy and pressure-volume product is known as enthalpy.

$$H = U + pV$$

on unit mass basis

$$h = u + pv, \quad \text{where} \quad h = \text{specific enthalpy}$$

or

$$dh = du + pdv + vdp$$

consider only '*pdv*' work, then

$$dh = du + pdv \qquad \text{[for constant pressure process } vdp = 0]$$

$$\boxed{dh = dQp = \text{const}}$$

where

$$dQ = du + pdv \text{ (from Ist law)}$$

3.8 APPLICATION OF FIRST LAW TO A CLOSED SYSTEM (NON FLOW SYSTEM)

According to first law of thermodynamics for a closed system, if the changes in kinetic energy and potential energy are negligible, the law can be reduced to the form

$$dU = dQ - dW$$

when *KE* and *PE* becomes zero, the work done will always be displacement work.

Most of the processes that are normally encountered in thermodynamic analysis of systems, can be identified as one or combination of some of the following processes:

(1) Constant volume (Isochoric) process
(2) Constant pressure (Isobaric) process
(3) Constant temperature (Isothermal) process
(4) Adiabatic process
[irreversible adiabatic, reversible adiabatic or isentropic process]
(5) Polytropic process

3.8.1 Constant Volume Process

Consider a gas confined in a rigid vessel of volume *V* as the system. For the rigid walls, the expansion work done (*pdv*) by the system is zero. During execution of process (heating or cooling), both pressure and temperatures are the variables and volume remains constant. The process is defined by Charle's law.

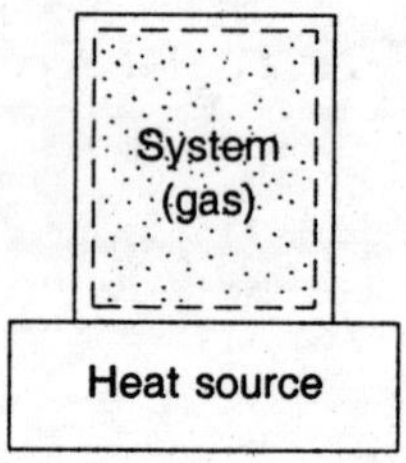

(a) System containing gas

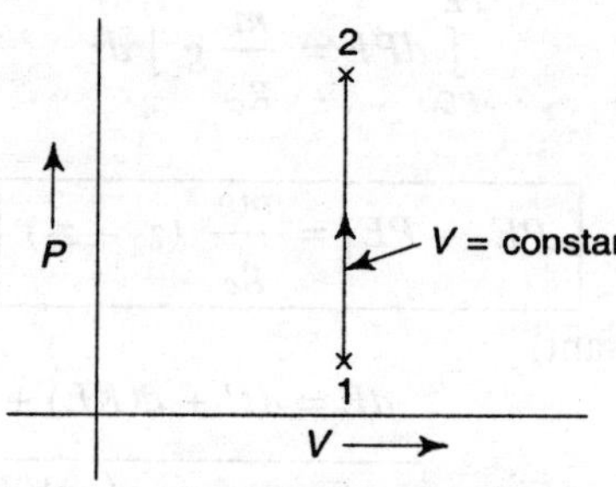

(b) P-V diagram

Fig. 3.5

At state 1, $P_1V_1 = mRT_1$ (1)

At state 2, $P_2V_2 = mRT_2$ (2)

½ gives

$$\frac{P_1}{P_2} = \frac{T_1}{T_2} \quad [V_1 = V_2]$$

Work done

From the Ist law, we can write

$$\text{word done} = W_{1\text{-}2} = \int_1^2 pdv = \int_1^2 p \cdot 0 = 0$$

$$\boxed{W_{1\text{-}2} = 0}$$

For a constant volume process in a closed system, work done is zero.

Heat transfer

When work done is zero, the heat interaction is equal to change in internal energy of the system. From Ist law,

$$Q_{1\text{-}2} = (U_2 - U_1) + W_{1\text{-}2}$$
$$= m(u_2 - u_1) \qquad [w_{1-2} = 0]$$

$\therefore \quad dQ = du = C_v\, dT$

$\therefore \quad C_v = \left.\dfrac{du}{dT}\right|_{v = \text{constant}}$

where C_v = specific heat at constant volume

or $dQ = du = C_v \cdot dT$

$$\int_1^2 dQ = \int_1^2 du = \int_1^2 C_v\, dT$$

$\therefore \quad u_2 - u_1 = C_v\,(T_2 - T_1)$, where $C_v = f(T)$

3.8.2 The Constant Pressure Process

Consider a gas enclosed in a cylinder as a system. The gas pressure in the cylinder is maintained constant by adding weight on the piston. The constant pressure process is defined by Boyle's law. Now, neglect friction between cylinder and piston. The volume and temperature will become the variables.

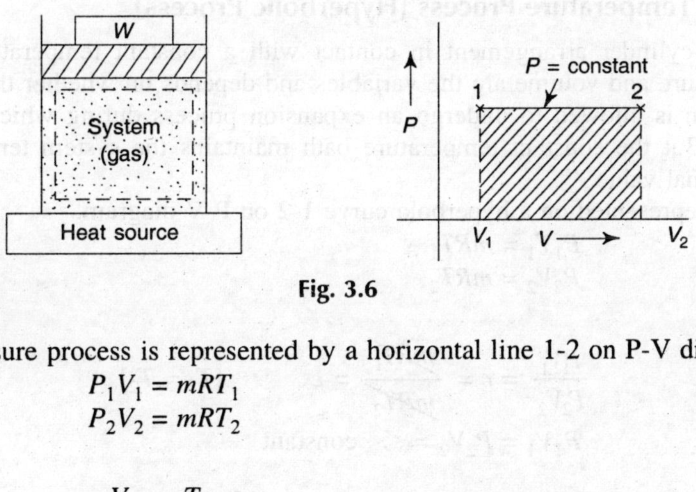

Fig. 3.6

The constant pressure process is represented by a horizontal line 1-2 on P-V diagram.

At state 1, $\quad P_1V_1 = mRT_1 \qquad (1)$

At state 2, $\quad P_2V_2 = mRT_2 \qquad (2)$

½ gives

$$\frac{V_1}{V_2} = \frac{T_1}{T_2} \qquad [\because\; P_1 = P_2]$$

Work done

The area under 1-2, on P-V diagram (hatched portion) gives work done during that process.

From Ist law,

$$W_{1\text{-}2} = \int_{V_1}^{V_2} pdv$$

$$= P \cdot (V_2 - V_1)$$

$$\boxed{W_{1\text{-}2} = P \cdot (V_2 - V_1)}$$

Heat transfer

For a closed system, Ist law can be written as

$$\begin{aligned} Q_{1\text{-}2} &= W_{1\text{-}2} + \Delta U \\ &= P(V_2 - V_1) + (U_2 - U_1) \\ &= (U_2 + P_2V_2) - (U_1 + P_1V_1) \\ Q_{1\text{-}2} &= H_2 - H_1 \quad \because \quad [H = U + PV) \end{aligned}$$

For a perfect gas,

$$Q_{1\text{-}2} = mcp\,(T_2 - T_1)$$

$$dh = dQ = C_p \cdot dT$$

$$\therefore \quad C_p = \left.\frac{dh}{dT}\right|_{P = \text{constant}}$$

or

$$dh = C_p \cdot dt$$

$$\int_1^2 dh = \int_1^2 C_p\, dT$$

i.e.

$$h_2 - h_1 = C_p\,(T_2 - T_1)$$

C_p = specific heat at constant pressure.

$$\begin{aligned} Q &= du + W \\ &= du + pdv \\ dQ &= du + d(PV) \\ dQ &= d(u + PV) \\ dQ &= dh \end{aligned}$$

3.8.3 Constant Temperature Process [Hyperbolic Process]

Consider a piston-cylinder arrangement in contact with a constant temperature bath. During this process, both pressure and volume are the variables and depends on whether the system is heated or cooled. The system is allowed to undergo an expansion process during which system temperature decreases by dT. But the constant temperature bath maintains the system temperature constant by bringing it to original value.

The process is represented on a hyperbola curve 1-2 on P-V diagram.

At 1, $\quad P_1V_1 = mRT_1 \quad$ (1)

At 2, $\quad P_2V_2 = mRT_2 \quad$ (2)

(1)/(2) gives

$$\frac{P_1V_1}{P_2V_2} = c = \frac{mRT_1}{mRT_2} = c \quad \because \quad [T_1 = T_2]$$

$$\therefore \quad P_1V_1 = P_2V_2 = \ldots \text{ constant}$$

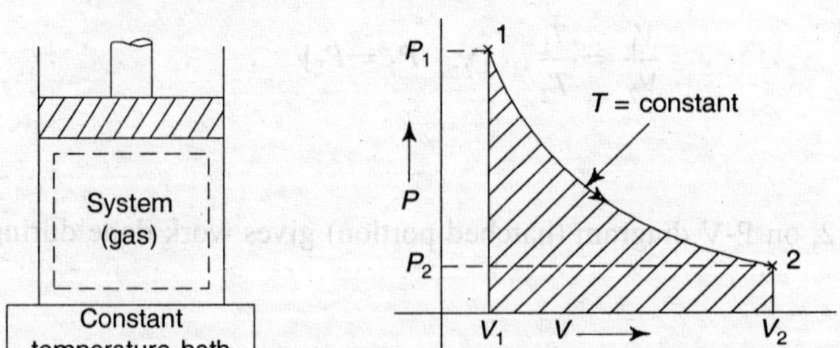

Fig. 3.7

Work done

$$P_1V_1 = P_2V_2 = c$$

$\therefore$ $$PV = c$$

Now $$W_{1\text{-}2} = \int_{V_1}^{V_2} pdv \qquad \left[PV = P_1V_1 = c, \quad P = \frac{P_1V_1}{V} = \frac{c}{V}\right]$$

$\therefore$ $$W_{1\text{-}2} = \int_{V_1}^{V_2} P_1V_1 \frac{dV}{V} = P_1V_1 \int_{V_1}^{V_2} \frac{dV}{V}$$

$$: P_1V_1 \ln V_2/V_1 \qquad \left[P_1V_1 = P_2V_2 \quad \therefore \quad \frac{V_2}{V_1} = \frac{P_1}{P_2}\right]$$

$$= P_1V_1 \ln\left(\frac{P_1}{P_2}\right) = P_2V_2 \ln\left(\frac{P_1}{P_2}\right)$$

Heat transfer

Since temperature is held constant, $\Delta u = 0$

We know that $$Q = W + \Delta u$$

$\therefore$ $$Q = W$$

$\therefore$ $$Q = W = \int_1^2 pdv \qquad \left[P = \frac{m^1 RT}{V}\right]$$

$$= \int_1^2 \frac{RT \cdot dV}{V} = RT \ln\left(\frac{V_2}{V_1}\right)$$

It is clear from the above discussion that the internal energy of the system does not change and heat transferred to the system is equal to the work done by the system. The working fluid receives heat during isothermal expansion and rejects heat during isothermal compression.

3.8.4 Reversible Adiabatic Process or Isentropic Process

It is defined as a process in which no heat transfer occurs across the boundaries of the system *i.e.* the system is enclosed by an insulated wall. Temperature, pressure and volume are the variables in these processes.

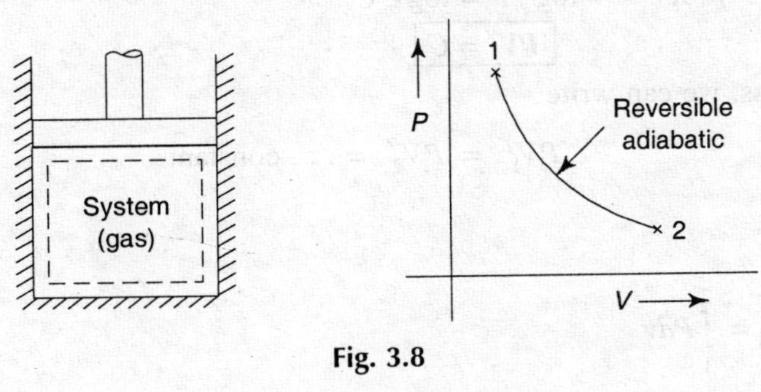

Fig. 3.8

Consider a gas contained in a cylinder as a system, in which cylinder walls are adiabatic walls (perfectly insulated) and there is no friction between piston and the cylinder. The expansion or compression of the gas in the above conditions is considered as an adiabatic process.

From first law of thermodynamics we can write

$$\delta q = \delta W + du$$

$$= pdv + C_v\, dT \qquad [du = C_v dT]$$

Since for adiabatic process $\delta q = 0$

$$pdv + C_v\, dT = 0 \qquad (1)$$

also we know that, $PV = \not{m}^1 RT$ (from perfect gas relation)

differentiate the above equation

$$\therefore \quad pdv + vdp = RdT$$

$$\therefore \quad \boxed{dT = \frac{PdV + VdP}{R}} \qquad (2)$$

Substitute the value of (2) in (1), we get

$$pdv + C_v \left[\frac{pdV + VdP}{R} \right] = 0$$

multiplying by $R = C_p - C_v$, we get

$$(C_p - C_v)\, pdv + C_v \cdot [pdv + vdp] = 0$$

$$\therefore \quad C_p \cdot pdv - C_v \not{p}dv + C_v p\not{d}v + C_v vdp = 0$$

$$C_p \cdot pdv + C_v \cdot vdp = 0$$

divide by $C_v.vp$, we get

$$\frac{C_P \cdot \not{p} dV}{C_V \cdot V \not{P}} + \frac{\not{C_v} \not{V} dP}{\not{C_v} \not{V} P} = 0$$

$$\therefore \quad \frac{C_P}{C_V} \frac{dV}{V} + \frac{dP}{P} = 0$$

Now put $\gamma = C_p/C_v$

$$\therefore \quad \gamma \frac{dV}{V} + \frac{dP}{P} = 0$$

On integration, we get

$$\gamma \log_e V + \log_e P = \log e\, C$$

$$\therefore \quad \boxed{PV^\gamma = C}$$

for an adiabatic process, we can write

$$P_1 V_1^\gamma = P_1 V_2^\gamma = \ldots \text{constant}$$

Work done

We know that, $W_{1\text{-}2} = \int_1^2 Pdv$

$$= \int_1^2 \left(\frac{c}{V^{\gamma}}\right) dv \qquad [\because\ PV^{\gamma} = c]$$

$$= \frac{c}{1-\gamma}\left(V_2^{1-\gamma} - V_1^{1-\gamma}\right)$$

$$= \frac{P_2V_2^{\gamma} \cdot V_2^{1-\gamma} - P_1V_1^{\gamma}V_1^{1-\gamma}}{(1-\gamma)} \qquad [\because\ P_1V_1^{\gamma} = P_1V_2^{\gamma} = c]$$

$$= \frac{P_2V_2 - P_1V_1}{1-\gamma}$$

$$\boxed{W_{1\text{-}2} = \frac{(P_1V_1 - P_2V_2)}{\gamma - 1}}$$

Relation between Temperature, Pressure and Volume

At state 1, $\quad P_1V_1 = mRT_1 \qquad (1)$

At state 2, $\quad P_2V_2 = mRT_2 \qquad (2)$

(1)/(2) gives

$$\frac{P_1V_1}{T_1} = \frac{P_2V_2}{T_2}$$

$$\therefore \quad \frac{T_2}{T_1} = \frac{P_2V_2}{P_1V_1}$$

$$\because \quad P_1V_1^{\gamma} = P_2V_2^{\gamma}$$

$$\left(\frac{V_2}{V_1}\right)^{\gamma} = \left(\frac{P_1}{P_2}\right)$$

$$\therefore \quad \left(\frac{V_2}{V_1}\right) = \left(\frac{P_1}{P_2}\right)^{1/\gamma}$$

$$\therefore \quad \frac{T_2}{T_1} = \left(\frac{P_2}{P_1}\right)\left(\frac{P_1}{P_2}\right)^{1/\gamma} = \frac{P_2 \cdot P_1^{1/\gamma}}{P_1 \cdot P_2^{1/\gamma}}$$

$$\therefore \quad \frac{T_2}{T_1} = P_2P_2^{-\frac{1}{\gamma}}\, P_1^{\frac{1}{\gamma}} \cdot P_1^{-1} = P_2^{\frac{\gamma-1}{\gamma}} \cdot P_1^{\frac{1-\gamma}{\gamma}}$$

$$\frac{T_2}{T_1} = \left(\frac{P_2}{P_1}\right)^{\frac{\gamma-1}{\gamma}}$$

But
$$\left(\frac{P_2}{P_1}\right) = \left(\frac{V_1}{V_2}\right)^{\gamma}$$

$\therefore$
$$\left(\frac{T_2}{T_1}\right) = \left(\frac{V_1}{V_2}\right)^{\gamma\cdot\frac{V-1}{\gamma}} = \left(\frac{V_1}{V_2}\right)^{\gamma-1}$$

$\therefore$
$$\boxed{\frac{T_2}{T_1} = \left(\frac{P_2}{P_1}\right)^{\frac{\gamma-1}{\gamma}} = \left(\frac{V_1}{V_2}\right)^{\gamma-1}}$$

Heat transfer

From the first law,

$$\delta Q = \delta W + du$$
$$\delta Q = 0, \quad \text{for adiabatic process}$$

$\therefore$ $\quad du = -dW = -W$

or $\quad W_{1\text{-}2} \doteq U_1 - U_2$

i.e. the work done is equal to change in internal energy of the system.

3.8.5 Polytropic process

It has been found experimentally that the pressure-volume relationship for compression and expansions of actual gases may be expressed by the equation $pv^n = c$, where 'n' is the index of process considered. The value of 'n' depends on the type of process.

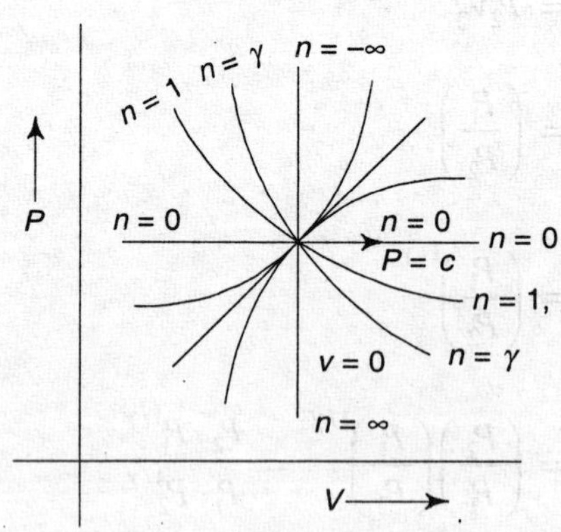

Fig. 3.9

The values of 'n' for different processes are as follows;

$n = 0$;	$pv^0 = c$,	$p = c$;	constant pressure process
$n = 1$;	$pv^1 = c$,	$pv = c$;	constant temperature process
$n = \gamma$;	$pv^\gamma = c$,	$pv^\gamma = c$;	reversible adiabatic process
$n = \infty$;	$pv^\infty = c$,	$v = c$;	constant volume process

Work done

We know that, $W_{1\text{-}2} = \int_1^2 pdv \quad [Pv^n = c]$

$$= \int_1^2 \frac{c}{V^n} dV$$

$$= c\left[\frac{V^{-n+1}}{-n+1}\right]_{V_1}^{V_2} = \frac{c}{1-n}\left[V_2^{1-n} - V_1^{1-n}\right]$$

$$= \frac{c \cdot V_2^{1-n} - c \cdot V_1^{1-n}}{1-n} \qquad \left[P_1V_1^n = P_2V_2^n = c\right]$$

$$= \frac{P_2V_2 \cdot V_2^{1-n} - P_1V_1 V_1^{1-n}}{1-n} = \frac{P_2V_2 - P_1V_1}{1-n}$$

$$W_{1\text{-}2} = \frac{P_1V_1 - P_2V_2}{n-1} = \frac{mR(T_1 - T_2)}{n-1}$$

Alternate solution

Since we know that, $P_1V_1^n = P_2V_2^n = Pv^n = c$

$$P = \frac{P_1V_1^n}{V^n}$$

also we know that, $W_{1\text{-}2} = \int_1^2 pdv$

$$W_{1\text{-}2} = \int_1^2 \frac{P_1V_1^n}{V^n} \cdot dv = P_1V_1^n \int_1^2 \frac{dV}{V^n}$$

[As state 1 remains the same $P_1V_1^n = c$]

After integration, we get

$$W_{1\text{-}2} = P_1V_1^n \left[\frac{V^{-n+1}}{-n+1}\right]_1^2$$

$$\therefore \qquad W_{1\text{-}2} = \frac{P_1V_1 - P_2V_2}{n-1}$$

Change in internal energy

$$U_2 - U_1 = mC_v\,(T_2 - T_1)$$

Heat transfer

From Ist law of *TD*, we know that,

$$\delta Q = \delta W + du$$

$$Q_{1\text{-}2} = W_{1\text{-}2} + (U_2 - U_1)$$

$$= \frac{P_1V_1 - P_2V_2}{n-1} + mC_v\,(T_2 - T_1)$$

$$= \frac{mR\,(T_1 - T_2)}{n-1} + mC_v\,(T_2 - T_1)$$

$$= \frac{mR\,(T_1 - T_2)}{(n-1)} + \frac{mR}{(\gamma-1)}\,(T_2 - T_1), \qquad [\because \; R = C_v\,(\gamma - 1)]$$

$$= mR\,(T_1 - T_2)\left[\frac{1}{n-1} - \frac{1}{\gamma-1}\right]$$

$$\therefore \qquad = \frac{mR\,(T_1 - T_2)\,(\gamma - n)}{(n-1)\,(\gamma-1)}$$

$$\left[\frac{1}{n-1} - \frac{1}{\gamma-1}\right] = \frac{\gamma - \not{1} - n + \not{1}}{(n-1)\,(\gamma-1)}$$

$$= \left(\frac{\gamma - n}{\gamma - 1}\right) \cdot \frac{mR\,(T_1 - T_2)}{(n-1)}$$

$$= \left(\frac{\gamma - n}{\gamma - 1}\right) \times \text{polytropic work done}$$

Polytropic index 'n'

We know that $\qquad P_1V_1^n = P_2V_2^n = c$

Taking log on both sides

$$\ln P_1 + n \ln V_1 = \ln P_2 + n \ln V_2$$

$$n\,(\ln V_1 - \ln V_2) = \ln P_2 - \ln P_1$$

$$n = \frac{\ln P_2 - \ln P_1}{\ln V_1 - \ln V_2}$$

$$\boxed{n = \frac{\ln\,(P_2/P_1)}{\ln\,(V_1/V_2)}}$$

Pressure, Volume and temperature relation

For a polytropic process, we know that,

$$P_1V_1^n = C; \quad P_2V_2^n = C$$

$$\therefore \quad \left(\frac{P_2}{P_1}\right) = \left(\frac{V_1}{V_2}\right)^n \tag{1}$$

from perfect gas relation

$$P_1V_1 = mRT_1$$

$$P_2V_2 = mRT_2$$

$$\therefore \quad \frac{P_2V_2}{P_1V_1} = \frac{\cancel{mR}\,T_2}{\cancel{mR}\,T_1}$$

$$\therefore \quad \frac{P_2}{P_1} = \left(\frac{V_1}{V_2}\right)\cdot\left(\frac{T_2}{T_1}\right) \tag{2}$$

form (1) and (2), we get

$$\left(\frac{V_1}{V_2}\right)^n = \left(\frac{V_1}{V_2}\right)\left(\frac{T_2}{T_1}\right)$$

$$\therefore \quad \left(\frac{T_2}{T_1}\right) = \left(\frac{V_1}{V_2}\right)^n \cdot \left(\frac{V_2}{V_1}\right)$$

$$\boxed{\left(\frac{T_2}{T_1}\right) = \left(\frac{V_1}{V_2}\right)^{n-1}} \tag{3}$$

from (1) and (3), we can write

$$\left(\frac{P_2}{P_1}\right) = \left(\frac{V_1}{V_2}\right)^n$$

$$\therefore \quad \left(\frac{V_1}{V_2}\right) = \left(\frac{P_2}{P_1}\right)^{1/n}$$

$$\text{and} \quad \left(\frac{T_2}{T_1}\right) = \left(\frac{V_1}{V_2}\right)^{n-1} = \left(\frac{P_2}{P_1}\right)^{\frac{n-1}{n}} \quad \left[\because \; \frac{V_1}{V_2} = \left(\frac{P_2}{P_1}\right)^{1/n}\right]$$

$$\therefore \quad \left(\frac{T_2}{T_1}\right) = \left(\frac{P_2}{P_1}\right)^{\frac{n-1}{n}}$$

$$\therefore \quad \boxed{\frac{T_2}{T_1} = \left(\frac{P_2}{P_1}\right)^{\frac{n-1}{n}} = \left(\frac{V_1}{V_2}\right)^{n-1}}$$

Constant Volume and Constant Pressure Specific Heats

These are thermodynamic properties of a substance. These two are defined by the relations

$$C_v = \left(\frac{du}{dT}\right)_v = \text{constant and}$$

$$C_p = \left(\frac{dh}{dT}\right)_p = \text{constant}$$

3.9 CORROLLARIES OF FIRST LAW

The first law of thermodynamics leads to several important corrollaries. They are:

(1) "Energy can neither be created nor be destroyed, but it can be converted from one form to another". (principle of conservation of energy). For an isolated system, it will not exchange energy either in the form of heat or work, therefore $dE = 0$, E = constant, therefore for an isolated system energy is conserved.

(2) Energy is a property of the system. Whenever a closed system undergoes a change of state, the change in energy of the system is the difference between heat supplied and work done.

Thus $dE = dQ - dW$

where dE = change in stored energy of the system

(3) Perpetual motion machine of first kind (PMMI) is impossible.

PMMI is an engine working in a cycle, delivers work without receiving heat from any reservoir.

(4) If changes in KE and PE are negligible, the first law of thermodynamics reduces to the form

$$dU = dQ - dW$$

3.10 PERPETUAL MOTION MACHINE OF FIRST KIND (PMMI)

PMMI is an engine or a cyclic device delivers work, without receiving heat from any source.

Therefore, if $\oint \delta Q = 0$, for a cycle, it is possible to obtain useful work $\oint \delta W$. But from Ist law of thermodynamics we can write

$$\oint \delta W = \oint \delta Q$$

if $\oint dQ = 0$, then

$$\oint dW = 0,$$

There PMMI is impossible.

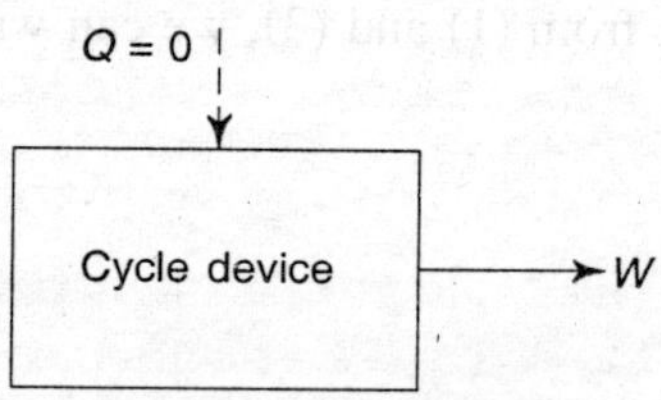

Fig. 3.10 PMMI

3.11 FIRST LAW APPLIED TO AN OPEN SYSTEM

In an open system, a substance, usually fluid, crosses the system boundary in one or more places. The fluid may enter and leave the system, in addition to possibility of heat and work transfer at one point or at several points. The fluid flows through the system either at steady state (mass inflow into the system is equal to the mass of the fluid leaves the system and change of energy within the system is equal to zero) or may flow intermittently. To analyse such systems, control volume concept is to be introduced. It is a region in the space upon which our attention is focussed for study. The boundary of control volume is known as control surface. Most of the engineering equipments are flow devices which need to be analysed by using the concept of control volume.

In a steady flow system, following assumptions are made:

(1) The state of the fluid at each point in the control volume does not vary with time.

(2) Net amount of heat and work entering the control volume for a specified time must be equal to the sum of net amount of heat and work leaving the control volume for the same interval of time, *i.e* the heat and work interact at a constant rate.

(3) The mass flow rate at inlet and outlet of the control volume is constant.

(4) The control volume does not move relative to the co-ordinate frame.

(5) Fluid is uniform in it's composition and will not change it's chemical composition.

When a system exists in steady state, at any particular location, the thermodynamic properties will have fixed values and will not change with time.

3.12 STEADY FLOW ENERGY EQUATION

Consider a steady flow system in which fluid enters and leaves the system as shown in figure. The properties are steady with time and at any location within the control volume, the mass or energy will not be accumulated.

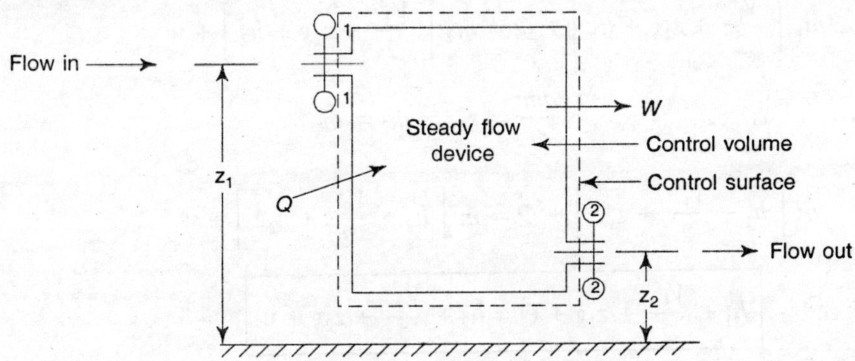

Fig. 3.11 Steady flow system

Let $A_1 A_2$ — c/s of stream, m^2

m_1, m_2 — mass flow rate, kg/s

P_1, P_2 — absolute pressure, N/m^2

v_1, v_2 — specific volume, m^3/kg

u_1, u_2 — specific internal energy, J/kg

V_1, V_2 — velocity, m/s

z_1, z_2 — elevation above reference plane, m

Q — $\dfrac{dQ}{dt}$ Net rate of heat transfer through control surface, J/sec

W — $\dfrac{dW}{dt}$ Net rate of work transfer through control surface, J/sec

t — time in sec

The principle of conservation of mass gives

$$= m_1 = m_2 \quad \left[\because \quad \frac{dm}{dt} = m_1 - m_2 \right]$$

i.e. there is no accumulation of the mass within the control volume. The principle of conservation of energy says that, the rate of flow of energy entering the control volume must be equal to that of leaving the control volume.

$$m_1 e_1 + m_1 P_1 v_1 + Q = m_2 e_2 + m_2 P_2 v_2 + W$$

e_1 & e_2 – energy at inlet and outlet of control volume
$P_1 v_1$ and $P_2 v_2$ – flow work at (1)-(1) and (2)-(2)
But total energy = $KE + PE$ + Int. energy

$$\therefore \quad e = \frac{V^2}{2} + zg + u$$

$$\therefore \quad \dot{m}_1 \left[\frac{V_1^2}{2} + z_1 g + u_1 + P_1 v_1 \right] + Q = \dot{m}_2 \left[\frac{V_2^2}{2} + z_2 g + u_2 + P_2 v_2 \right] + w$$

But
$$h = u + pv$$

$$\therefore \quad \dot{m}_1 \left[\frac{V_1^2}{2} + z_1 g + h_1 \right] + Q = \dot{m}_2 \left[\frac{V_2^2}{2} + z_2 g + h_2 \right] + w$$

for steady flow,
$$\frac{d\dot{m}}{dt} = \dot{m}_1 = \dot{m}_2 = \dot{m}$$

$$\therefore \quad \dot{m} \left[h_1 + \frac{V_1^2}{2} + z_1 g \right] + Q = \dot{m} \left[h_2 + \frac{V_2^2}{2} + z_2 g \right] + w$$

$$\therefore \quad \boxed{h_1 + \frac{V_1^2}{2} + z_1 g + Q = h_2 + \frac{V_2^2}{2} + z_2 g + w}$$

This equation is known as a steady flow energy equation (SFEE) in terms of per unit time and per unit mass.

SFEE; Involving two fluid streams at the inlet and outlet of control volume

Mass balance

$$\dot{m}_1 + \dot{m}_2 = \dot{m}_3 + \dot{m}_4$$

i.e.
$$\frac{A_1 V_1}{v_1} + \frac{A_2 V_2}{v_2} = \frac{A_3 V_3}{v_3} + \frac{A_4 V_4}{v_4}$$

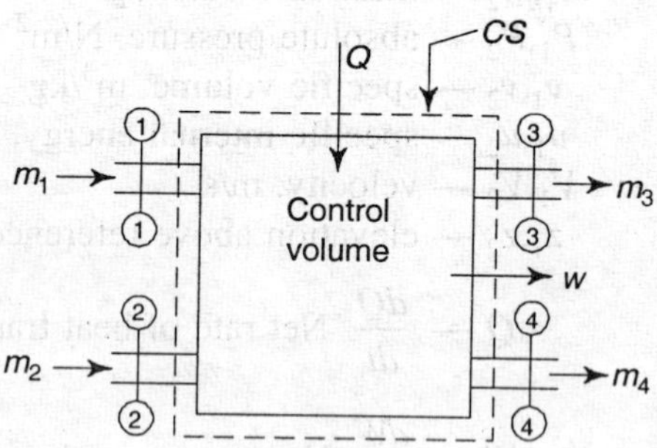

Fig. 3.12

Energy balance

$$\dot{m}_1 \left[h_1 + \frac{V_1^2}{2} + z_1 g \right]$$

$$+ \dot{m}_2 \left[h_2 + \frac{V_2^2}{2} + z_2 g \right] + Q$$

$$= \dot{m}_3 \left[h_3 + \frac{V_3^2}{2} + z_3 g \right] + \dot{m}_4 \left[h_4 + \frac{V_4^2}{2} + z_4 g \right] + W$$

3.13 SFEE APPLIED TO ENGINEERING SYSTEMS

SFEE is very useful for analysing most of the engineering systems. Here, a discussion is made on work developing and work absorbing engineering systems.

3.13.1 Work Absorbing Systems

1. Compressor

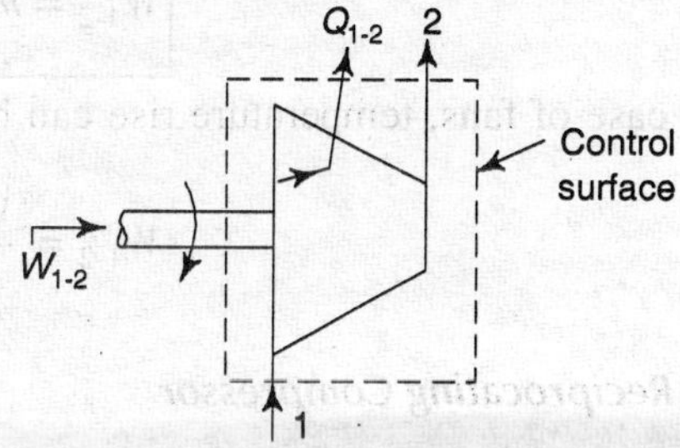

Fig. 3.13 Compressor

A compressor can be considered as a turbine operating in reverse. It is a work absorbing device, which increases pressure of a fluid by doing work upon it.

In a compressor

- *KE* and *PE* terms can be usually ignored.
- Work is done on the system to compress the fluid, hence *W* is negative.
- The system rejects heat, hence '*Q*' is negative.
- For an adiabatic device, $Q = 0$.

From SFEE, we know that,

$$m\left(h_1 + \frac{V_1^2}{2} + gz_1\right) + Q = m\left(h_2 + \frac{V_2^2}{2} + gz_2\right) + W$$

for air compressor, it becomes

$$m\left(h_1 + \cancelto{0}{\frac{V_1^2}{2}} + \cancelto{0}{gz_1}\right) - \cancelto{0}{Q_{1-2}} = m\left(h_2 + \cancelto{0}{\frac{V_2^2}{2}} + \cancelto{0}{gz_2}\right) - W_{1-2}$$

$$\therefore \qquad \boxed{W_{1-2} = m\,(h_2 - h_1)}$$

The work done on the compressor per unit mass of the fluid is equal to increase in the enthalpy of the fluid.

2. Blower

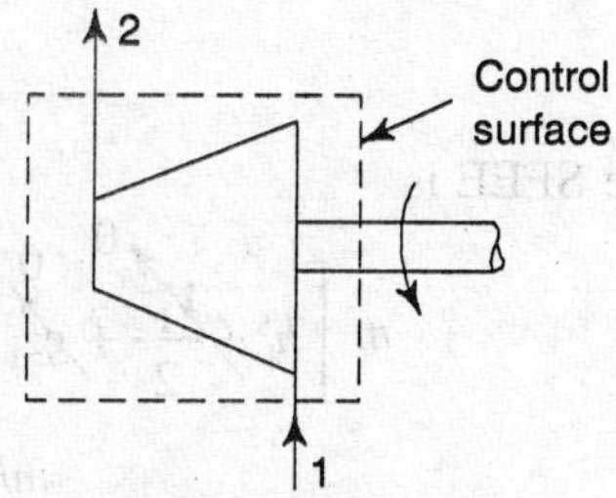

Fig .3.14. Blower

Blower is a device which delivers air at high flow rates and comparatively at low pressure.

In a blower

- Work is done on the system to compress the fluid, hence *W* is negative.
- PE term is usually ignored.
- Since fluid compression is assumed to be adiabatic, $Q = 0$.

$\therefore$ SFEE becomes

$$m\left[h_1 + \frac{V_1^2}{2} + \cancelto{0}{gz_1}\right] - \cancelto{0}{Q_{1-2}} = m\left[h_2 + \frac{V_2^2}{2} + \cancelto{0}{gz_2}\right] - W_{1-2}$$

$$W_{1-2} = m\left[\left(h_1 + \frac{V_1^2}{2}\right) - \left(h_2 + \frac{V_2^2}{2}\right)\right]$$

$$= m\left[\left(u_1 + P_1v_1 + \frac{V_1^2}{2}\right) - \left(u_2 + P_2v_2 + \frac{V_2^2}{2}\right)\right]$$

$$= m\left[\left(u_1 + \frac{V_1^2}{2}\right) - \left(u_2 + \frac{V_2^2}{2}\right)\right]$$

$[\because \quad P_1 v_1 \approx P_2 v_2 \text{ and } v_2 \geq v_1]$

$$\therefore \qquad \boxed{W_{1\text{-}2} = m\left[(u_1 - u_2) + \left(\frac{V_1^2}{2} - \frac{V_2^2}{2}\right)\right]}$$

In case of fans, temperature rise can be ignored and heat loss is small

$$\therefore \qquad W_{1\text{-}2} = \frac{V_2^2}{2}$$

3. Reciprocating Compressor

A reciprocating compressor delivers small quantity of fluid and increases fluid pressure as high as 1000 bar. It can be treated as steady flow system in which control volume includes receiver. The receiver ensures uniform flow rate in the control volume by reducing fluctuations of flow.

In a reciprocating compressor

- $\Delta KE + \Delta PE$ can be ignored.
- W is negative.
- 'Q' is negative.

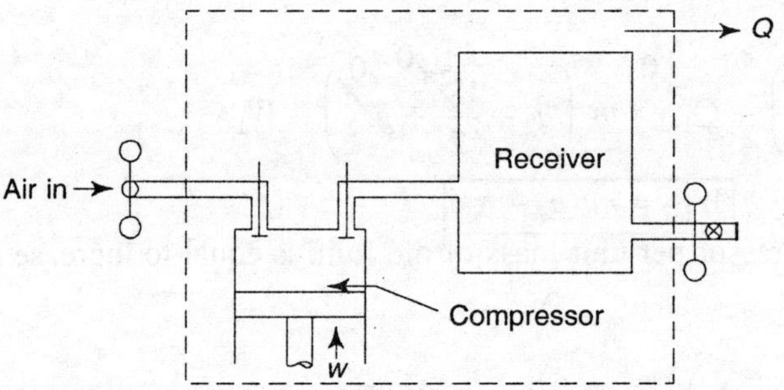

Fig. 3.15 Reciprocating Compressor

The SFEE is

$$m\left[h_1 + \cancelto{0}{\frac{V_1^2}{2}} + \cancelto{0}{gz_1}\right] - Q_{1\text{-}2} = m\left[h_2 + \cancelto{0}{\frac{V_2^2}{2}} + \cancelto{0}{gz_2}\right] - W_{1\text{-}2}$$

$$mh_1 - Q_{1\text{-}2} = mh_2 - W_{1\text{-}2}$$

$$\therefore \qquad \boxed{W_{1\text{-}2} = m(h_2 - h_1) + Q_{1\text{-}2}}$$

4. Centrifugal Pump

A pump is used to deliver liquid at higher heads. It converts mechanical energy into potential energy of liquid.

While using SFEE in a pump, following assumptions are made:

- Work is done on the system and is considered as negative.
- Heat transfer is zero i.e. $Q = 0$
- As water temperature remains same, change in internal energy can be neglected i.e. $\Delta U = 0$

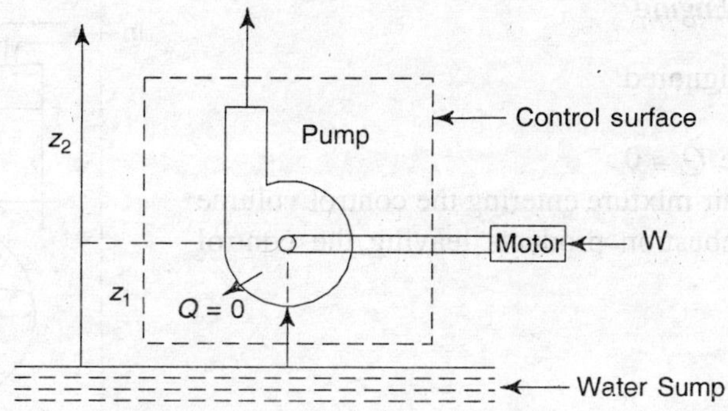

Fig. 3.16 Centrifugal Pump

$\therefore$ The SFEE becomes

$$m\left[h_1 + \frac{V_1^2}{2} + z_1 g\right] - \cancelto{0}{Q_{1\text{-}2}} = m\left[h_2 + \frac{V_2^2}{2} + z_2 g\right] - W_{1\text{-}2}$$

$$m\left[\cancelto{0}{u_1} + P_1 v_1 + \frac{v_1^2}{2} + z_1 g\right] = m\left[\cancelto{0}{u_2} + P_2 v_2 + \frac{v_2^2}{2} + z_2 g\right] - W_{1\text{-}2}$$

$\therefore$
$$W_{1\text{-}2} = m\left[\left(P_2 v_2 + \frac{v_2^2}{2} + z_2 g\right) - \left(P_1 v_1 + \frac{v_1^2}{2} + z_1 g\right)\right]$$

$\therefore$
$$\boxed{W_{1\text{-}2} = m\left\{(P_2 v_2 - P_1 v_1) + \frac{V_2^2 - V_1^2}{2} + g(z_2 - z_1)\right\}}$$

3.13.2 Work Developing Devices

1. Steam or Gas Turbine

A turbine is a device in which the working fluid (steam or gas) expands from high pressure to low pressure to develop power. While using SFEE for a turbine, following assumptions are made:

- ΔKE and ΔPE can be neglected
- Work done by the system, hence W is positive
- The expansion of fluid in the turbine is assumed to be adiabatic, hence $Q = 0$
- There SFEE becomes

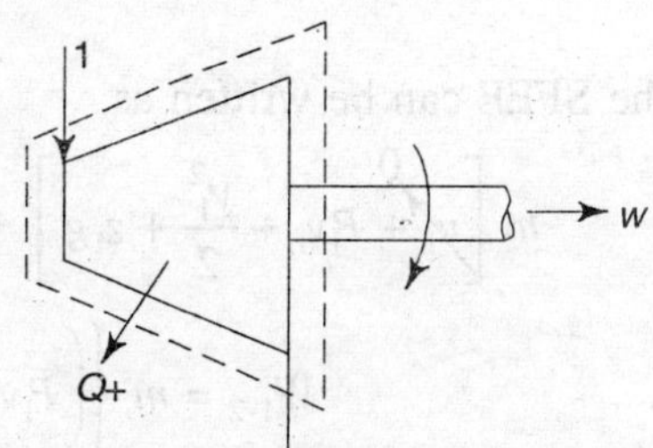

Fig. 3.17 Turbine

$$m\left[h_1 + \cancelto{0}{\frac{V_1^2}{2}} + \cancelto{0}{z_1 g}\right] + \cancelto{0}{Q_{1\text{-}2}} = m\left[h_2 + \cancelto{0}{\frac{V_2^2}{2}} + \cancelto{0}{z_2 g}\right] + W_{1\text{-}2}$$

$\therefore$
$$\boxed{W_{1\text{-}2} = m(h_1 - h_2)}$$

$\therefore$ In an adiabatic turbine, the work done per unit mass of the fluid is equal to decrease in enthalpy of the fluid.

2. Internal Combustion Engine

- ΔKE and ΔPE are ignored
- W is positive
- for adiabatic device $Q = 0$

Let m_1 = mass of fuel-air mixture entering the control volume

m_2 = mass of combustion products leaving the control volume

∴ The SFEE becomes

$$W_{1\text{-}2} = m_1h_1 - m_2h_2$$

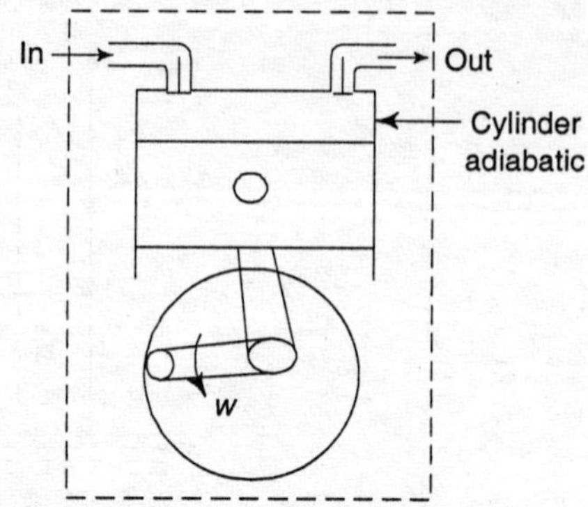

Fig. 3.18 I.C. engine

3. Hydraulic Turbine

While using SFEE, following assumptions are made

- workdone by the system, W is positive
- water temperature remains same, $\Delta u = 0$
- adiabatic device, $Q = 0$

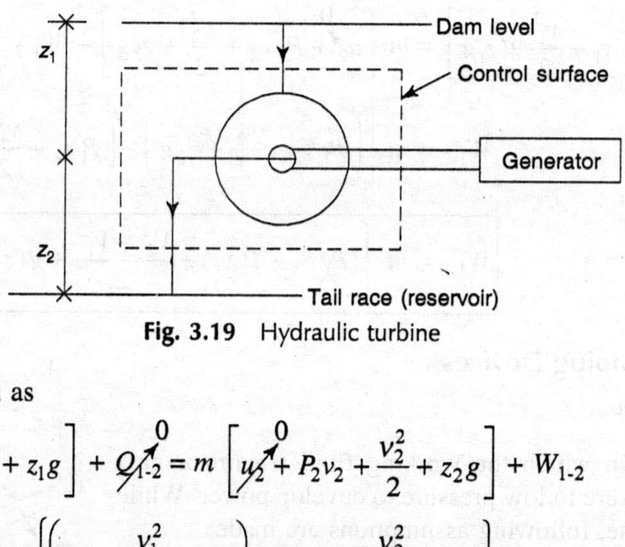

Fig. 3.19 Hydraulic turbine

The SFEE can be written as

$$m\left[\overset{0}{\cancel{u_1}} + P_1v_1 + \frac{V_1^2}{2} + z_1g\right] + \overset{0}{\cancel{Q_{1\text{-}2}}} = m\left[\overset{0}{\cancel{u_2}} + P_2v_2 + \frac{V_2^2}{2} + z_2g\right] + W_{1\text{-}2}$$

$$\therefore \quad W_{1\text{-}2} = m\left\{\left(P_1v_1 + \frac{V_1^2}{2} + z_1g\right) - (P_2v_2) + \frac{V_2^2}{2} + z_2g\right\}$$

$$\boxed{W_{1\text{-}2} = m\left\{(P_1v_1 - P_2v_2) + \left(\frac{V_1^2 - V_2^2}{2}\right) + g(z_1 - z_2)\right\}}$$

3.14 SFEE FOR OTHER DEVICES

1. Nozzle or Diffuser

This is used to increase the flow velocity. It increases kinetic energy of fluid by decreasing pressure energy.

In a nozzle

- Work done is zero

- ΔPE can be neglected
- Due to perfect insulation, $Q = 0$
- Enthalpy decreases & KE increases

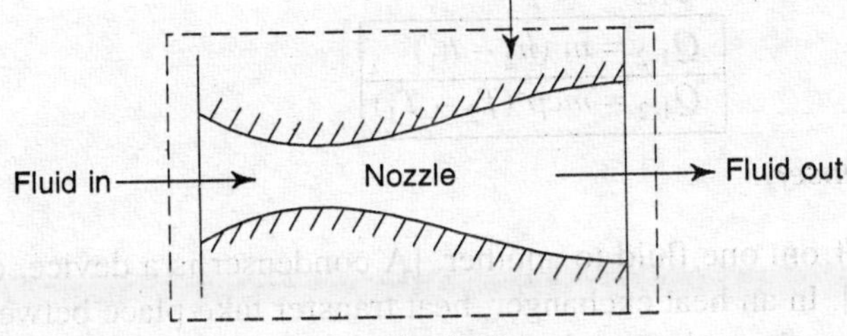

Fig. 3.20

$\therefore$ The SFEE is

$$m\left(h_1 + \frac{V_1^2}{2} + \overset{0}{z_1 g}\right) + \overset{0}{Q_{1\text{-}2}} = m\left(h_2 + \frac{V_2^2}{2} + \overset{0}{z_2 g}\right) + \overset{0}{W_{1\text{-}2}}$$

$$\therefore \quad h_1 + \frac{V_1^2}{2} = h_2 + \frac{V_2^2}{2}$$

$$\boxed{V_2^2 - V_1^2 = z(h_1 - h_2) \text{ m/s}}$$

When the initial velocity or approach velocity of fluid is small, then

$$\boxed{V_2 = \sqrt{2(h_1 - h_2)}}$$

Diffuser: Diffuser increases fluid pressure at the expense of it's kinetic energy. In this case the direction of flow is reversed.

2. Boiler

The function of a boiler is to generate steam at constant pressure. The steam can be generated by supplying heat externally.

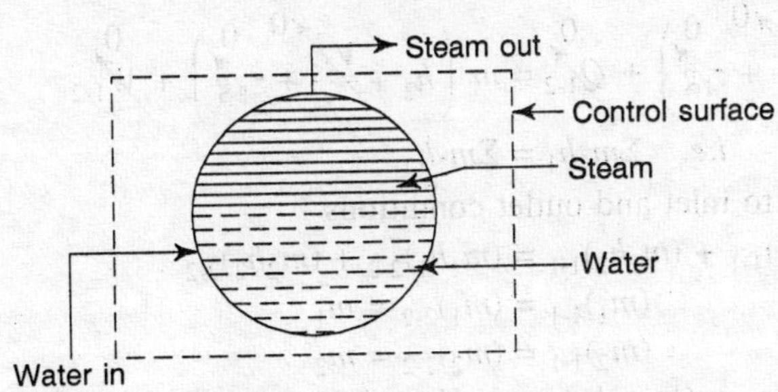

Fig. 3.21 Boiler

In a boiler

- Boiler is a closed vessel, in which there is no work interaction, hence w is zero.
- Change in potential and kinetic energy can be neglected.

Apply the SFEE,

$$m\left(h_1 + \cancelto{0}{\frac{V_1^2}{2}} + \cancelto{0}{z_1 g}\right) + \cancelto{0}{Q_{1-2}} = m\left(h_2 + \cancelto{0}{\frac{V_2^2}{2}} + \cancelto{0}{z_2 g}\right) + \cancelto{0}{W_{1-2}}$$

$$mh_1 + Q_{1-2} = mh_2$$

$$\therefore \quad \boxed{Q_{1-2} = m\,(h_2 - h_1)}$$

$$\boxed{Q_{1-2} = mcp\,(T_2 - T_1)}$$

3. Heat Exchanger (Condenser)

It transfers energy as heat from one fluid to another. [A condenser is a device, condenses vapour into liquid at constant pressure]. In an heat exchanger, heat transfer take place between hot and cold fluids indirectly, as one fluid flows through the tubes and other fluid outside the tubes in the shell.

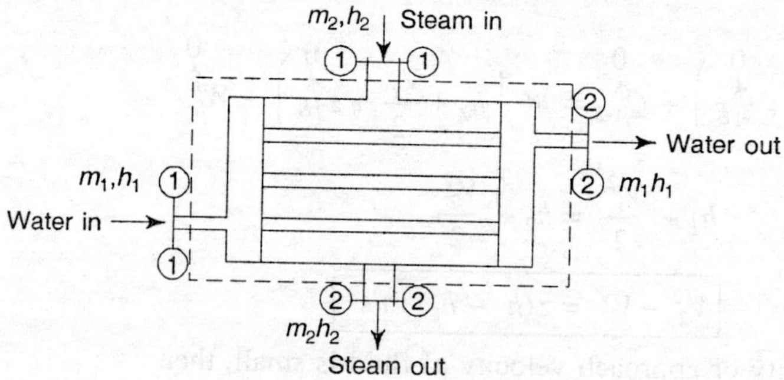

Fig. 3.22 Heat Exchanger

In a heat exchanger

- No work interaction, W is zero
- As there is perfect insulation, Q is zero
- ΔKE and ΔPE are neglected.

Let m_1 and h_1 – mass and enthalpy of water
m_2 and h_2 – mass and enthalpy of steam and condensate

Now SFEE is

$$m\left(h_1 + \cancelto{0}{\frac{V_1^2}{2}} + \cancelto{0}{z_1 g}\right) + \cancelto{0}{Q_{1-2}} = m\left(h_2 + \cancelto{0}{\frac{V_2^2}{2}} + \cancelto{0}{z_2 g}\right) + \cancelto{0}{W_{1-2}}$$

$$\therefore \quad i.e. \quad \Sigma m_1 h_1 = \Sigma m_2 h_2$$

where (1) and (2) refers to inlet and outlet conditions

$$(m_1 h_1)_{1-1} + (m_2 h_2)_{1-1} = (m_1 h_1)_{2-2} + (m_2 h_2)_{2-2}$$

Let

$$(m_1)_{1-1} = (m_1)_{2-2} = m_1$$

$$(m_2)_{1-1} = (m_2)_{2-2} = m_2$$

$$\therefore \quad m_1(h_2 - h_1)_1 = m_2(h_1 - h_2)_2$$

For an evaporator

From SFEE,

$$mh_1 + Q_{1-2} = mh_2$$

$$\therefore \quad \boxed{Q_{1-2} = m(h_2 - h_1)}$$

4. Throttling Process

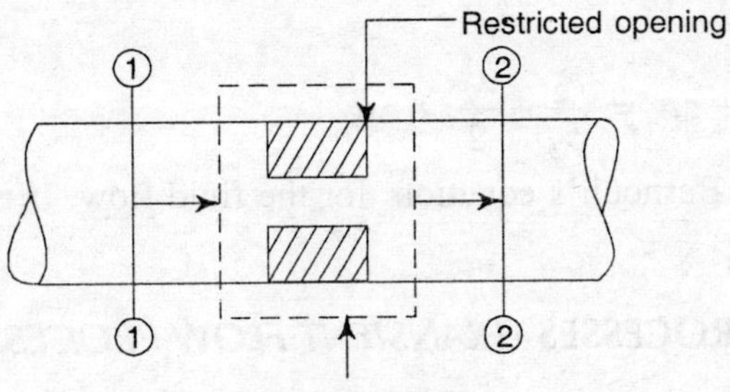

Fig. 3.23

Consider the fluid flow through a restricted passage, like an orifice or porous plug, in which fluid expands from high pressure to low pressure without doing any work or any considerable change in kinetic energy. As the fluid flows through restricted passage, the flow assumes to be adiabatic as there is neither sufficient time nor large enough area for heat transfer.

In a throttling process

- No heat and work interactions.
- ΔKE and ΔPE can be ignored.

$$\therefore \quad \cancel{m}\left(h_1 + \cancelto{0}{\frac{V_1^2}{2}} + \cancelto{0}{z_1 g}\right) + \cancelto{0}{Q_{1\text{-}2}} = \cancel{m}\left(h_2 + \cancelto{0}{\frac{V_2^2}{2}} + \cancelto{0}{z_2 g}\right) + \cancelto{0}{W_{1\text{-}2}}$$

$$h_1 = h_2$$

i.e Enthalpy of fluid before and after throttling is same.

5. Flow Through Pipe

For the flow through a pipe,

- Work and heat interaction is zero.
- As there is no change in temperature of fluid flowing through pipe, there is no change in internal energy.

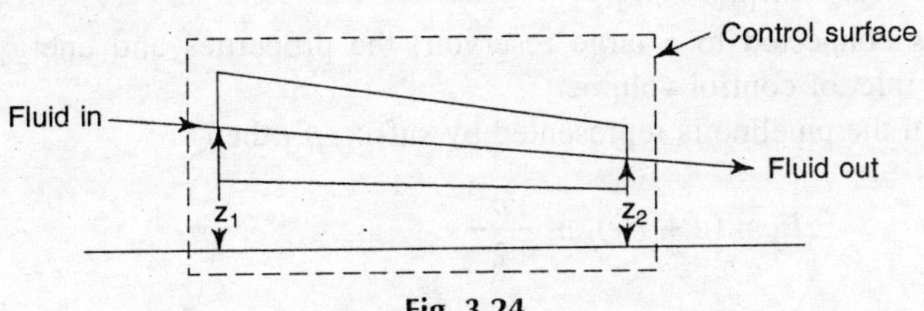

Fig. 3.24

The SFEE is

$$\cancel{m}\left[\cancelto{0}{u_1} + P_1 v_1 + \frac{v_1^2}{2} + z_1 g\right] + \cancelto{0}{Q_{1\text{-}2}} = \cancel{m}\left[\cancelto{0}{u_2} + P_2 v_2 + \frac{v_2^2}{2} + z_2 g\right] + \cancelto{0}{W_{1\text{-}2}}$$

$$\therefore \quad P_1 v_1 + \frac{v_1^2}{2} + z_1 g = P_2 v_2 + \frac{v_2^2}{2} + z_2 g$$

$\because \quad \left[v = \frac{1}{\rho} \right]$

$$\frac{P_1}{\rho_1} + \frac{V_1^2}{2} + z_1 g = \frac{P_2}{\rho_2} + \frac{V_2^2}{2} + z_2 g$$

This equation is same as the Bernouli's equation for the fluid flow. If the fluid is incompressible like water, then $\rho_1 = \rho_2 = \rho$.

3.15 UNSTEADY FLOW PROCESSES (*TRANSIENT FLOW PROCESSES*)

So far in the discussion made for flow processes, it has been assumed that the properties do not change with time. It was also assumed that, the mass flow rate of the fluid through the control volume was constant. But in the transient flow processes, properties in the control volume changes with time. There is accumulation of mass and energy inside the control volume and hence the rate of fluid flow through the control volume is not constant. The analysis of these variable flow processes can be done either by control volume or control mass (system technique) technique. Here a brief discussion is made on tank filling and emptying process.

3.15.1 Charging of a Tank (Tank Filling)

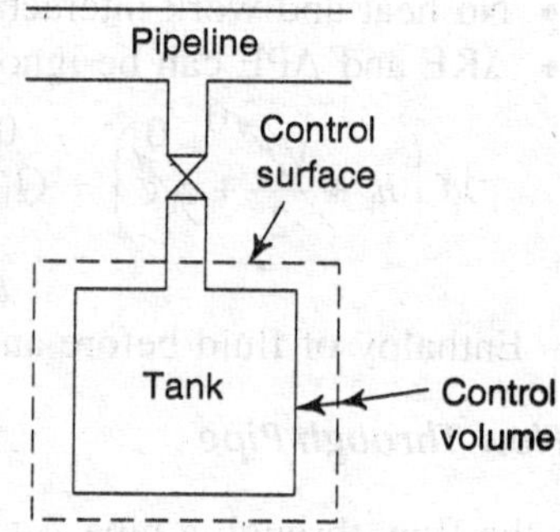

Fig. 3.25

Consider a process in which the tank is to be filled with gas from the pipeline which is maintained at constant pressure and temperature.

Let m_1 be the mass of gas at the initial state. Let P_1, t_1, V_1, h_1 and u_1 are the values at the initial state.

When the valve is opened, gas from the pipeline enters into the tank till the mass of gas becomes m_2 at state, P_2, t_2, V_2, h_2 and u_2.

During charging process, let $(m_2 - m_1)$ be the mass of the gas entering the tank, and energy of gas entering the tank is $(m_2 - m_1) E$.

Making energy balance, we get

(Energy entering into control volume – Energy leaving control volume) in time dt = Change in energy in control volume in time dt

$$(m_2 - m_1) E + Q_{1\text{-}2} = (m_2 u_2 - m_1 u_1) + W_{1\text{-}2} \qquad (1)$$

Since the pipeline is connected to a large reservoir, the properties and energy of gas entering remains constant at the inlet of control volume.

Let the state of gas in the pipeline is represented by suffix 'p', then

$$E_p = (u + Pv)_p + \frac{Vp^2}{2}$$

$$E_p = hp + \frac{Vp^2}{2} \qquad (2)$$

Substitute (2) in (1), we get

$$(m_2 - m_1)\left(hp + \frac{Vp^2}{2}\right) + Q_{1\text{-}2} = (m_2 u_2 - m_1 u_1) + W_{1\text{-}2}$$

$[\because \quad E = E_p]$

If we ignore KE and PE changes and there is no work and heat (insulated) transfer, then $(m_2 - m_1)(hp) = m_2u_2 - m_1u_1$

If the tank is empty at initial state,

$$m_1 = 0$$

$$\therefore \quad \cancel{m_2}\, hp = \cancel{m_2}\, u_2$$

$$\boxed{hp = u_2}$$

Enthalpy of gas entering the tank is equal to final internal energy of gas in the tank.

If an ideal gas is filled adiabatically into an empty tank, then

$$Cp \cdot Tp = C_vT_2$$

$$\therefore \quad \boxed{T_2 = \gamma\, Tp}$$

3.15.2 Discharging of a Tank

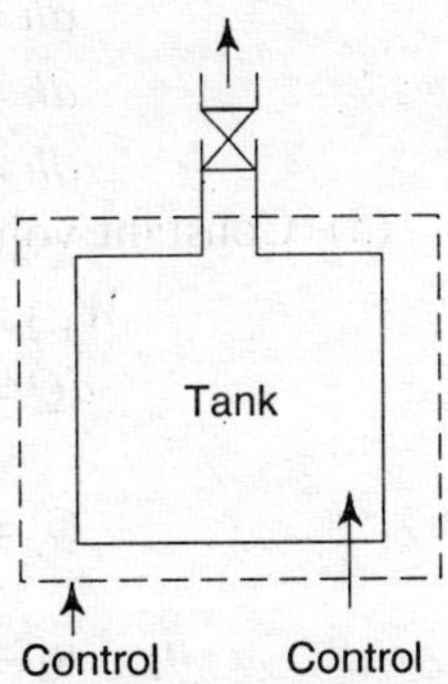

Fig. 3.26

While discharging a tank, fluid (gas) flows from tank to the surroundings, which is larger than tank being emptied. Hence energy per unit mass of the fluid at the outlet of the control volume remains constant.

Initially the tank has mass 'm_1' after discharging the tank, the mass left in the tank is 'm_2'.

Making energy balance

(Energy entering – Energy leaving) control volume in time dt = Change in energy in control volume in time dt

$$\therefore \quad (m_1 - m_2)\, E - Q_{1\text{-}2} = m_1u_1 - m_2u_2$$

We know that $E = E_p = hp + \dfrac{Vp^2}{2}$

$$\therefore \quad (m_1 - m_2)\left(hp + \frac{Vp^2}{2}\right) - Q_{1\text{-}2} = m_1u_1 - m_2u_2$$

Neglect ΔKE and heat transfer, the equation can be written as

$$(m_1 - m_2)\, hp = m_1u_1 - m_2u_2$$

When the tank is completely evacuated, then

$$m_2 = 0,$$

$$\therefore \quad m_1hp = m_1u_1$$

$$\boxed{hp = u_1}$$

Enthalpy of the fluid leaving the tank is equal to internal energy of the fluid at initial state.

3.16 FORMULAE USED IN SOLVING PROBLEMS

(1) First law of thermodynamics for a closed system; $\oint dQ = \oint dW$ under going a cycle

i.e. $Q_{1\text{-}2} + Q_{2\text{-}3} + Q_{3\text{-}4} + Q_{4\text{-}1} = W_{1\text{-}2} + W_{2\text{-}3} + W_{3\text{-}4} + W_{4\text{-}1}$

(2) First law for a closed system which executes a process or change of state

$$Q - W = \Delta E = E_2 - E_1 \text{ or } Q = W + \Delta U$$

(3) E = Internal energy + KE + PE

$$E = U + KE + PE$$

$$\therefore \quad dQ = dW + dE = dw + dU + d(KE) + d(PE)$$

$$dQ = dW + dU + d\left(\frac{1}{2}\frac{mv^2}{gc}\right) + d\left(\frac{mg}{gc}\cdot 2\right)$$

$$Q_{1\text{-}2} = W_{1\text{-}2} + u_2 - u_1 + m\,\frac{(v_2^2 - v_1^2)}{2gc} + \frac{mg}{gc}\,(z_2 - z_1)$$

(4) Enthalpy $H = U + PV$ or

$$h = u + PV$$

$$dh = du + pdv + vdp \qquad \text{(consider only } pdv)$$

$$dh = du + pdv$$

$$dh = dQ \cdot p = \text{const.} \qquad [dq = du + pdv]$$

(5) Constant volume process (for a closed system)

$$W_{1\text{-}2} = 0$$

$$dQ = du = C_v\,dT$$

$$C_v = \left.\frac{du}{dT}\right|_v = \text{constant}$$

$$u_2 - u_1 = C_v\,(T_2 - T_1), \text{ where } C_v = f(T)$$

(6) *Constant Pressure Process*

$$W_{1\text{-}2} = P(V_2 - V_1)$$

$$Q_{1\text{-}2} = H_2 - H_1 \; [H = U + PV]$$

$$Q_{1\text{-}2} = mcp\,(T_2 - T_1) \qquad \text{[for a perfect gas]}$$

$$C_p = \left.\frac{dh}{dT}\right|_{p = \text{const.}}$$

(7) *Constant Temperature Process*

$$P_1V_1 = P_2V_2 = C$$

$$WD = \int_{v_1}^{v_2} Pdv = P_1V_1 \ln\frac{V_2}{V_1} = P_1V_1 \ln\left(\frac{P_1}{P_2}\right)$$

$$= P_2V_2 \ln(P_1/P_2)$$

$$Q = W + \Delta U \nearrow 0$$

$$Q = W = \int_1^2 pdv = RT \ln\left(\frac{V_2}{V_1}\right)$$

(8) *Rev. adiabatic or Isentropic process*

$$PV^\gamma = c \quad i.e. \quad P_1V_1^\gamma = P_2V_2^\gamma = c$$

$$WD = \int_1^2 pdv = \frac{P_1V_1 - P_2V_2}{\gamma - 1}$$

$$\left(\frac{T_2}{T_1}\right) = \left(\frac{P_2}{P_1}\right)^{\frac{\gamma-1}{\gamma}} = \left(\frac{V_1}{V_2}\right)^{\gamma - 1}$$

$$\delta Q \overset{0}{\nearrow} = \delta W + du$$

$$du = -\delta W$$

$$W_{1\text{-}2} = U_1 - U_2$$

(9) *Polytropic process*

$$WD = \frac{P_1V_1 - P_2V_2}{n-1} = \frac{mR\,(T_1 - T_2)}{n-1}$$

Change in Internal energy

$$U_2 - U_1 = m\,C_v\,(T_2 - T_1)$$

Heat transfer

$$Q_{1\text{-}2} = \left(\frac{\gamma - n}{\gamma - 1}\right) \times \frac{mR\,(T_1 - T_2)}{(n-1)}$$

$$n = \frac{\ln\,(P_2/P_1)}{\ln\,(V_1/V_2)}$$

$$\left(\frac{T_2}{T_1}\right) = \left(\frac{P_2}{P_1}\right)^{\frac{n-1}{n}} = \left(\frac{V_1}{V_2}\right)^{n-1}$$

(10) $h_1 + \frac{V_1^2}{2} + z_1 g + Q = h_2 + \frac{V_2^2}{2} + z_2 g + W$

(11) For compressor; $w_{1\text{-}2} = m\,(h_2 - h_1)$

(12) Blower; $W_{1\text{-}2} = m\left[(u_1 - u_2) + \left(\frac{V_1^2}{2} - \frac{V_2^2}{2}\right)\right]$

(13) Reciprocating compressor $w_{1\text{-}2} = m(h_2 - h_1) + Q_{1\text{-}2}$

(14) Centrifugal pump

$$W_{1\text{-}2} = m\left\{(P_2v_2 - P_1v_1) + \frac{V_2^2 - V_1^2}{2} + g\,(z_2 - z_1)\right\}$$

(15) Turbine $W_{1\text{-}2} = m(h_1 - h_2)$

(16) IC engine $W_{1\text{-}2} = m_1h_1 - m_2h_2$

(17) Hydraulic turbine

$$W_{1\text{-}2} = m\left\{(P_1v_1 - P_2v_2) + \left(\frac{V_1^2 - V_2^2}{2}\right) + g\,(z_1 - z_2)\right\}$$

(18) Nozzle or diffuser

$$V_2 = \sqrt{2(h_1 - h_2)}$$

(19) Boiler $\quad Q_{1\text{-}2} = m\,(h_2 - h_1) = mcp\,(T_2 - T_1)$

(20) Heat exchanger

$$m_1(h_2 - h_1)_1 = m_2(h_1 - h_2)_2$$

for evaporator

$$Q_{1\text{-}2} = m(h_2 - h_1)$$

(21) Throttling process

$$h_1 = h_2$$

(22) Flow through pipe

$$\frac{P_1}{\rho_1} + \frac{V_1^2}{2} + gz_1 = \frac{P_2}{\rho_2} + \frac{V_2^2}{2} + gz_2 \text{ (Bernouli's equation)}$$

(23) Tank filling $hp = u_2$

$$T_2 = \gamma T_P.$$

(24) Discharging a tank

$$hp = u_1$$

3.17 NUMERICAL PROBLEMS

Problem 1: *The fluid contained in a tank is stirred by a paddle wheel. The work input to the paddle wheel is 5075 kJ and heat rejected from the tank is 1450 kJ. Determine change in internal energy by considering tank and fluid as a system.*

Solution : From the Ist law, we can write

$$Q = \Delta U + W$$

$$-1450 = U_2 - U_1 - 5075$$

$\therefore$ change in internal energy $= \Delta U = 3625$ kJ

$$\boxed{\Delta V = 3625 \text{ kJ}}$$

Ans.

Problem 2 : *In a system, the work and heat interactions for a process 'A' are 22 kJ and 18 kJ and for process 'B' between the same end conditions, the heat input is 11 kJ. Determine change in internal energy and work done in process 'B'. If a cycle is formed with processes 'A' and 'B', then prove that first law is obeyed.*

Solution :

We know that, Ist law of thermodynamics for a closed system, for process 'A'

$$Q = \Delta U + W$$

$$18 = \Delta U + 22$$

$$\Delta U = 18 - 22 = -4 \text{ kJ}$$

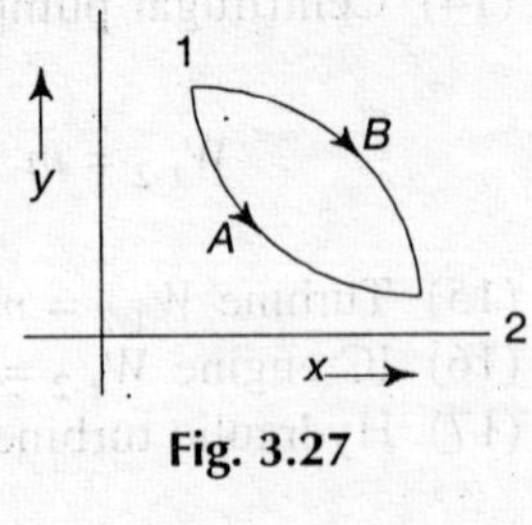

Fig. 3.27

Since internal energy is a property of the system. It is a point function. Hence ΔU for path 'B' is also same.

∴ for process '*B*'

$$Q = W + \Delta U$$

$$11 = W - 4 \qquad [(\Delta U) \text{ path } B = -4 \text{ kJ}]$$

$$W = 15 \text{ kJ}$$

When a cycle is formed with '*A*' and '*B*', then

$$\oint W = 22 - 15 = 7 \text{ kJ}$$

$$\oint Q = 18 - 11 = 7 \text{ kJ}$$

∴ $$\oint Q = \oint W$$ ***Ans.***

∴ first law is obeyed.

Problem 3: *The table given below shows the values of heat transfer and work done during a cyclic process. Determine work interaction for the process 4-1 and state whether it is done upon the system or by the system?*

Process	*Q kJ*	*W kJ*
1-2	850	75
2-3	−80	−65
3-4	−700	70
4-1	210	?

Solution: From first law, we can write

$$\oint dQ = \oint dW$$

$$Q_{1-2} + Q_{2-3} + Q_{3-4} + Q_{4-1} = W_{1-2} + W_{2-3} + W_{3-4} + W_{4-1}$$

$$850 - 80 - 700 + 210 = 75 - 65 + 70 + w_{4-1}$$

$$280 = 80 + W_{4-1}$$

∴ $$\boxed{W_{4-1} = +200 \text{ kJ}}$$ ***Ans.***

The positive sign indicates that, the work is done by the system.

Problem 4: *The table given below shows the values of heat and work interactions for a system which undergoes a cyclic change. Complete the table and determine the net work done by the system.*

Process	*W(kJ)*	*Q(kJ)*	*v(kJ)*
1-2	650	–	−515
2-3	125	750	–
3-4	−150	–	–
4-1	0	60	–

Solution: From first law, we can write,

for process 1-2,

$$Q = W + \Delta U$$

$$Q = 650 - 515$$

$$Q : 135 \text{ kJ}$$

for the process 2-3, $$750 = 125 + \Delta U_{2-3}$$

∴ $$\Delta U_{2-3} = 750 - 125$$

$$\Delta U_{2-3} : 625 \text{ kJ}$$

for the process 3-4

$$Q : \Delta U - 150$$

for the process 4-1

$$Q : 0 + \Delta U$$
$$60 : \Delta U$$
$$\Delta U_{4\text{-}1} : 60 \text{ kJ}$$

We know that,

$$\Delta U_{net} = (\Delta U)_{1\text{-}2} + (\Delta U)_{2\text{-}3} + (\Delta U)_{3\text{-}4} + (\Delta U)_{4\text{-}1} = 0$$

$\therefore$ $$\Delta U_{net} = -515 + 625 + (\Delta U)_{3\text{-}4} + 60 = 0$$

$\therefore$ $$\boxed{(\Delta U)_{3\text{-}4} = -170 \text{ kJ}}$$

$\therefore$ for the process 3-4; $$Q = \Delta U - 150$$
$$Q : -170 - 150$$
$$\boxed{Q : -320 \text{ kJ}}$$

$\therefore$ The net work done by the system

$$= W_{net} = W_{1\text{-}2} + W_{2\text{-}3} + W_{3\text{-}4} + W_{4\text{-}1}$$
$$= 650 + 125 - 150 + 0$$
$$: 625 \text{ kJ}$$

$\therefore$ The net heat transfer

$$Q_{net} = Q_{1\text{-}2} + Q_{2\text{-}3} + Q_{3\text{-}4} + Q_{4\text{-}1}$$
$$: 135 + 750 - 320 + 60$$
$$= 625 \text{ kJ}$$

$$\Sigma Q_{cycle} = \Sigma w_{cycle} : \text{The system obeys first law.}$$

Problem 5: *The gas is compressed from an initial state of 0.35 m³ and 105 kpa to a final state of 0.14 m³ and to the same pressure. Determine change in internal energy of gas which transfers 38 kJ of heat.*

Solution: We know that, from the first law

$$Q = W + \Delta U$$
$$Q_{1\text{-}2} = U_2 - U_1 + W_{1\text{-}2}$$
$$W_{1\text{-}2} = \int_1^2 p dv = P(V_2 - V_1)$$
$$= 105(0.14 - 0.35)$$
$$W_{1\text{-}2} = -22.05 \text{ kJ}$$

$\therefore$ $$Q_{1\text{-}2} = -38 \text{ kJ}$$ [$\because$ gas transfers or rejects heat]

$\therefore$ $$(\Delta U) = (Q - W)_{1\text{-}2}$$
$$= -38 + 22.05$$
$$(\Delta U)_{1\text{-}2} = U_2 - U_1 = -15.95 \text{ kJ}$$

Negative sign indicates that, the internal energy decreases by an amount of 15.95 kJ during the process.

Problem 6: *In a system, 86 kJ of heat absorbed by it to do 30 kJ of work, during a change of state 1-3 along path 1-2-3 (a) Determine the quantity of heat flows into the system along path 1-4-3 and* *during this change of state, the work done by the system is 12 kJ, (b) If the system returns from 3 to 1 along curved path, 21 kJ of work will be done on the system. Does the system absorb or liberate heat and how much? (c) If $u_1 = 0$, and $u_4 = 42$ kJ, find heat absorbed in the process '1-4' and '4-3'.*

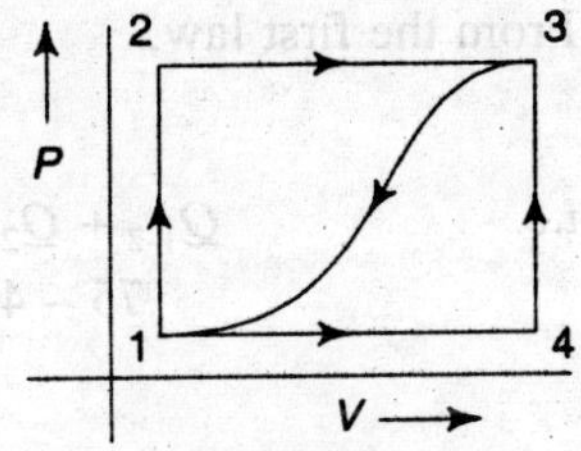

Fig. 3.28

Solution: Given data:

$$Q_{1\text{-}2\text{-}3} = 86 \text{ kJ}$$
$$W_{1\text{-}2\text{-}3} = 30 \text{ kJ}$$
$$W_{1\text{-}4\text{-}3} = 12 \text{ kJ}$$
$$Q_{1\text{-}4\text{-}3} = ?$$

From first law, we know that,

$$Q_{1\text{-}2\text{-}3} = U_3 - U_1 + W_{1\text{-}2\text{-}3}$$
$$86 = U_3 - U_1 + 30$$

also, $$U_3 - U_1 = 56 \text{ kJ}$$

(a) $Q_{1\text{-}4\text{-}3} = U_3 - U_1 + W_{1\text{-}4\text{-}3}$

$$= 56 + 12$$
$$\boxed{Q_{1\text{-}4\text{-}3} = 68 \text{ kJ}}$$ *Ans.*

(b) When $W_{3\text{-}1} = -21$ kJ (work done on the system)

$\therefore$ $$Q_{3\text{-}1} = U_1 - U_3 + W_{3\text{-}1}$$
$$= -56 - 21$$
$$\boxed{Q_{3\text{-}1} = -77 \text{ kJ}}$$

The system liberates 77 kJ of heat

(c) When $U_1 = 0$, $U_4 = 42$ kJ

$$Q_{1\text{-}4} = ? \; Q_{4\text{-}3} = ?$$
$$W_{1\text{-}4\text{-}3} = W_{1\text{-}4} + W_{4\text{-}3} = W_{1\text{-}4} = 12 \text{ kJ}$$

[$\because$ $W_{4\text{-}3}$ = const. volume process, $W_{4\text{-}3} = 0$]

$$Q_{1\text{-}4} = U_4 - U_1 + W_{1\text{-}4}$$
$$Q_{1\text{-}4} = 42 - 0 + 12 = \boxed{56 \text{ kJ}}$$ *Ans.*

also

$$Q_{1\text{-}4\text{-}3} = 68 \text{ kJ} = Q_{1\text{-}4} + Q_{4\text{-}3}$$
$$68 = 56 + Q_{4\text{-}3}$$
$$\boxed{Q_{4\text{-}3} = 12 \text{ kJ}}$$ *Ans.*

Problem 7: *In an engine, for one cycle the working fluid engages in two work interactions: 15 kJ to the fluid and 44 kJ from the fluid and three heat interactions; 75 kJ to the fluid and 40 KJ from the fluid. Determine the magnitude and direction of third heat transfer.*

Solution: Consider fluid contained in the engine cylinder as system;

$W_{1\text{-}2} = -15$ kJ (fluid absorbs work)

$W_{2\text{-}3} = +44$ kJ (work done by the system)

$Q_{1\text{-}2} = 75$ kJ (heat enters into the system)

$Q_{2\text{-}3} = -40$ kJ (system rejects heat)

$Q_{3\text{-}4} = ?$

From the first law,

$$\oint \delta Q = \oint \delta W$$

i.e. $$Q_{1-2} + Q_{2-3} + Q_{3-4} = W_{1-2} + W_{2-3}$$

$$75 - 40 + Q_{3-4} = -15 + 44$$

$$\boxed{Q_{3-4} : -6 \text{ kJ}}$$ **Ans.**

Q_{3-4} : heat rejected from the fluid

Problem 8 : *0.015 m^3 gas at constant pressure of 2060 kN/m^2 expands to a pressure of 210 kN/m^2 by following the law $PV^{1.35} = c$. Determine the work done by the gas during expansion process.*

Solution : For a polytropic expansion, we know that,

$$W_{1-2} = \frac{P_1V_1 - P_2V_2}{n-1}$$

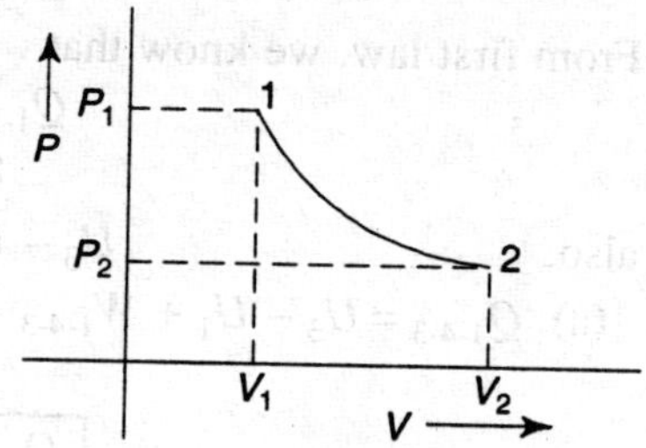

Fig. 3.29

here
$$P_1 = 2060 \text{ kN/m}^2$$
$$P_2 = 210 \text{ kN/m}^2$$
$$V_1 = 0.015 \text{ m}^3,\ n = 1.35$$
$$V_2 = ?$$
$$W_{1-2} = ?$$
$$P_1V_1^n = P_2V_2^n = c$$
$$V_2 = V_1 \left(\frac{P_1}{P_2}\right)^{1/n}$$
$$V_2 = 0.015 \left(\frac{2060}{210}\right)^{1/1.35}$$
$$\boxed{V_2 = 0.0812 \text{ m}^3}$$

$\therefore$ $$W_{1-2} = \frac{2060 \times 10^3 \times 0.015 - 210 \times 10^3 \times 0.0812}{1.35 - 1}$$

$$W_{1-2} = \frac{30900 - 17052}{0.35}$$ $$\left[\frac{N}{m^2} \times m^3 - \frac{N}{m^2} \times m^3 \quad Nm : J\right]$$

$$= 39565.71 \text{ J}$$

$$\boxed{W_{1-2} = 39.56 \text{ kJ}}$$

Problem 9: *A gas is compressed hyperbolically from a pressure and volume of 105 kN/m^2 and 0.055 m^3 to a volume of 0.006 m^3. Determine the final pressure and work done on the gas.*

Solution: For hyperbolic compression,

$$P_1V_1 = P_2V_2 = c$$

$\therefore$ $$P_2 = \frac{P_1V_1}{V_2} = 105 \times \frac{0.055}{0.006}$$

$$\boxed{P_2 = 962.5 \text{ kN/m}^2}$$

We know that, $W_{1\text{-}2} = P_1 V_1 \ln \dfrac{V_2}{V_1}$ $\quad \left| \dfrac{N}{m^2} \times m^3 \right.$

$$= 105 \times 10^3 \, (0.055) \ln \frac{0.006}{0.055} \quad \Big| \; Nm : J$$

$$= -\,12.794 \times 10^3 \text{ Nm}$$

$\therefore$ $\quad -12.794$ kJ

$\therefore$ The work done on the gas

$$= -12.794$$ **Ans.**

Problem 10: *5.5 kg of water is contained in a vessel of mass 1.8 kg. If the initial temperature of vessel plus water is 15°C and final temperature is 85°C, find heat transfer required to accomplish this change. Take specific heat of water to be 4.19 × 10^3 J/kg k and for vessel to be 390 J/kg k*

Solution:

We know that,

$$(Q)_{\text{vessel}} = m \cdot c \, (\Delta T)$$

$$1.8 \times 390 \, (85 - 15) = 49.14 \text{ kJ}$$

$$(Q)_{\text{water}} = 5.5 \, (4.19) \, (85 - 15)$$

$$= 1613.15 \text{ kJ}$$

$\therefore$ $\quad (Q)_{\text{water + vessel}} = 49.14 + 1613.15 = 1662.29$ kJ **Ans.**

Problem 11: *In a piston cylinder machine air at 1.0132 bar and 20°C has it's initial volume 1000 cm^3. Assuming gas pressure remains constant, heat is added to air until it's temperature reaches 250°C. Calculate the heat added, the work the gas does on the piston and change in internal energy of the gas.*

Solution: We know that, from the ideal gas relation

$$PV = mRT$$

$$m = \frac{PV}{RT} = \frac{(1.0132 \times 10^5)\,(1000 \times 10^{-6})}{287\,(20 + 273)}$$

$$= 1.205 \times 10^{-3} \text{ kg}$$

$$\left[\begin{array}{l} R = \dfrac{\bar{R}}{m} : \dfrac{8314 \text{ J/kg mol k}}{28.97 \text{ kg/kg mol}} \\ \quad = 287 \text{ J/kg k} \\ R = \text{gas constant} \\ m = \text{molecular weight of air} \end{array}\right]$$

The heat transfer is given by

$$Q = mC \cdot (\Delta T)$$

$$: 1.205 \times 10^{-3} \cdot 1.005 \times 10^3 \, (250 - 20)$$

$$Q = 278.53 \text{ J}$$

The work done on the piston is

$$W = \int p\,dv$$

$$W = P(V_2 - V_1) \quad \text{[for constant pressure process]}$$

Also $\dfrac{P_1 V_1 : mRT_1}{P_2 V_2 : mRT_2}$ $\quad \therefore \quad V_2 = V_1 \left(\dfrac{T_2}{T_1}\right)$:

$$= 1000 \left(\frac{250 + 273}{20 + 273}\right)$$

$$= 1784.98 \text{ cm}^3$$

$$\therefore \quad W = 1.0132 \times 10^5 \, (1784.98 - 1000)\, 10^{-6}$$

$$W = 79.53 \text{ J}$$

Similarly the work done on the gas is negative of this quantity

$$W = -79.53 \text{ J}$$

Consider air as a system, then

$$Q = W + \Delta U$$

$$\Delta U = Q - W$$

$$= 278.53 + 79.53 = 358.06 \text{ J.}$$

Ans.

Problem 12: *Twenty five people attend a dinner party in a hall which measures 9 m × 8 m × 2.5 m height. Each person gives up 120 watts of heat. Assuming that the party hall is completely sealed off and insulated, calculate the air temperature rise occurring within 15 minutes. Assume that each person occupies a volume of 0.08 m³.*

Solution: The party hall is considered as a thermodynamic system. The people are adding heat to the air in a constant volume process.

∴ The volume of air in the hall

$$= \text{volume of hall} - \text{volume occupied by 25 people}$$

$$= 9 \times 8 \times 2.5 - 25 \times 0.08$$

$$= 178 \text{ m}^3$$

The mass of air in the hall

$$m = \frac{PV}{RT} = \frac{1.03 \times 10^5 \times 178}{287 \times (273 + 21)}$$

$$m = 217.28 \text{ kg}$$

Assume initial pressure = P = 1.03 bar

Initial temperature = 21°C

From first law of thermodynamics

$$Q = \Delta U + W$$

$$Q = \Delta U$$

[W = 0, for constant volume process)

$$\therefore \quad \Delta U = Q = 120 \times 25 = 3000 \text{ watts}$$

For 15 minutes

$$\Delta U = 3000 \times 15 \times 60 = 2700 \text{ kJ}$$

$\therefore$ The temperature rise during 15 minutes

$$= \frac{\Delta U}{MCv}$$

C_v for air: 0.743 kJ/kg k

$$= \frac{2700}{217.28 \times 0.743} = 16.72°\text{C} \qquad \textit{Ans.}$$

Problem 13: *1 g of Nitrogen contained in a cylinder undergoes the following sequence of quasi-static processes:*

1. *An adiabatic expansion which follows the law $PV^{1.3} = c$ and the volume doubles.*
2. *A constant pressure cooling process which reduces volume to its initial value.*
3. *With piston in the firmly locked position, heat is supplied to rise the pressure to its initial value.*

The fluid is initially at 160°C and 5 atm. pressure. Calculate the net work done by the fluid and show the processes on P-V diagram.

Solution: Consider fluid contained in a cylinder as system

Then $P_1 = 5 \text{ atm} = 5{,}066 \times 10^5 \ N/m^2$

$T_1 = 160 + 273 = 433$ k

$m = 1 \text{ g} = 10^{-3}$ kg

$$\left[\begin{aligned} 1 \text{ atm} &= 1.01325 \text{ bar} \\ &= 1.01325 \times 10^5 \text{ N/m}^2 \end{aligned}\right]$$

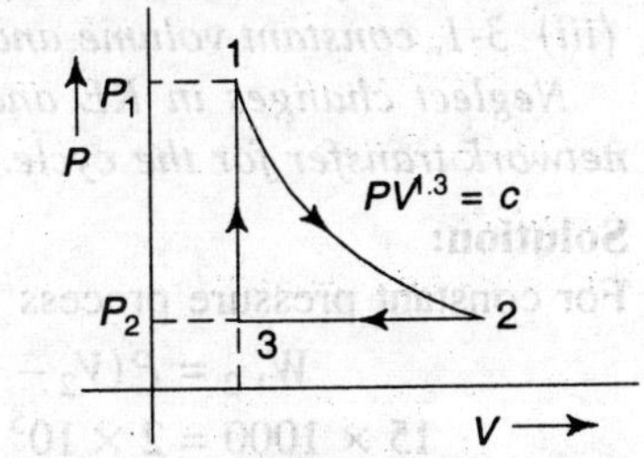

Fig. 3.30

For nitrogen

$$R = \text{gas constant} = \frac{\bar{R}}{M}$$

$$= \frac{8314}{28} = 297 \text{ J/kgk}$$

$$\therefore \quad V_1 = \frac{mRT_1}{P_1} = \frac{(0.001)\ 297\ (433)}{5.066 \times 10^5}$$

$$V_1 = 2.538 \times 10^{-4} \text{ m}^3$$

Also given that,

$$V_2 = 2V_1 = 2 \times 2.538 \times 10^{-4}$$

$$V_2 = 5.077 \times 10^{-4} \text{ m}^3$$

$$P_1 V_1^{\,n} = P_2 V_2^{\,n} = c \qquad \{n = 1.3\}$$

$$\therefore \quad P_2 = P_1 \left(\frac{V_1}{V_2}\right)^n = 5.066 \times 10^5 \left(\frac{2.538 \times 10^{-4}}{5.077 \times 10^{-4}}\right)^{1.3}$$

$$P_2 = 2.05 \times 10^5 \text{ N/m}^2$$

For the process 1-2, Rev. adiabatic expansion

$$\therefore \quad W_{1\text{-}2} = \frac{P_2V_2 - P_1V_1}{1-n}$$

$$= \frac{(2.05 \times 10^5)\,(5.077 \times 10^{-4}) - (5.066 \times 10^5)\,(2.538 \times 10^{-4})}{1-1.3}$$

$$W_{1\text{-}2} = +\ 81.65 \text{ J}$$

For the constant pressure process 2-3

$$W_{2\text{-}3} = \int_{V_2}^{V_3} pdv = \int_{V_2}^{V_3} p_2 dv$$

$$= 2.05 \times 10^5 (V_3 - V_2)$$

$$= 2.05 \times 10^5 (2.538 \times 10^{-4} - 5.077 \times 10^{-4})$$

$$= -52.04 \text{ J}$$

During a constant volume process 3-1, the workdone is zero [$\because$ $V_3 = V_1$]

$\therefore$ $W_{net} = W_{1\text{-}2} + W_{2\text{-}3} + W_{3\text{-}4}$

$= +81.65 - 52.04 + 0 = +29.61$ J *Ans.*

$\therefore$ The positive sign indicates that there is net work output from all three processes.

Problem 14: *A gas occupies 0.3 m³ at 2 bar. It executes a cycle consisting of processes*

(i) 1-2 constant pressure process with work interaction of 15 kJ.

(ii) 2-3 compression process which follows the law PV = constant and $U_3 = U_2$.

(iii) 3-1, constant volume and change in internal energy $U_1 - U_3$ is – 40 kJ

Neglect changes in KE and PE. Draw P-V diagram for the following processes and determine network transfer for the cycle. Also show that first law is obeyed by the cycle.

Solution:

For constant pressure process 1-2

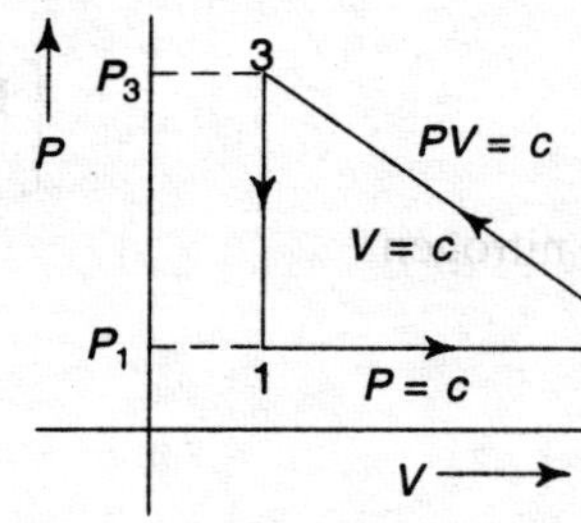

Fig. 3.31

$$W_{1\text{-}2} = P(V_2 - V_1) : P_1(V_2 - V_1)$$

$$15 \times 1000 = 2 \times 10^5 (V_2 - 0.3)$$

$\therefore$ $\boxed{V_2 = 0.375 \text{ m}^3}$

From first law,

$$Q_{1\text{-}2} = W_{1\text{-}2} + \Delta U_{1\text{-}2}$$

$= 15 + U_2 - U_1$ [$\because$ $U_3 = U_2$]

$= 15 + U_3 - U_1$ $\left[\begin{array}{l} U_1 - U_3 = -40 \\ U_3 - U_1 = +40 \end{array}\right]$

$= 15 + 40 = 55$ kJ

Now for the process 2-3, $PV = c$

[$\because$ $U_3 = U_2,\ \Delta U = 0$]

$$Q_{2\text{-}3} = W_{2\text{-}3} + \Delta U_{2\text{-}3}$$

$$Q_{2\text{-}3} = W_{2\text{-}3} + U_3 - U_2$$

$$= W_{2\text{-}3} + \cancel{U_3} - \cancel{U_3}$$

$\therefore$ $Q_{2\text{-}3} = W_{2\text{-}3}$

$\therefore$ $Q_{2\text{-}3} = P_2V_2 \ln\left(\dfrac{V_3}{V_2}\right)$, also

$$P_2V_2 = P_3V_3 = c \qquad\qquad V_3 = V_1$$

$$P_3 = P_2\left(\frac{V_2}{V_3}\right) \qquad\qquad P_2 = P_1$$

$$P_3 = 2\left[\frac{0.375}{0.3}\right]$$

$$\boxed{P_3 = 2.5 \text{ bar}}$$

$$\therefore \quad Q_{2\text{-}3} = 2 \times 10^5 \times 0.375 \times \ln\left(\frac{0.3}{0.375}\right)$$

$$= -\ 16735.76 \text{ J}$$

$$Q_{2\text{-}3} = -16.735 \text{ kJ} = W_{2\text{-}3}$$

For the constant volume process 3-1;

$$Q_{3\text{-}1} = W_{3\text{-}1} + \Delta U_{3\text{-}1}$$

$$= W_{3\text{-}1} + U_1 - U_3$$

$$Q_{3\text{-}1} = W_{3\text{-}1} - 40$$

$$W_{3\text{-}1} = -\int_3^1 vdp = V_3\,(P_1 - P_3)$$

$$= -0.3\,(2 - 2.5) = 15 \text{ kJ}$$

$$\therefore \quad Q_{3\text{-}1} = 15 - 40 = -25 \text{ kJ}$$

W_{net} for the cycle $= W_{net} = W_{1\text{-}2} + W_{2\text{-}3} + W_{3\text{-}1}$

$$= 15 + (-16.74) + 15$$

$$= 13.26 \text{ kJ} \qquad \textit{Ans.}$$

Net heat transfer for the cycle

$$= Q_{net} = Q_{1\text{-}2} + Q_{2\text{-}3} + Q_{3\text{-}1}$$

$$= 55 + (-16.74) + (-25) = 13.26 \text{ kJ} \qquad \textit{Ans.}$$

$$\therefore \quad \Sigma Q_{cycle} = \Sigma W_{cycle} \qquad \textit{Ans.}$$

Problem 15: *A cylinder piston arrangement contains unit mass of liquid water at a pressure of 1 × 10⁵ pa. If heat is added at constant pressure till the water is completely evaporated, determine the amount of heat and work interaction of the system.*

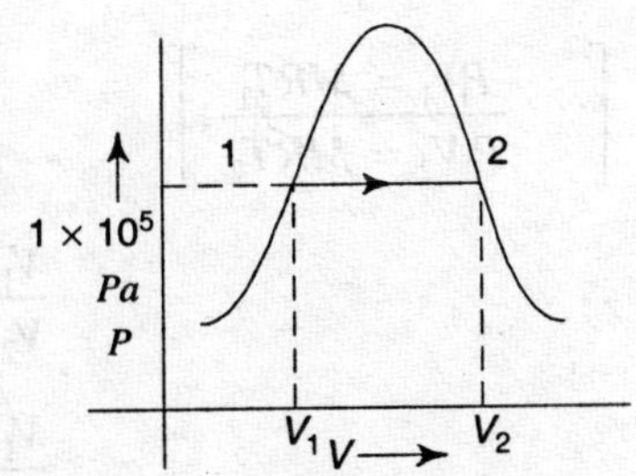

Fig. 3.32

Solution:

At pressure 1×10^5 pa

$$h_f = h_1 = 417.46 \text{ kJ/kg}$$

$$h_g = h_2 = 2675.47 \text{ kJ/kg}$$

$$v_f = v_1 = 0.001043 \text{ m}^3\text{/kg}$$

$$v_g = v_2 = 1.6949 \text{ m}^3\text{/kg}$$

[from steam tables
hf and *vf* = sp. enthalpy and sp. volume of saturated liquid
hg and *vg* = sp. enthalpy and sp. volume of saturated vapour]

We know that for constant pressure process

$$W = \int pdv = P\,(V_g - V_f)$$
$$= 1 \times 10^5\ (1.6949 - 0.001043)$$
$$= 169.38 \text{ kJ} \qquad \textit{Ans.}$$
$$q = \Delta h = h_g - h_f$$
$$= 2675.47 - 417.46$$
$$= 2258.01 \text{ kJ} \qquad \textit{Ans.}$$

It means 2258.01 kJ of heat is transferred to the system and it does 169.38 kJ of work

Problem 16: *In an engine, the charge is at 105 kpa and 310 k at the beginning of compression. It reaches 2.5 MPa after compression by following the law $PV^\gamma = c$. Calculate the temperature at the end of compression, compression ratio and work done.*

Solution:

For a reversible adiabatic process, consider $\gamma = 1.4$

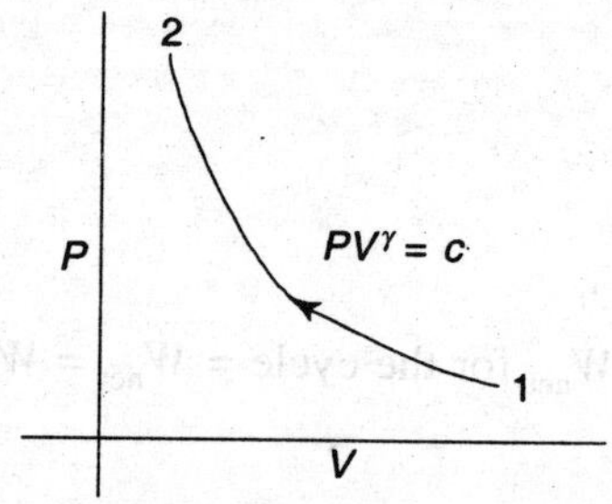

Fig. 3.33

$$\therefore \qquad T_2 = T_1 \left(\frac{P_2}{P_1}\right)^{\frac{\gamma-1}{\gamma}}$$

$$= 310 \left(\frac{2.5 \times 10^6}{105 \times 10^3}\right)^{\frac{1.4-1}{1.4}} = 766.83 \text{ k}$$

$$\boxed{T_2 = 766.83 \text{ k}} \qquad \textit{Ans.}$$

Compression ratio $\displaystyle = \frac{V_1}{V_2} = \left(\frac{P_2}{P_1}\right)\left(\frac{T_1}{T_2}\right)$

$$\left[\because \begin{array}{l} P_1V_1 = \cancel{M}\cancel{R}T_1 \\ P_2V_2 = \cancel{M}\cancel{R}T_2 \end{array} :\right]$$

$$\therefore \qquad \frac{V_1}{V_2} = \frac{2.5 \times 10^6 \times 310}{105 \times 10^3 \times 766.83}$$

$$\frac{V_1}{V_2} = 9.62 \qquad \textit{Ans.}$$

$$W_{1\text{-}2} = \frac{P_1V_1 - P_2V_2}{\gamma - 1} = \frac{R(T_1 - T_2)}{\gamma - 1}$$

$$W_{1\text{-}2} = \frac{0.2927\ (310 - 766.83)}{1.4 - 1}$$

$$\boxed{W_{1\text{-}2} = -\ 334.28 \text{ kJ/kg of air}} \qquad \textit{Ans.}$$

Problem 17: *The initial state of air is 100 kpa and 30°C. After compression the final volume is $\frac{1}{15}$th of original volume and compression follows polytropic law $PV^{1.25} = c$. Determine pressure and temperature at the final state, work done and energy transfer as heat.*

Solution: We know that,

$$P_1V_1^{1.25} = P_2V_2^{1.25} = c$$

and

$$\frac{P_2}{P_1} = \left(\frac{V_1}{V_2}\right)^n$$

$$\therefore \quad P_2 = 100 \times 10^3\ (15)^{1.25} = 2.95 \text{ MPa}$$ *Ans.*

Also from Ideal gas equation, we can write

$$\frac{P_1V_1 = MRT_1}{P_2V_2 = MRT_2}$$

$$T_2 = T_1 \cdot \left(\frac{P_2}{P_1}\right)\left(\frac{V_1}{V_2}\right)$$

$$T_2 = 303\left(\frac{2.95 \times 10^6}{100 \times 10^3}\right)\left(\frac{1}{15}\right)$$

$$\boxed{T_2 = 595.9 \text{ k}}$$ *Ans.*

or Alternate Method

we know that,

$$\left(\frac{T_2}{T_1}\right) = \left(\frac{V_1}{V_2}\right)^{n-1}$$

$$T_2 = 303\ (15)^{1.25-1}$$

$$\boxed{T_2 = 596.3} \approx 595.9 \text{ k}$$

for polytropic process 1-2

$$W_{1\text{-}2} = \frac{P_1V_1 - P_2V_2}{n-1} = \frac{R(T_1 - T_2)}{n-1}$$

$\bar{R} = 8.314$ kJ/mol k

$$W_{1\text{-}2} = \frac{0.287(303 - 596.3)}{1.25 - 1}$$

$R = C_p - C_v$: 0.287 kJ/kgk for air

$$= -336.70 \text{ kJ/kg of air}$$

The energy transfer as heat can be calculated as

$$Q_{1\text{-}2} = \left(\frac{\gamma - n}{\gamma - 1}\right) \times W_{1\text{-}2}$$

$$= \left(\frac{1.4 - 1.25}{1.4 - 1}\right) \times (-336.70)$$

$$\boxed{Q_{1\text{-}2} = -126.26 \text{ kJ/kg of air}}$$ *Ans.*

Alternate method

$$Q = \frac{R(T_2 - T_1)}{\gamma - 1} + W$$

$$= \frac{8.314\,(596.3 - 303)}{1.4 - 1} - 9.75 = -3.65 \text{ kJ/mol}$$

or

$$Q_{1\text{-}2} = \frac{0.287\,(596.3 - 303)}{1.4 - 1} - (336.7) = -126.26 \text{ kJ/kg of air} \qquad \textit{Ans.}$$

Problem 18: *A cylinder with air comprises the system. The cycle is completed as follows*

(i) 82,000 Nm work is done by the piston on the air during compression and 45 kJ of heat is rejected to the surrounding

(ii) During expansion, 1,00,000 Nm of work is done by the air on the piston. Determine quantity of heat added to the system. *(VTU, Aug 2001)*

Solution:

For compression process 1–2,

$$w_{1-2} = -82{,}000 \text{ Nm}$$
$$= -82 \text{ kJ}$$
$$Q_{1-2} = -45 \text{ kJ}$$

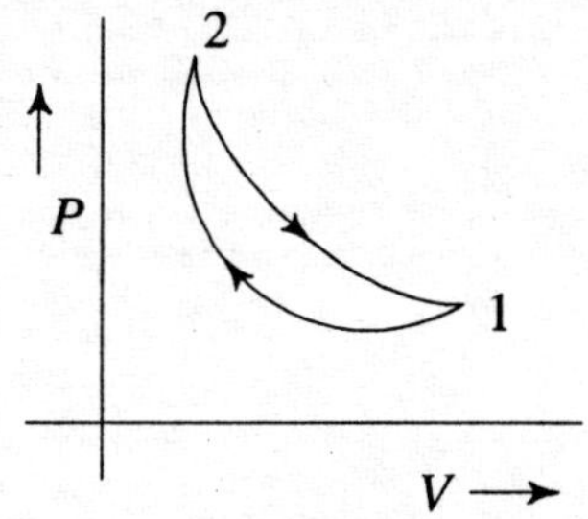

Fig. 3.34

from first law of T.D.

$$Q_{1-2} = U_2 - U_1 + w_{1-2}$$
$$-45 = U_2 - U_1 + (-82)$$
$$U_2 - U_1 = 37 \text{ kJ}$$

for expansion process 2 – 1,

$$w_{2-1} = 1{,}00{,}000 \text{ Nm} = 100 \text{ kJ.}$$
$$\Delta U = U_1 - U_2 = -(U_2 - U_1) = -37 \text{ kJ}$$
$$\therefore \quad Q_{2-1} = U_1 - U_2 + w_{2-1}$$
$$= -37 + 100$$
$$\boxed{Q_{2-1} = 63 \text{ kJ}} \qquad \textit{Ans.}$$

∴ The heat added to the System is 63 kJ *Ans.*

Problem 19: *The air is compressed from an initial state of 105 kpa and 30°C to a final state of 515 kpa and 30°C by the following processes:*

(i) Constant volume heat addition followed by cooling at constant pressure

(ii) Process 1-2 is considered to be isothermal compression.

(iii) Adiabatic compression which is followed by cooling at constant volume.

Determine and compare heat and work interactions and change in internal energy for each process.

Solution:

(i) For the path 1-*a*-2

1*a* – constant volume heat addition

*a*2 – constant pressure cooling

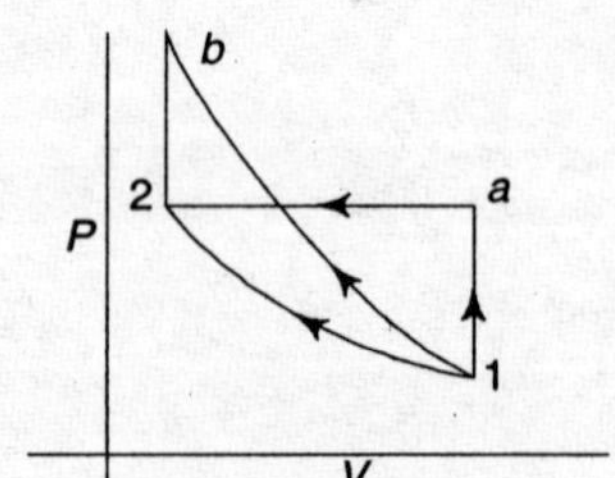

Fig. 3.35

$$W_{1\text{-}a} = \int_1^a p.dv = 0$$

from Ist law of Thermodynamics

$$\Delta U_{1\text{-}a} = U_a - U_1 = Q - W = C_v\,(T_a - T_1)$$

$$\therefore \quad \frac{P_1 V_1 = M RT_1}{P_a V_a : M RT_a}$$

$$\therefore \qquad T_a = \frac{P_a T_1}{P_1} : \frac{P_2 \cdot T_1}{P_1} \qquad [\because \quad P_a = P_2]$$

$$T_a = \frac{515 \times 10^3 \times 303}{105 \times 10^3} = 1486.14 \text{ k}$$

$$\boxed{T_a = 1486.14 \text{ k}}$$

For ideal gas, we can write

$[\because \quad V_a = V_1]$ $\qquad du = C_v \, dT$

$$\therefore \qquad U_a - U_1 = 0.718 \; (1486.14 - 303)$$

$$U_a - U_1 = 849.49 \text{ kJ/kg °k}$$

$[\because \quad C_v = 0.718$ kJ/kg k for air]

for the process a-2

We know that, $\qquad dQ = dh = C_p dT$

$$\therefore \qquad Q_{a\text{-}2} = C_p \; (T_2 - T_a)$$

$$= 1.005 \; (303 - 1486.14)$$

$$Q_{a\text{-}2} = -1189.05 \text{ kJ/kg}$$

$$U_{a\text{-}2} = U_2 - U_a : Cv \; (T_2 - T_a) : 0.718 \; (303 - 1486.14)$$

$$U_2 - U_a = -849.49 \text{ kJ/kg}$$

$$\therefore \qquad Q = W + \Delta U$$

$$W = Q - \Delta U$$

$$= -1189.05 + 849.49$$

$$W_{a\text{-}2} = -339.56 \text{ kJ/kg}$$

$$\therefore \qquad (\Delta U)_{1\text{-}a\text{-}2} = (\Delta U)_{1\text{-}a} + (\Delta U)_{a\text{-}2}$$

$$= 849.49 \; - 849.49 = 0$$

$$(Q)_{1\text{-}a\text{-}2} = (Q)_{1\text{-}a} + (Q)_{a\text{-}2}$$

$$= 849.49 - 1189.05$$

$$Q_{1\text{-}a\text{-}2} = -339.56 \text{ kJ/kg}$$

$$W_{1\text{-}a\text{-}2} = (W)_{1\text{-}a} + (W)_{a\text{-}2}$$

$$= 0 - 339.56$$

$$\boxed{W_{1\text{-}a\text{-}2} = -339.56 \text{ kJ/kg}}$$

$$Q_{1\text{-}a} = (\Delta U)_{1\text{-}a}$$

$$Q_{1a} = \cancelto{0}{W_{1a}} + (\Delta U)_{1a}$$

$$Q_{1a} = (\Delta U)_{1a}$$

$(\Delta U)_{1-a-2} = 0$ *Ans.*

$(Q)_{1-a-2} = -339.56$ kJ/kg *Ans.*

$(W)_{1-a-2} = 339.56$ kJ/kg *Ans.*

(ii) Now, consider the process 1-2

Isothermal compression

From Ist law

$$\delta Q = \delta W + du$$

$$Q_{1\text{-}2} = W_{1\text{-}2} + (U_2 - U_1)$$

$$= P_1 V_1 \ln \left(\frac{V_2}{V_1} \right) + mC_v \; (T_2 - T_1)$$

For isothermal $T_2 = T_1$

$$Q_{1\text{-}2} = P_1 V_1 \ln\left(\frac{V_2}{V_1}\right) = W_{1\text{-}2}$$

$$\therefore \quad Q_{1\text{-}2} = W_{1\text{-}2} = \text{RT} \cdot \ln\left(\frac{V_2}{V_1}\right)$$

Now $\dfrac{P_1V_1 = MR\cancel{T_1}}{P_2V_2 = MR\cancel{T_2}}$ $[\because \; T_1 = T_2]$

$$\therefore \quad P_1V_1 = P_2V_2 = c \qquad \text{[Isothermal]}$$

$$\therefore \quad \left(\frac{V_2}{V_1}\right) = \left(\frac{P_1}{P_2}\right)$$

$$\therefore \quad Q_{1\text{-}2} = W_{1\text{-}2} = RT \ln\left(\frac{P_1}{P_2}\right)$$

$$= 0.287\ (303) \ln\left(\frac{105}{515}\right)$$

$$\boxed{Q_{1\text{-}2} = -138.28 \text{ kJ/kg}} \quad \textit{Ans.}$$

$$\boxed{W_{1\text{-}2} = -138.28 \text{ kJ/kg}} \quad \textit{Ans.}$$

$$\boxed{(\Delta U)_{1\text{-}2} = U_2 - U_1 = 0} \quad \textit{Ans.}$$

(iii) For the process 1-*b*-2
1-*b*; adiabatic compression
b-2; heat rejection at constant volume
for 1-*b*; $PV^\gamma = c$

$$\therefore \quad \left(\frac{T_b}{T_1}\right) = \left(\frac{V_1}{V_b}\right)^{\gamma - 1} = \left(\frac{P_2}{P_1}\right)^{\frac{\gamma-1}{\gamma}}$$

$$\therefore \quad T_b = T_1 \left(\frac{P_2}{P_1}\right)^{\gamma - 1/\gamma}$$

$$= 303 \left(\frac{515}{105}\right)^{\frac{1.4-1}{1.4}}$$

$$\boxed{T_b = 477.25 \text{ k}}$$

$$Q_{1\text{-}b} = 0$$

$$\therefore \quad Q_{1\text{-}b} = W_{1\text{-}b} + (\Delta U)_{W1\text{-}b}$$

$$\therefore \quad (\Delta U)_{1\text{-}b} = (-W_{1-b})$$

$$U_b - U_1 = C_v\,(T_b - T_1) = 0.718\ (477.25 - 303)$$

$$= \boxed{U_b - U_1 = 125.11 \text{ kJ/kg}}$$

$\therefore \quad \boxed{W_{1\text{-}b} = -125.11 \text{ kJ/kg}}$

For process b-2, $W_{b\text{-}2} = 0$

$$\therefore \quad Q_{b\text{-}2} = (\Delta U)_{b\text{-}2}$$

$$(\Delta U)_{b\text{-}2} = U_2 - U_b = Cv\,(T_2 - T_b)$$

$$= 0.718\,(303 - 477.25)$$

$$\boxed{U_2 - U_b = -125.11 \text{ kJ/kg}}$$

$$\therefore \quad \boxed{Q_{b\text{-}2} = -125.11 \text{ kJ/kg}}$$

For the complete process 1-b-2

$$(Q)_{1\text{-}b\text{-}2} = Q_{1\text{-}b} + Q_{b\text{-}2}$$

$$= 0 - 125.11$$

$$\boxed{Q_{1\text{-}b\text{-}2} = -125.11 \text{ kJ/kg}}$$ *Ans.*

$$(W)_{1\text{-}b\text{-}2} = W_{1\text{-}b} + W_{b\text{-}2}$$

$$= -125.11 + 0$$

$$\boxed{W_{1\text{-}b\text{-}2} = -125.11 \text{ kJ/kg}}$$ *Ans.*

$$(\Delta U)_{1\text{-}b\text{-}2} = (\Delta U)_{1\text{-}b} + (\Delta U)_{b\text{-}2}$$

$$= 125.11 - 125.11 = 0$$

Quantity	*Path 1-a-2*	*Path 1-2*	*Path 1-b-2*
ΔU(kJ/kg)	0	0	0
Q (kJ/kg)	–339.56	–138.28	–125.11
W(kJ/kg)	–339.56	–138.28	–125.11

ΔU represent change in property which is independent of path followed by the system. Hence ΔU is same for all processes. But Q and W depends on the path of the system. As initial and final temperatures are same for all the paths, Q becomes equal to W in all these cases. This obeys first law of thermodynamics.

Problem 20: *2.5 kg of substance expands from initial state (515 kpa and 0.25 m^3) to the final state of 105 kpa by following the law $PV^{1.25} = C$. The internal energy of the substance is related by the equation $u = 3.55\ pv + 85$, where u is in kJ/kg, p is in kpa, v is in m^3/kg*

(a) For quasi-static expansion, determine heat and work interactions & change in internal energy.

(b) For the same above conditions, if heat transfer is 30 kJ, find work transfer

Solution:

(a)

$$u = 3.55\ PV + 85$$

$$(\Delta U)_{1\text{-}2} = u_2 - u_1 = 3.55\,(P_2V_2 - P_1V_1)$$

and

$$(\Delta U)_{1\text{-}2} = U_2 - U_1 = 3.55\,(P_2V_2 - P_1V_1)$$

$$\because \quad P_1V_1^{1.25} = P_2V_2^{1.25} = C$$

$$V_2 = V_1\left(\frac{P_1}{P_2}\right)^{1/1.25}$$

$$= 0.25\left(\frac{515 \times 10^3}{105 \times 10^3}\right)^{\frac{1}{1.25}}$$

$$\boxed{V_2 = 0.892 \text{ m}^3}$$

$$\therefore \quad \Delta U = 3.55\,(105 \times 10^3 \times 0.892 - 515 \times 10^3 \times 0.25)$$

$$\boxed{\Delta U = -124.56 \text{ kJ}}$$ **Ans.**

for a quasi-static process

$$W_{1\text{-}2} = \int p dv = \frac{P_1V_1 - P_2V_2}{n-1}$$

$$= \frac{515 \times 10^3 \times 0.25 - 105 \times 10^3 \times 0.892}{1.2 - 1}$$

$$= 175.45 \text{ kJ}$$ **Ans.**

$$\therefore \quad Q_{1\text{-}2} = \Delta U_{1\text{-}2} + W_{1\text{-}2}$$

$$= -124.56 + 175.45$$

$$\boxed{Q_{1\text{-}2} = 50.89 \text{ kJ}}$$ **Ans.**

(b) In this case between same initial and final states $Q_{1\text{-}2} = 30$ kJ.
Since end states are same ΔU would be same

$$Q = \Delta U + W$$

$$30 = -124.56 + W$$

$$W_{1\text{-}2} = 154.56 \text{ kJ}$$ **Ans.**

Note: In case (b) work is not equal to $\int p dv$ as the process is not quasi-static.

Problem 21: *Unit mass of working substance is contained in a piston-cylinder machine. When the piston moves outward, the system temperature decreases from 75 to 25°C and rejects 25 kJ of heat. The specific internal energy is given by* $u = \left[A + \frac{t}{1.25}\right]$ *kJ. where 'A'= constant, t = temperature in °C. Determine the work done by the system.*

Solution: The given equation is

$$u = \left[A + \frac{t}{1.25}\right] \text{ kJ}$$

At state 1,
$$u_1 = \left[A + \frac{t_1}{1.25}\right] = A + \frac{75}{1.25} = (A + 60) \text{ kJ/kg}$$

At state 2,
$$u_2 = \left[A + \frac{t_2}{1.25}\right] = A + \frac{25}{1.25} = (A + 20) \text{ kJ/kg}$$

$$\therefore \quad du = u_2 - u_1 = (A + 20) - (A + 60) = -40 \text{ kJ/kg}$$

$$\therefore \quad dU = mdu = 1 \times (-40) = -40 \text{ kJ}$$

From first law of thermodynamics

$$\delta Q_{1\text{-}2} = \delta W_{1\text{-}2} + dU_{1\text{-}2}$$

$$-25 = \delta W_{1\text{-}2} + (-40)$$
$$\delta W_{1\text{-}2} = -25 + 40 = 15 \text{ kJ}$$
$$W_{1\text{-}2} = 15 \text{ kJ} \quad \text{[Work is done by the system].}$$

Problem 22: *An engine cylinder of diameter 22.5 cms has a stroke length of 37.5 cms. The swept volume is 4 times the clearance volume. The pressure of gases at the beginning of expansion stroke is 1570 kpa. Find the work done during expansion stroke assuming the process as reversible adiabatic. Assume $\gamma = 1.4$* (VTU March 2001)

Given:

$$d = 22.5 \text{ cms} = 0.225 \text{ m}$$
$$L = 37.5 \text{ cms} = 0.375 \text{ m}$$
$$V_2 = 4V_c = 4V_1$$
$$P_1 = 1570 \text{ kpa}$$
$$\gamma = 1.4$$
$$W_{1\text{-}2} = ?$$

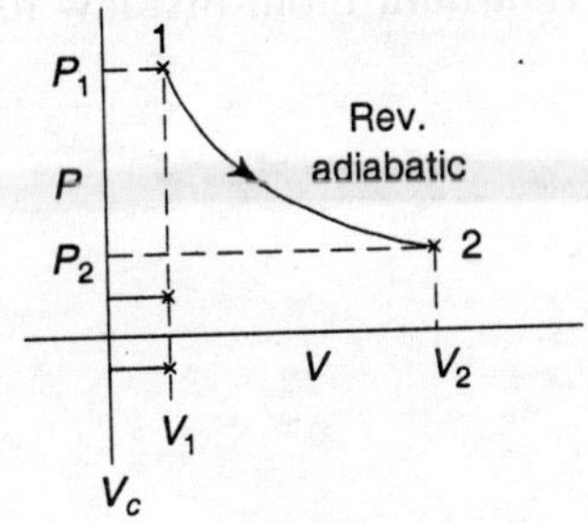

Fig. 3.36

Solution: Swept volume

$$V_2 = \frac{\pi d^2}{4} \times L$$

$$V_2 = \frac{\pi}{4}(0.225)^2 \times 0.375$$

$$V_2 = 0.0149 \text{ m}^3$$

also,
$$V_2 = 4V_1$$

$$\therefore \quad V_1 = \frac{V_2}{4} = \frac{0.0149}{4} = 3.727 \times 10^{-3} \text{ m}^3$$

For the process 1-2, isentropic expansion

we can write,
$$\left(\frac{P_2}{P_1}\right)^{\frac{\gamma-1}{\gamma}} = \left(\frac{V_1}{V_2}\right)^{\gamma-1}$$

$$\therefore \quad \left(\frac{P_2}{P_1}\right) = \left(\frac{V_1}{V_2}\right)^{\gamma \times \left(\frac{\gamma-1}{\gamma-1}\right)} = \left(\frac{V_1}{V_2}\right)^{\gamma}$$

$$\therefore \quad P_2 = 1570 \times \left(\frac{3.727 \times 10^{-3}}{0.0149}\right)^{1.4}$$

$$\boxed{P_2 = 225.60 \text{ kpa}}$$

For rev. adiabatic (isentropic) process

$$W_{1\text{-}2} = \frac{P_1V_1 - P_2V_2}{\gamma - 1}$$

$$= \frac{1570 \times 10^3(3.727 \times 10^{-3}) - 225.6 \times 10^3(0.0149)}{1.4 - 1}$$

$$= W_{1\text{-}2} = 6224 \text{ J}$$

$$\boxed{W_{1\text{-}2} = 6.224 \text{ kJ}}$$ *Ans.*

Problem 23: *A gas of mass 1.5 kg undergoes a quasi-static expansion by following the equation P = a + bV, where 'a' and 'b' are constants. The initial and final pressures are 10 bar and 2 bar respectively and the corresponding volumes are 0.2 m³ and 1.2 m³. The specific internal energy of the gas is given by u = 1.5 PV – 85 kJ/kg, where 'P' is in bar and 'V' is in m³/kg. Calculate the net heat transfer and internal energy of the gas during expansion*

Solution: From first law for a quasi-static process

$$W_{1\text{-}2} = \int_1^2 pdv = \int_1^2 (a + bV)dV = \int_1^2 adV + b \cdot V.\, dV$$

$$= V + \left.\frac{bV^2}{2}\right|_0^2 = \left(aV_2 + \frac{bV_2^2}{2}\right) - \left(aV_1 + \frac{bV_1^2}{2}\right)$$

$$= a(V_2 - V_1) + b\left(\frac{V_2^2 - V_1^2}{2}\right)$$

$$= a(1.2 - 0.2) + b\left(\frac{1.2^2 - 0.2^2}{2}\right)$$

$$W_{1\text{-}2} = a(1) + 0.76$$

To find constants 'a' and 'b'

We know that, $P = a + bV$

At state 1; $10 = a + b(0.2)$

At state 2; $2 = a + b\,(1.2)$

Solving for 'a' and 'b'

$$b = -8$$

$$a = 11.6$$

$$\therefore \quad W_{1\text{-}2} = 11.6 + 0.7(-8)$$

$$\boxed{W_{1\text{-}2} = 6 \text{ kJ}}$$ *Ans.*

The specific internal energy of the given gas is

$$u = 1.5\, PV - 85$$

$$u_1 = 1.5\, P_1V_1 - 85$$

$$= 1.5\,(10 \times 10^5)\left(\frac{0.2}{1.5}\right) - 85 = 199.91 \text{ kJ/kg}$$

$$199.91 \times 1.5 = 299.87 \text{ kJ}$$

Similarly $u_2 = 1.5\,(2 \times 10^5)\left(\frac{1.2}{1.5}\right) - 85$

$$= 239.915 \text{ kJ/kg}$$

$$239.915 \times 1.5 = 359.87 \text{ kJ}$$

$$u_2 - u_1 = 239.91 - 199.91$$

$$= 40 \text{ kJ/kg} = 40 \times 1.5$$

$$u_2 - u_1 = 60 \text{ kJ}$$ *Ans.*

$\therefore$ $Q_{1\text{-}2} = (\Delta U)_{1\text{-}2} + W_{1\text{-}2}$

$$= 60 + 6 = 66 \text{ kJ.}$$ *Ans.*

Problem 24: *A domestic refrigerator is loaded with food and the door is closed. During certain period, the refrigerator consumes 1.5 kW hr. of energy and the internal energy of the system drops by 3800 kJ. Find the net heat transfer for the system.*

Solution: The work consumed by the refrigerator

$$W = 1.5 \text{ kW hr} = 1.5\ \frac{\text{kJ}}{\text{s}}\ \text{hr.} \qquad [1 \text{ W} = 1 \text{ J/s}]$$

$$= 1.5 \times 60 \times 60\ \frac{\text{kJ}}{\cancel{\text{hr}}}\ \cancel{\text{hr}}$$

$$= 5400 \text{ kJ}$$

As internal energy of the system decreases ΔU is to be negative.

$\therefore$ $\delta Q = du + \Delta W$ from Ist law

$$Q_{1\text{-}2} = \Delta U + W_{1\text{-}2}$$

$$= -3800 - 5400$$

[$W_{1\text{-}2}$: work consumed –ve sign is to be used]

$$Q_{1\text{-}2} = -9200 \text{ kJ}$$ *Ans.*

The refrigerator rejects 9200 kJ of energy in the form of heat.

Problem 25: *One kg of air having an initial volume of 0.3 m^3 is heated at constant pressure of 3.5 bar until the volume is doubled. Calculate*

(i) Initial and final temperature of air,

(ii) Work done, and

(iii) heat added.

Take C_p = 1.005 kJ/kg k, R = 0.2927 kJ/kgk *(VTU Aug 2000)*

Solution: For a constant pressure process

$\frac{N}{\cancel{m^2}} \times \cancel{m^3}\ m$ $\qquad W_{1\text{-}2} = P(V_2 - V_1)$

$\frac{J}{1000}$ kJ $\qquad = \frac{3.5(0.6 - 0.3)}{1000} \times 10^5$

$$\boxed{W_{1\text{-}2} = 105 \text{ kJ}}$$ *Ans.*

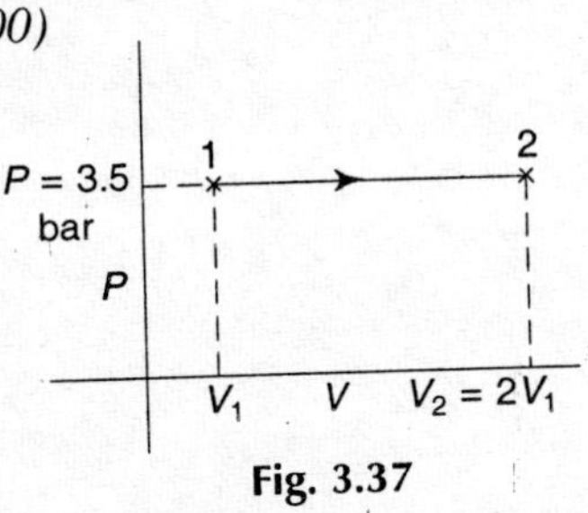

Fig. 3.37

Heat added during the process 1-2

$$Q_{1\text{-}2} = mC_p\,(T_2 - T_1)$$

also,

$$W_{1\text{-}2} = mR\,(T_2 - T_1)$$

$$\therefore \quad T_2 - T_1 = \frac{W_{1-2}}{1 \times R} = \frac{105}{0.2927}$$

$$T_2 - T_1 = 358.72 \text{ k}$$

$$\therefore \quad Q_{1\text{-}2} = 1 \times 1.005 \times (358.72)$$

$$= 360.51 \text{ kJ/kg}$$

Ans.

Then

$$\frac{P_1 V_1 = MRT_1}{P_2 V_2 = MRT_2}$$

$$\therefore \quad \frac{V_1}{V_2} = \frac{T_1}{T_2}$$

$$\therefore \quad T_2 = \left(\frac{V_2}{V_1}\right) \cdot T_1 = \frac{2V_1}{V_1}\,T_1 = 2T_1$$

$$T_2 - T_1 = 358.72 \text{ k}$$

$$2T_1 - T_1 = 358.72$$

$$\boxed{T_1 = 358.72 \text{ k}}$$

Ans.

$$\therefore \quad T_2 = 2 \times 358.72$$

$$\boxed{T_2 = 717.44 \text{ k}}$$

Ans.

Problem 26: *An air standard diesel cycle has atmospheric air at initial pressure 1 bar and volume 3.5 m³. The compression ratio of the cycle is 14. After adiabatic compression, heat addition at constant pressure doubles the volume. Then adiabatic expansion reduces the volume to its original value and further heat rejection take place at constant volume and thus completes the cycle. Determine net work transfer during the cycle.*

Solution:

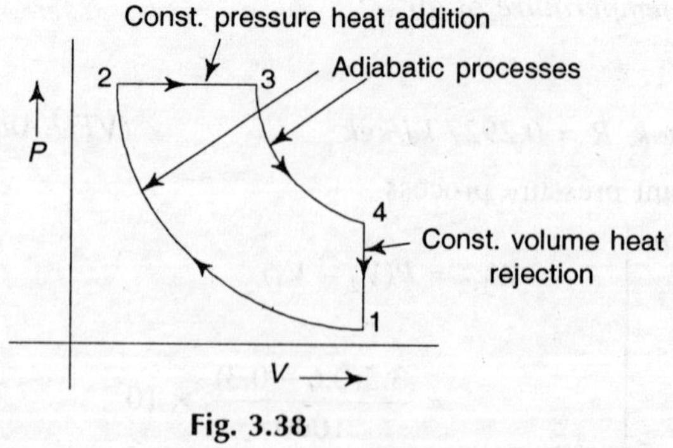

Fig. 3.38

Let $P_1 = 1 \text{ bar}, V_1 = 3.5 \text{ m}^3$

Compression ratio $= \dfrac{V_1}{V_2} = 14$

$$V_2 = \frac{3.5}{14} = 0.25 \text{ m}^3$$

For the process 1-2; adiabatic compression

$$\left(\frac{P_2}{P_1}\right)^{\frac{\gamma-1}{\gamma}} = \left(\frac{V_1}{V_2}\right)^{\gamma-1}$$

$$\therefore \quad P_2 = P_1\left(\frac{V_1}{V_2}\right)^{\gamma} = 1 \times 10^5 \left[\frac{3.5}{0.25}\right]^{1.4}$$

$$P_2 = 40.23 \times 10^5 \text{ N/m}^2$$

$$\boxed{P_2 = 40.23 \text{ bar}}$$

for the process 3-4, adiabatic expansion

$$\left(\frac{P_4}{P_3}\right) = \left(\frac{V_3}{V_4}\right)^{\gamma}$$

$$\therefore \quad P_4 = 40.23\left(\frac{0.5}{3.5}\right)^{1.4}$$

$$\boxed{P_4 = 2.63 \text{ bar}}$$

$P_3 = P_2$
$V_3 = 2 \times V_2$
$V_3 = 0.5$
$V_4 = V_1$

Work done can be calculated as follows

for process 1-2; $W_{1\text{-}2} = \dfrac{P_1V_1 - P_2V_2}{\gamma - 1}$

$$= \frac{1 \times 10^5 \times 3.5 - 40.23 \times 10^5 \times 0.25}{1.4 - 1}$$

$$\boxed{W_{1\text{-}2} = -1639.375 \text{ kJ}}$$

for process 2-3; $W_{2\text{-}3} = P_2(V_3 - V_2)$

$$= 40.23 \times 10^5 (0.5 - 0.25)$$

$$\boxed{W_{2\text{-}3} = 1005.75 \text{ kJ}}$$

for process 3-4; $W_{3\text{-}4} = \dfrac{P_3V_3 - P_4V_4}{\gamma - 1}$

$$= \frac{40.23 \times 10^5 \times 0.5 - 2.63 \times 10^5 \times 3.5}{1.4 - 1}$$

$$\boxed{W_{3\text{-}4} = 2727.5 \text{ kJ}}$$

for process 4-1 = $W_{4-1} = 0$ [$\because$ it is a constant volume process]

$\therefore$ Net work done for the entire cycle

$$= W_{net} = W_{1-2} + W_{2-3} + W_{3-4} + W_{4-1}$$
$$= -1639.375 + 1005.75 + 2727.5 + 0$$

$$\boxed{W_{net} = 2093.875 \text{ kJ}}$$ ***Ans.***

Problem 27 : *The specific heat of a fluid involved in a constant pressure non-flow process is given by $C_p = a + \frac{b}{t+85}$ J/kg °C, where t is in °C, a =2, b = 38. The pressure of the system is 125 kPa and volume changes from 0.35 to 0.4 m^3. The temperature of the system changes from 10 to 100°C. Determine (i) heat and work interactions and (ii) change in internal energy.* *(VTU, Jan 2003)*

Solution : We can write, for the process 1-2

$$Q_{1-2} = \int_{T_1}^{T_2} Cp\, dT = \int_{T_1}^{T_2} \left(a + \frac{b}{t+85} \right) dT$$

$$= \int_{T_1}^{T_2} \left(2 + \frac{38}{t+85} \right) dT = \int_{10}^{100} 2dT + \int_{10}^{100} \frac{38}{t+85} dT$$

$$= 2(100 - 10) + 38 \ln (t+85)_{10}^{100}$$

$$= 2\,(100 - 10) + 38 \ln \left(\frac{100+85}{10+85} \right)$$

$$= 180 + 38 \ln 185 - 38 \ln 95$$
$$= 205.32 \text{ J/kg}$$

$$Q_{1-2} = 0.205 \text{ kJ/kg}$$ ***Ans.***

$$W_{1-2} = P(V_2 - V_1)$$
$$= 125 \times 10^3 (0.4 - 0.35)$$
$$= 6.25 \text{ kJ/kg}$$ ***Ans.***

from Ist law

$$Q = W + dU$$

For 1 kg of fluid

$$Q_{1-2} = W_{1-2} + \Delta U_{1-2}$$
$$\Delta U_{1-2} = 0.205 - 6.25$$
$$= -6.045 \text{ kJ/kg}$$ ***Ans.***

Problems on SFEE

Problem 28: *In a steady flow system, fluid enters the control volume at 100 kpa and leaves at 990 kpa with a mass flow rate of 1 kg/s. The velocities at inlet and exist of control volume are 45 m/s and 25 m/s. The heat interaction during the process is 35 × 10³ kJ/hr. The rise in enthalpy of fluid is 15 kJ/kg. By neglecting change in potential energy, determine power developed by the system.*

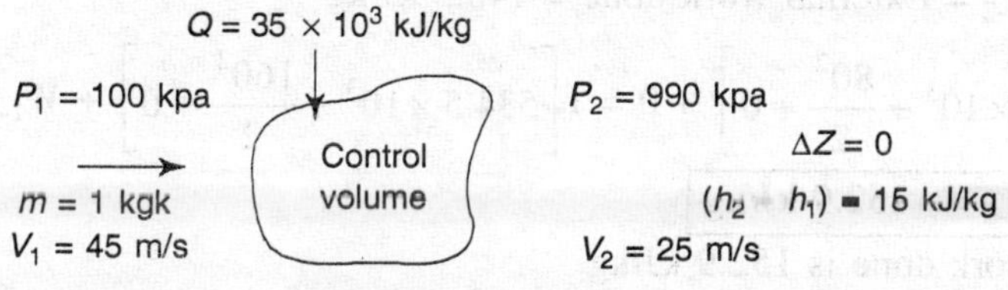

Fig. 3.39

Solution:
The SFEE can be written as

$$\dot{m}\left[h_1 + \frac{V_1^2}{2} + z_1 g + q\right] = \dot{m}\left[h_2 + \frac{V_2^2}{2} + z_2 g + w\right]$$

$$Q = \frac{35 \times 10^3 \times 1000}{3600} = 9722.22 \text{ J/sec}$$

$$\therefore \quad W = (h_1 - h_2) + \left(\frac{V_1^2 - V_2^2}{2}\right) + Q$$

$$\therefore \quad W = (-15 \times 10^3) + \left(\frac{45^2 - 25^2}{2}\right) + 9722.22$$

$$= -4577.78 \text{ J/s} = -4.577 \text{ kW}$$ *Ans.*

Problem 29: *In a rotary compressor air enters at 16°C and at a pressure of 100 × 10³ N/m², the enthalpy of which is being 391.2 kJ/kg. The corresponding properties of air at the exit are 245°C, 600 × 10³ N/m² and 534.5 kJ/kg respectively. By assuming no heat interaction in the compressor, determine*

(i) Work done per unit mass, by neglecting change in kinetic energy

(ii) Work done per unit mass, when velocity at inlet and exit of compressor are 85 and 160 m/s.

Solution:
The inlet conditions are:

$$T_1 = 16°\text{C}$$
$$P_1 = 100 \times 10^3 \text{ N/m}^2$$
$$h_1 = 391.2 \text{ kJ/kg}$$

The outlet conditions are

$$T_2 = 245°\text{C}$$
$$P_2 = 600 \times 10^3 \text{ N.m}^2$$
$$h_2 = 534.5 \text{ kJ/kg}$$

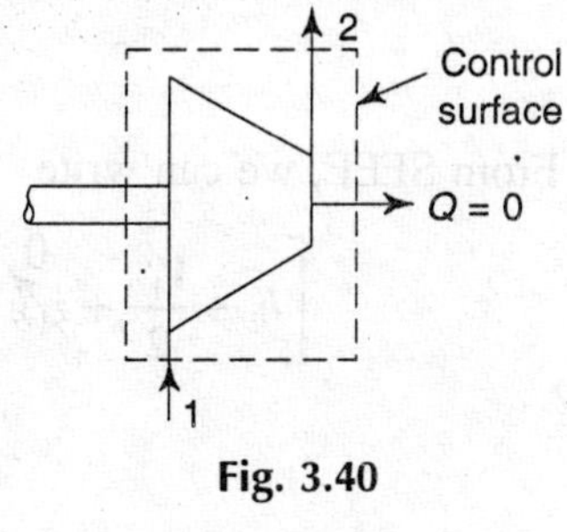

Fig. 3.40

(i) $$m\left[h_1 + \frac{V_1^2}{2} + z_1 g\right] + Q = m\left[h_2 + \frac{V_2^2}{2} + z_2 g\right] + W$$

$$1[391.2 \times 10^3 + 0 + 0] + 0 = 1[534.5 \times 10^3 + 0] + W_{1\text{-}2}$$

$$\therefore \quad W_{1\text{-}2} = -143.3 \text{ kJ/kg}$$

$$W_{1\text{-}2} = \text{External work done} = 143.3 \text{ kJ/kg}$$

(ii) $$1\left[391.2\times 10^3 + \frac{80^2}{2} + 0\right] + 0 = 1\left[534.5\times 10^3 + \frac{160^2}{2} + 0\right] + W_{1\text{-}2}$$

$$\boxed{W_{1\text{-}2} = -152.9 \text{ kJ/kg}}$$

∴ External work done is 152.9 kJ/kg.

[**Note:** The compressor absorbs $W_{1\text{-}2}$ amount of work, hence negative sign is used].

Problem 30: *In a rotary compressor air flow steadily at a rate of 1.5 kg/s. The air enters with a velocity of 80 m/s and leaves at 4.5 m/s. The specific volume at inlet is 0.9 m³/kg and at the exit is 0.4 m³/kg. As the air passes through the compressor, the specific enthalpy of air is increased by 110 kJ/kg and losses heat of 20 kJ/kg. Determine*

(a) The inlet and exit areas of the compressor in m².

(b) The power required to drive the compressor in kW.

Solution:

$$m = 1.5 \text{ kg/s}$$
$$v_1 = 0.9 \text{ m}^3\text{/kg}$$
$$V_1 = 80 \text{ m/s}$$
$$v_2 = 0.4 \text{ m}^3\text{/kg}$$
$$V_2 = 45 \text{ m/s}$$
$$h_1 - h_2 = -110 \text{ kJ/kg}$$
$$Q_{1\text{-}2} = -20 \text{ kJ/kg}$$

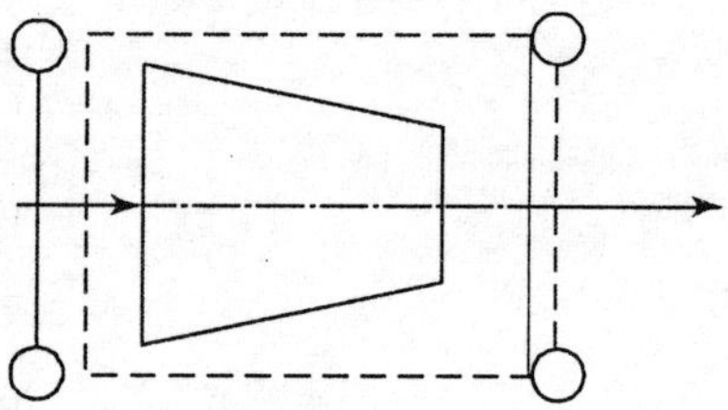

Fig. 3.41

From continuity equation,

$$m = \frac{A_1 V_1}{v_1} = \frac{A_2 V_2}{v_2}$$

$$1.5 = \frac{A_1(80)}{0.9}$$

Similarly $$1.5 = \frac{A_2(45)}{0.4}$$

$$\boxed{A_1 = 0.0168 \text{ m}^2}$$

$$\boxed{A_2 = 0.0133 \text{ m}^2}$$

From SFEE, we can write

$$\left[h_1 + \frac{V_1^2}{2} + \overset{0}{\cancel{z_1 g}}\right] + Q = \left[h_2 + \frac{V_2^2}{2} + \overset{0}{\cancel{z_2 g}}\right] + W$$

$$W_{1\text{-}2} = (h_1 - h_2) + \left(\frac{V_1^2 - V_2^2}{2}\right) + Q_{1\text{-}2}$$

$$= -110 \times 10^3 + \frac{(80^2 - 45^2)}{2} + (-20 \times 10^3)$$

$$W_{1\text{-}2} = -127812.5 \ \frac{\text{J}}{\text{kg}} = -127812.5 \times 1.5 \ \left[\frac{\text{J}}{\text{kg}} \times \frac{\text{kg}}{\text{sec}}\right]$$

$$= -191718.75 \ \frac{\text{J}}{\text{sec}} = -191.71 \ \frac{\text{kJ}}{\text{sec}} = \text{kW}$$

∴ The power required to drive the compressor is 191.71 kW.

Problem 31: *A centrifugal air compressor compresses 5.5 m³/min of air from 80 kpa to 675 kpa. The initial and final specific volumes of air are 0.35 m³/kg and 0.1 m³/kg. If the diameters at the inlet and outlet are 10 cms and 6.5 cms, determine (i) change in flow work, (ii) mass flow rate, and (iii) velocity change.* *(VTU, Aug. 2000)*

Solution: Consider, compressor as control volume

$$v_1 = 0.35 \text{ m}^3/\text{kg}$$
$$P_1 = 80 \text{ kpa}$$

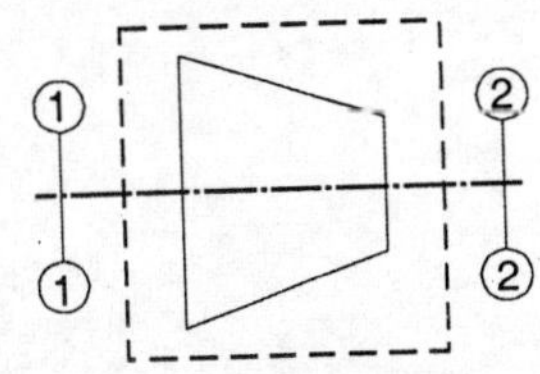

Fig. 3.42

Mass flow rate of air at inlet

$$m_1 = \frac{V_1}{v_1} = \frac{5.5}{0.35} \quad \left| \quad \frac{\text{m}^3/\text{min}}{\text{m}^3/\text{kg}} : \frac{\text{kg}}{\text{min}} \right.$$

$$m_1 = 15.71 \text{ kg/min}$$

At section (2) – (2)
From continuity eqn.

$$m = \frac{A_1 V_1}{v_1} = \frac{A_2 V_2}{v_2}$$

$$\therefore \quad m_1 V_1 = m_2 V_2 = V_2 \qquad [\because \ m_1 = m_2]$$
$$V_2 = 15.71 \times 0.1$$
$$V_2 = 1.571 \text{ m}^3/\text{min}$$

$$C_1 = \frac{V_1}{A_1} = \frac{5.75}{\frac{\pi d_1^2}{4} \times 60} : \frac{5.5}{\frac{\pi}{4} \cdot (0.1)^2 \times 60}$$

$$C_1 = 11.67 \text{ m/s}$$

$$C_2 = \frac{V_2}{A_2} = \frac{1.571}{\frac{\pi}{4} \cdot (0.061)^2 \times 60} = 7.890 \text{ m/s}$$

(i) Change in flow work $= \Delta W = \dfrac{P_2 V_2 - P_1 V_1}{60}$

$$= \frac{675 \times 1.571 - 80 \times 5.5}{60} \quad \left| \quad \frac{\text{kN}}{\text{m}^2} \times \frac{\text{m}^3}{\text{min}} = \frac{\text{kJ}}{\text{min}} : \div 60 = \frac{\text{kJ}}{\text{sec}} \right.$$

$$= 10.34 \text{ kW}$$

(ii) Mass flow rate $m_1 = m_2 = m = 15.71$ kg/min
(iii) Velocity change $C_2 - C_1 = 7.89 - 11.67 = -3.78$ m/s (decreases)
[**Note** C_1 and C_2 denotes velocities at inlet and outlet]

Problem 32: *In a steady flow device, the work done by each kg of fluid is 150 kJ. At the entry, the fluid properties are $v_1 = 0.40$ m³/kg, $P_1 = 550$ kpa and $V_1 = 15$ m/s and $V_2 = 0.65$ m³/kg, $P_2 = 105$ kpa and $V_2 = 275$ m/s are the fluid properties at the exit. The inlet is 30 m above the floor and exit section is at the floor level. The heat loss from the fluid is 10×10^3 J/kg. Calculate change in internal energy through the device*

Solution:

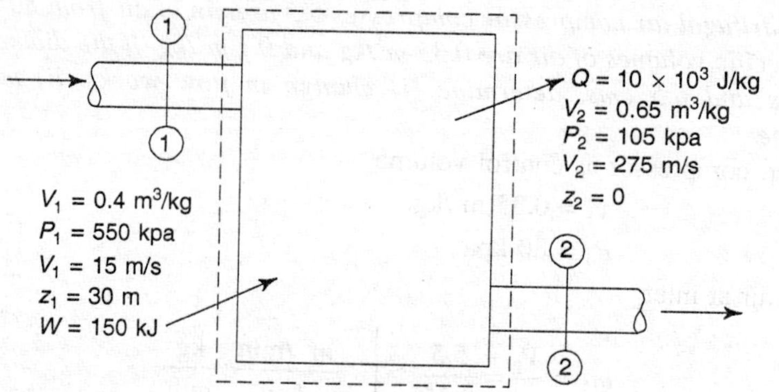

Fig. 3.43

The SFEE is

$$\left[u_1 + P_1V_1 + \frac{V_1^2}{2} + z_1g\right] + Q = \left[u_2 + P_2V_2 + \frac{V_2^2}{2} + z_2g\right] + W$$

$$u_1 - u_2 = (P_2V_2 - P_1V_1) + \frac{V_2^2 - V_1^2}{2} + (z_2 - z_1)g + W - Q$$

$$= (105 \times 10^3 \times 0.65 - 550 \times 10^3 \times 0.4) + \frac{275^2 - 15^2}{2}$$

$$+ (0 - 30) \cdot 9.81 + 150 \cdot 10^3 - (-10 \times 10^3)$$

$$-151750 + 37700 - 294.3 + 150 \times 10^3 + 10 \times 10^3$$

$$\boxed{u_1 - u_2 = +45.65 \text{ kJ/kg}}$$ *Ans.*

∴ The specific internal energy decreases by 45.65 kJ/kg of fluid.

Problem 33: *The gas leaving the turbine of an aircraft engine flows steadily into the engine jet-pipe at a temperature of 900°C, a pressure of 192 × 10³ N/m² and a velocity of 300 m/s relative to the pipe. The gas leaves the jet pipe at a temperature of 820°C and a pressure of 105 × 10³ N/m². By neglecting heat transfer, calculate velocity of gas leaving the jet pipe relative to the engine. The enthalpy of gas at different temperatures is*

t°C	820	900
h, J/kg	862.6 × 10³	951.7 × 10³

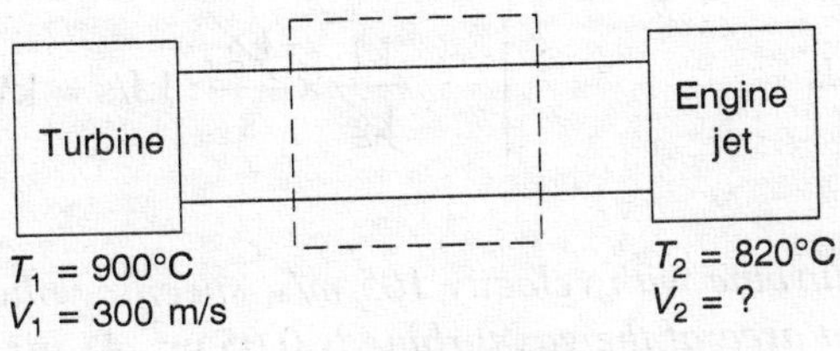

Fig. 3.44

Solution: Consider the engine as a system, the SFEE can be written as

$$\left[h_1 + \frac{V_1^2}{2} + z_1 g\right] + Q = \left[h_2 + \frac{V_2^2}{2} + z_2 g\right] + W$$

$$951.7 \times 10^3 + \frac{300^2}{2} + 0 + 0 = 862.6 \times 10^3 + \frac{V_2^2}{2} + 0 + 0$$

$$996700 - 862.6 \times 10^3 = \frac{V_2^2}{2}$$

$$\boxed{V_2 = 517.8 \text{ m/s}}$$ ***Ans.***

Problem 34: *The fluid flows through an open system with flow rate of 5 kg/s. It enters the system at a pressure of 600 kN/m², a velocity 200 m/s, internal energy 2100 kJ/kg and specific volume 0.45 m³/kg. The fluid leaves the system at a pressure of 150 kN/m², velocity = 145 m/s, internal energy 1600 kJ/kg and specific volume 1.5 m³/kg. The heat loss through the system is 40 × 10³ J/kg to the surroundings. By neglecting potential energy change, calculate power of the system.*

Solution: The SFEE is

$$(u_1 + P_1V_1) + \frac{V_1^2}{2} + \overset{0}{\cancel{z_1 g}} + Q = u_2 + P_2V_2 + \frac{V_2^2}{2} + \overset{0}{\cancel{z_2 g}} + W$$

$$\therefore \quad W = (u_1 - u_2) + (P_1V_1 - P_2V_2) + \left(\frac{V_1^2 - V_2^2}{2}\right) + Q$$

$$W = (2100 - 1600) + (600 \times 0.45 - 150 \times 1.5) + \frac{200^2 - 145^2}{2} - 40$$

$$= (2100 \times 10^3 - 1600 \times 10^3) + (600 \times 0.45 - 150 \times 1.5)\, 10^3$$

$$+ \frac{200^2 - 141^2}{2} - 40 \times 10^3$$

$$\boxed{W = 514.50 \text{ kJ/kg}}$$ ***Ans.***

Since '*W*' is + ve, work is done by the system

∴ Power output from the system

$$= 514.50 \times 5 \text{ kg/s} \qquad \Big| \quad \frac{\text{kJ}}{\cancel{\text{kg}}} \times \frac{\cancel{\text{kg}}}{\text{s}} : \text{kJ/s} = \text{kW}$$

$$\boxed{\text{Power} = 2572.5 \text{ kW}}$$

Ans.

Problem 35: *Air enters a gas turbine with velocity 105 m/s, specific volume 0.8 m³/kg and leaves at 135 m/s and 1.5 m³/kg. The inlet area of the gas turbine is 0.05 m². As air passes through the turbine, the specific enthalpy decreases by 145 kJ/kg and air loses 27 kJ/kg of heat. Determine*

(i) Mass flow rate of air

(ii) Exit area of the turbine

(iii) Power developed by the turbine

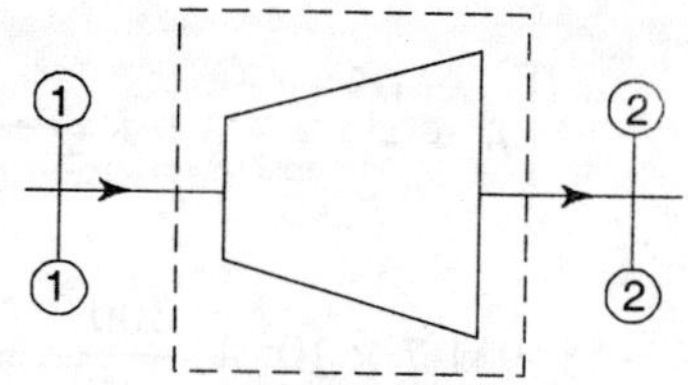

Fig. 3.45

Solution:

From continuity eqn, we can write

$$m = \frac{A_1 V_1}{V_1} = \frac{A_2 V_2}{V_2}$$

$$= \frac{0.05 \times 10^5}{0.8} = 6.56 \text{ kg/s} \qquad \textit{Ans.}$$

$$6.56 = \frac{A_2 \times 135}{1.5}; \; A_2 = 0.728 \text{ m}^2 \qquad \textit{Ans.}$$

From SFEE we can write

$$\left[h_1 + \frac{V_1^2}{2} + \cancelto{0}{z_1 g}\right] + Q = \left[h_2 + \frac{V_2^2}{2} + \cancelto{0}{z_2 g}\right] + W$$

$$\therefore \qquad W = (h_1 - h_2) + \left(\frac{V_1^2 - V_2^2}{2}\right) + Q$$

$$= 145 \times 10^3 + \frac{105^2 \times 135^2}{2} + (-27 \times 10^{+3})$$

$$= 114.4 \frac{\text{kJ}}{\text{kg}} = 114.4 \times 6.56 = 750.46 \text{ kW.} \qquad \textit{Ans.}$$

Problem 36: *In a nozzle, the ideal gas expands from a pressure of 20 bar to 3 bar and the process is reversible adiabatic. The inlet conditions are 500°C and 35 m/s. Determine area and velocity at the outlet section of the nozzle, if the flow rate is 5 kg/s. Take R = 190 J/kg k and γ = 1.35.*

Solution: The SFEE for unit mass

$$u_1 + P_1 V_1 + \frac{V_1^2}{2} + z_1 g + Q = u_2 + P_2 V_2 + \frac{V_2^2}{2} + z_2 g + W$$

$$W = 0, \quad \text{No shaft work}$$

$$\Delta PE = 0, \quad \text{Process is adiabatic } Q = 0;$$

$$\therefore \quad u_1 + P_1V_1 + \frac{V_1^2}{2} = u_2 + P_2V_2 + \frac{V_2^2}{2}$$

$$\therefore \quad \frac{V_2^2 - V_1^2}{2} = h_1 - h_2 = C_p\,(T_1 - T_2)$$

$$C_p = \frac{\gamma}{\gamma - 1} \times R = \frac{1.35}{1.35 - 1} \times 190 = 733 \text{ J/kg k}$$

also

$$\frac{T_2}{T_1} = \left(\frac{P_2}{P_1}\right)^{\gamma - 1/\gamma}$$

$$\therefore \quad T_2 = (500 + 273)\left(\frac{3}{20}\right)^{\frac{1.35-1}{1.35}}$$

$$\boxed{T_2 = 472.91 \text{ k}}$$

$$\therefore \quad \frac{V_2^2 - V_1^2}{2} = 733\,(500 - 472.91) = 19856.97$$

$$\therefore \quad V_2^2 = (19856.97 \times 2) + V_1^2 = 39713.94 + 1225$$

$$V_2 = 202.33 \text{ m/s}$$

We know that, $PV = mRT$ for ideal gas

$$\therefore \quad v_2 = \frac{RT_2}{P_2} = \frac{190 \times 472.91}{3 \times 10^5} = 0.299 \text{ m}^3/\text{kg}$$

from continuity equation, we can write

$$\dot{m} = \frac{A_2V_2}{v_2}$$

$$\therefore \quad A_2 = \frac{mv_2}{V_2} = \frac{4 \times 0.299}{202.33} = 7.38 \times 10^{-3}\text{m}^2$$

Problem 37: *In a machine, fluid flows steadily at the rate of 16 kg/min. The fluid enters the machine at a pressure of 105 kpa, velocity 5 m/s, and specific volume 0.5 m^3/kg and leaves at 650 kpa, 10 m/s and 0.25 m^3/kg. The internal energy of the fluid at exit is 150 kJ/kg and is greater than that at entrance and heat loss to the surroundings is 7000 kJ/min. Neglecting potential energy changes calculate the shaft work, and the ratio of inlet pipe diameter to outlet pipe diameter.*

Solution: From continuity equation,
We can write,

$$m = \frac{A_1V_1}{v_1} = \frac{A_2V_2}{v_2}$$

$$\therefore \quad \frac{A_1}{A_2} = \frac{V_2}{V_1} \cdot \frac{v_1}{v_2} = \frac{10}{5} \times \frac{0.5}{0.25} = 4$$

Ratio of pipe diameters

$$\frac{D_1}{D_2} = \sqrt{4} = 2 \qquad \left| \quad \frac{A_1 = \frac{\pi D_1^2}{4}}{A_2 = \frac{\pi D_2^2}{4}} : \frac{D_1^2}{D_2^2} : 4 \right.$$

$$\therefore \quad \boxed{\frac{D_1}{D_2} = 2}$$

$$Q = -7000 \text{ kJ/min}$$

$$= \frac{-7000}{16} \times \frac{\text{kJ}}{\text{kg}} \qquad \left| \quad m = 16 \frac{\text{kg}}{\text{min}} : \frac{\frac{\text{kJ}}{\text{min}}}{\frac{\text{kg}}{\text{min}}} \right.$$

$$\boxed{Q = -437.5 \text{ kJ/kg}}$$ *Ans.*

The SFEE can be written as

$$u_1 + P_1 v_1 + \frac{v_1^2}{2} + \overset{0}{g z_1} + Q = u_2 + P_2 v_2 + \frac{v_2^2}{2} + \overset{0}{g z_2} + W$$

$$W = (u_1 - u_2) + (P_1 V_1 - P_2 V_2) + \frac{v_1^2 - v_2^2}{2} + Q$$

$$= -150 \times 10^3 + [105 \times 10^3 \times 0.5 - 650 \times 10^3 \times 0.25]$$

$$+ \frac{5^2 - 10^2}{2} - 437.5 \times 10^3$$

$$W = -150 \times 10^3 - 110000 - 37.5 - 437.5 \times 10^3$$

$$W = -697.53 \text{ kJ/kg}$$

$$W = -697.53 \times 16 \frac{\text{kJ}}{\text{min}} \qquad \left| \quad \frac{\text{kJ}}{\text{kg}} \times \frac{\text{kg}}{\text{min}} \right.$$

$$= \frac{-697.53 \times 16}{60} \times \frac{\text{kJ}}{\text{s}}$$

$$\boxed{W = -186.00 \text{ kJ/s}}$$ *Ams.*

Problem 38: *A gasoline engine operates steadily and delivers work output of 35 kW and transfers heat to the cooling water at the rate of 135 × 10³ kJ/hr. The sum of all other heat transfers to the surrounding atmosphere and the engine foundation is 30 × 10³ kJ/hr. The fuel-air mixture flows through the system at the rate of 130 kg/hr. Determine enthalpy change of fuel-air mixture by neglecting changes in kinetic and potential energy.*

Solution:

$$W = 35 \text{ kJ/s} = 35 \times 60 \times 60 \text{ kJ/hr}$$

Total heat transfer from the system

$$Q_{total} = (Q)_{\text{to cooling water}} + (Q)_{\text{surroundings}}$$
$$= 135 \times 10^3 + 30 \times 10^3$$
$$= -165 \times 10^3 \text{ kJ/hr}$$

[Negative sign is due to the fact that engine dissipates heat to the cooling water and surroundings].

∴ The SFEE can be written as

$$m\left[h_1 + \frac{V_1^2}{2} + z_1 g\right] + Q = m\left[h_2 + \frac{V_2^2}{2} + z_2 g\right] + W$$

as

$$\Delta PE = 0$$
$$\Delta KE = 0$$
$$mh_1 + Q = mh_2 + W$$

∴

$$m\,(h_1 - h_2) = W - Q$$

∴

$$h_1 - h_2 = \frac{W - Q}{m} : \frac{(35 \times 60 \times 60) + 165 \times 10^3}{130}$$
$$h_1 - h_2 = 2238.46$$
$$\boxed{h_1 - h_2 = 2238.46 \text{ kJ/kg}}$$

As

$$h_1 > h_2 \qquad \left| \quad \frac{\text{kJ}/\cancel{\text{hr}}}{\frac{\text{kg}}{\cancel{\text{hr}}}} = \text{kJ/kg} \right.$$

The enthalpy of fuel air mixture has decreased.

Problem 39: *A steam turbine operating under steady flow conditions receives 4500 kg of steam per hour. The steam enters the turbine at a velocity of 45 m/s and at the elevation of 4 m and specific enthalpy of 2850 kJ/kg. The steam leaves the turbine at a velocity of 9.5 m/s and at an elevation of 1 m with specific enthalpy of 2260 kJ/kg. The heat losses from the turbine to the surroundings is 16780 kJ/hr. Determine power developed by turbine.* *(VTU, March 2001)*

Solution:

The SFEE can be written as per unit mass

$$\left[h_1 + \frac{V_1^2}{2} + z_1 g\right] + Q = \left[h_2 + \frac{V_2^2}{2} + z_2 g\right] + W$$

$$W = (h_1 - h_2) + \frac{V_1^2 - V_2^2}{2} + g\,(z_1 - z_2) + Q$$

$$= (2850 - 2260)10^3 + \left(\frac{45^2 - 9.5^2}{2}\right) + 9.81\,(4 - 1) - 16780 \times 10^3$$

$$= 590000 + 967.375 + 29.43 - \frac{16780 \times 10^3}{4500} \qquad \frac{\text{J}}{\cancel{\text{hr}}} \Big/ \frac{\text{kg}}{\cancel{\text{hr}}}$$

$$W = 587.26 \text{ kJ/kg} \qquad \frac{\text{kJ}}{\text{kg}} \times \frac{\text{kg}}{\text{hr}}$$

Power developed by the turbine

$$P = \frac{587.26 \times 4500}{60 \times 60} \qquad \frac{\text{kJ}}{\text{hr}} \div 3600$$

$$\boxed{P = 734.07 \text{ kW}} \qquad \frac{\text{kJ}}{\text{S}} = \text{kW}$$

Problem 40: *A water heater operates under steady flow conditions receives 4.5 kg/s of water at 70°C with enthalpy of 315 kJ/kg. The water heating is done by mixing it with steam which is being supplied at temperature of 100°C and enthalpy 2665 kJ/kg. The mixture leaves the heater as liquid water at temperature 100°C and enthalpy 420 kJ/kg. Determine the amount of steam supplied to the heater per hour?*

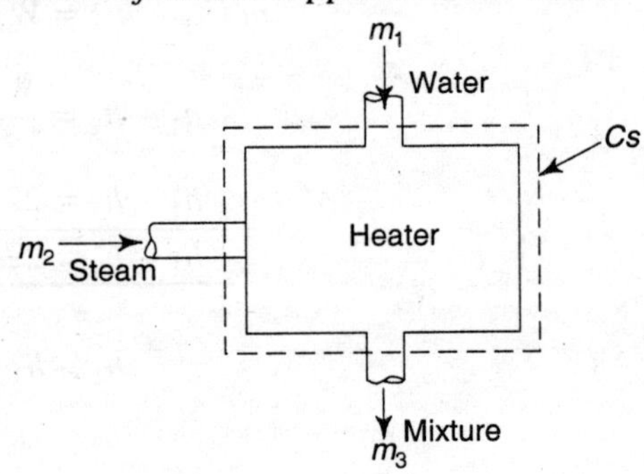

Fig. 3.46

Solution: Consider heater as the control volume, m_1 quantity of water enters into control volume, m_2 quantity of steam is supplied to heat the water.

Let m_3 = quantity of mixture leaving the control volume

∴ The mass balance gives

$$m_1 + m_2 = m_3$$

The SFEE can be written as

$$m_1\left[h_1 + \frac{V_1^2}{2} + z_1 g\right] + m_2\left[h_2 + \frac{V_2^2}{2} + z_2 g\right] +$$

$$Q = m_3\left[h_3 + \frac{V_3^2}{2} + z_3 g\right] + W$$

In the above equation

$$W = 0$$

$$\Delta PE = 0$$

$$\Delta KE = 0$$

$$Q = 0$$

$$\therefore \quad m_1 h_1 + m_2 h_2 = m_3 h_3$$

$$4.5 \times 315 + m_2\,(2665) = (4.5 + m_2)\,(420)$$

$$1417.5 + 2665\, m_2 = 18.90 + 420\, m_2$$

$$-472.5 = -2245 m_2$$

$$\therefore \quad \boxed{\begin{aligned} m_2 &= 0.21 \text{ kg/s} \\ m_2 &= 757.68 \text{ kg/hr} \end{aligned}}$$

Ans.

Problem 41: *Air passes through a heat exchanger with velocity of 25 m/s and raises its temperature from 20°C to 850°C. Then the air at same velocity enters the turbine and expands so that temperature decreases to 585°C. On leaving the turbine, air is taken at a velocity of 55 m/s to a nozzle, where temperature of air decreases to 500°C. If the flow rate is 2.5 kg/s, calculate (1) Heat transfer in the heat exchanger, (ii) Power output from the turbine with heat loss Q = 2000 × 10³ J/kg, (iii) Exit velocity from the nozzle assuming no heat loss.*

Take enthalpy of air as h = C_p . t, where C_p = 1.005 kJ/kg k and t is temp-erature.

Solution:

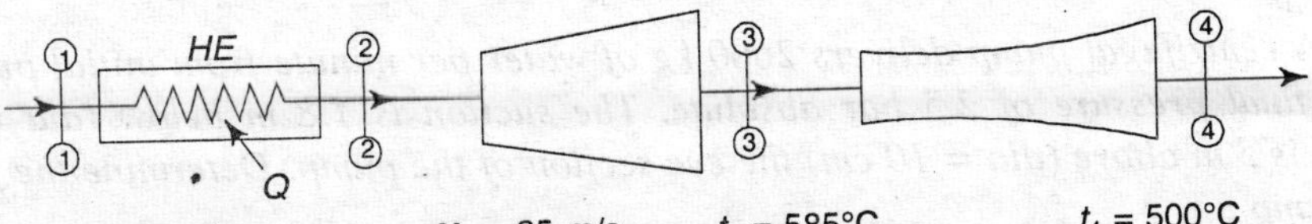

Fig. 3.47

For heat exchanger, SFEE is

$$m\left[h_1 + \frac{V_1^2}{2} + z_1 g\right] + Q = m\left[h_2 + \frac{V_2^2}{2} + z_2 g\right] + W$$

$$\left[\begin{array}{l} \because V_1 = V_2 = 25 \text{ m/s} \\ \Delta PE = 0,\ W = 0 \end{array}\right]$$

$$mh_1 \pm Q = mh_2$$

$$\therefore \quad Q = m\,(h_2 - h_1)$$

$$= m \cdot C_p\,(t_2 - t_1)$$

$$= 2.5 \times 1.005\,(850 - 20)$$

$$= 2085.3 \text{ kJ/s} \qquad \textit{Ans.}$$

For the turbines SFEE is

$$m\left[h_2 + \frac{V_2^2}{2}\right] + Q = m\left[h_3 + \frac{V_3^2}{2}\right] + W_T$$

$$W_T = m\,[h_2 - h_3] + m\left[\frac{V_2^2 - V_3^2}{2}\right] - Q$$

$$W_T = m \cdot C_p\,(t_2 - t_3) + m\left[\frac{V_2^2 - V_3^2}{2}\right] - Q$$

$$W_T = 2.5 \times 1.005 \times (850 - 585) + 2.5\left(\frac{25^2 - 55^2}{2}\right) + 665.8125$$

$$- 3000 - 2000 \qquad \textit{Ans.}$$

$$\boxed{W_T = 660.81 \text{ kJ/s}}$$

For the nozzle, SFEE per unit mass can be written as

$$m\left[h_3 + \frac{V_3^2}{2}\right] = m\left[h_4 + \frac{V_4^2}{2}\right]$$

$$\frac{V_4^2 - V_3^2}{2} = (h_3 - h_4) = C_p\,(t_3 - t_4)$$

$$V_4^2 - V_3^2 = 2 \times 1005 \times (585 - 500)$$

$$= 170850$$

$$V_4^2 = 170850 + 55^2$$

$$V_4 = 416.98 \text{ m/s}$$

Ans.

The exit velocity of air from the nozzle is 416.98 m/s.

Problem 42: *A centrifugal pump delivers 2600 kg of water per minute from initial pressure of 1 bar absolute to a final pressure of 2.5 bar absolute. The suction is 1.8 m below (dia = 12 cms) and discharge side is 5 m above (dia = 10 cm) the eye section of the pump. Determine the power required to drive the pump.*

Solution: For a centrifugal pump

$$\Delta U = 0,\; Q = 0$$

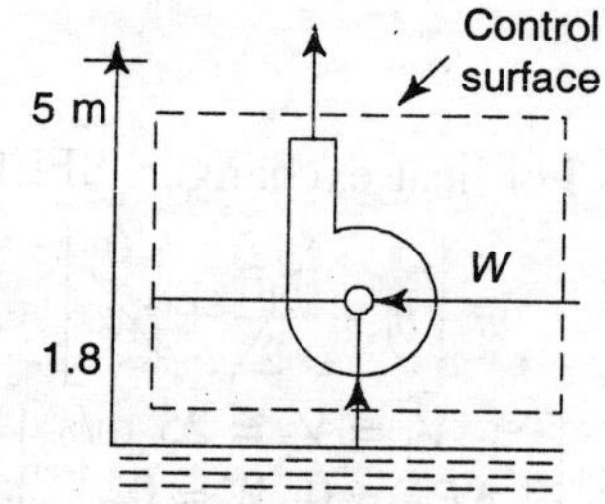

Fig. 3.48

The SFEE on unit mass basis

$$P_1 V_1 + \frac{V_1^2}{2} + gz_1 = P_2 V_2 + \frac{V_2^2}{2} + gz_2 + W$$

$$\frac{P_1}{\rho_1} + \frac{V_1^2}{2} + gz_1 = \frac{P_2}{\rho_2} + \frac{V_2^2}{2} + gz_2 + W \qquad \Bigg|\; V = \frac{1}{\rho}$$

For water $\rho_1 = \rho_2 = 1000 \text{ kg/m}^3$

also

$$m = \frac{A_1 V_1}{v_1} = A_1 V_1 \rho_1 = A_2 V_2 \rho_2$$

$$\frac{2600}{60} = \frac{\pi (0.12)^2}{4} \times V_1 \times 1000 = \frac{\pi}{4}\,(0.1)^2 \times V_2 \times 1000$$

$$V_1 = 3.83 \text{ m/s};\; V_2 = 5.51 \text{ m/s}$$

Taking suction level as datum for elevation head

$$z_1 = 0,\; z_2 = 1.8 + 5 = 6.8 \text{ m}$$

$$\therefore \qquad W = \frac{P_1 - P_2}{\rho} + \frac{V_1^2 - V_2^2}{2} + g(z_1 - z_2)$$

$$(\rho_1 = \rho_2 = \rho)$$

$$= \frac{1 \times 10^5 - 2.5 \times 10^5}{1000} + \frac{3.83^2 - 5.51^2}{2} + 9.81\,(0 - 6.8)$$

$$-\,150 - 7.845 - 66.708$$

$$W = -224.553 \text{ J/kg}$$

$\therefore$ The power required to drive the pump is $-224.53 \times \dfrac{2600}{60}$

$$\boxed{P = 9.72 \text{ kW}} \quad \Bigg| \quad \frac{J}{kg} \times \frac{kg}{sec} : \frac{J}{s} \div 1000 \text{ kJ/s : kW}$$

Problem 43: *In a video theatre during winter, comfortable conditions are achieved by circulation of hot fluid through a piping system. At the inlet of pipe fluid has enthalpy of 230 kJ/kg and at the exit, it has an enthalpy of 180 kJ/kg. The exit section is 10 m above the ground. If 30×10^3 kJ/hr of heat is added to the theatre, calculate quantity of fluid circulated through pipe. Neglect ΔKE and assume that there are no pumps in the system.*

Solution: The SFEE per kg of fluid is

$$h_1 + \frac{V_1^2}{2} + z_1 g + Q = h_2 + \frac{V_2^2}{2} + z_2 g + W$$

In the piping system, given that

$$\Delta KE = 0 \quad \therefore \quad h_1 + z_1 g + Q = h_2 + z_2 g$$

$$W = 0 \quad \therefore \quad Q = (h_2 - h_1) + g(z_2 - z_1)$$

[Assuming entrance section of the pipe at ground level]

$$= (180 - 230) \times 10^3 + 9.81\,(10 - 0)$$

$$= Q = -49.90 \frac{kJ}{kg}$$

[Negative sign indicates that fluid in the pipe rejects heat]

$$\text{Mass of fluid circulated} = \frac{30 \times 10^3}{49.90} \left[\frac{kJ/hr}{kJ/kg}\right] \quad 601.20 : \frac{kg}{hr}$$

$$= 0.167 \frac{kg}{sec}$$

Problem 44: *A turbo jet engine is at a speed of 265 m/s in the atmosphere at –15°C. The fuel air ratio for the engine is 0.018 with calorific value of the fuel used as 44000 kJ/kg. The fuel liberates only 90% of energy due to incomplete combustion. The heat rejected by the engine is 25 kJ/kg of air and at the exit of nozzle, temperature measured is found to be 625°C and enthalpy of air and gases are 250 kJ/kg and 850 kJ/kg respectively. Calculate velocity of exhaust jet by neglecting potential energy changes.*

Solution:

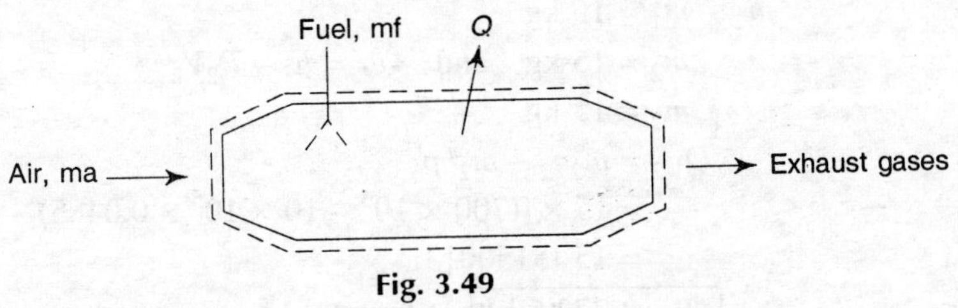

Fig. 3.49

Let suffix g, a and f refer to gas, air and fuel respectively

From mass balances we can write

$$mg = ma + mf = 1 + 0.018 = 1.018 \ \frac{\text{kg}}{\text{kg of air}}$$

The SFEE can be written as

$$ma\left(ha + \frac{Va^2}{2}\right) + mf\,Ef + Q = mg\left(hg + \frac{Vg^2}{2}\right)$$

where $mfEf$ = heat energy liberated by fuel

$$\Delta PE = 0$$

$$W = 0$$

$$\therefore \quad 1\left[250 \times 10^3 + \frac{265^2}{2}\right] + (0.018 \times 0.9 \times 44{,}000 \times 10^3) - 25 \times 10^3$$

$$= 1.018\left[850 \times 10^3 + \frac{Vg^2}{2}\right]$$

$$285112.5 + 712800 - 25 \times 10^3 = 865300 + 0.509\ Vg^2$$

$$\boxed{Vg = 459.8 \text{ m/s}}$$ *Ans.*

Problem 45: *A household empty gas cylinder is filled by 15 kg gas supply of enthalpy 600 kJ/kg. After filling, the gas in the cylinder has the properties P = 10 bar, h = 700 kJ/kg and specific volume 0.0485 m³/kg. Evaluate the heat received by the cylinder from the surrounding.*

Solution: In case of charging of a cylinder energy equation can be written as [assume cylinder as the control volume]

$$[\text{Energy entering} - \text{Energy leaving}]_{\text{control volume in time } dt}$$

$$= \text{change in energy in control volume in time } dt$$

$$(m_2 - m_1)\,E + Q_{1\text{-}2} = m_2u_2 - m_1u_1 + W_{1\text{-}2}$$

$$(m_2 - m_1)\left(hp + \frac{Vp^2}{2}\right) + Q_{1\text{-}2} = m_2u_2 - m_1u_1 + W_{1\text{-}2}$$

Suffix 'p' refers to state of gas in the intake pipe

Since $\Delta KE = 0$, and $W_{1\text{-}2} = 0$

$$\therefore \quad (m_2 - m_1)\,hp + Q_{1\text{-}2} = m_2u_2 - m_1u_1$$

Initially, the cylinder is completely empty

$$\therefore \quad m_1 = 0$$

$$m_2hp + Q_{1\text{-}2} = m_2u_2$$

Given:

$$m_2 - m_1 = 15 \text{ kg}$$

$$m_2 = 15 \text{ kg} \quad \text{and} \quad u_2 = h_2 - P_2V_2$$

$$\therefore \quad m_2 = 15 \text{ kg}$$

$$\therefore \quad Q_{1\text{-}2} = m_2u_2 - m_2hp$$

$$= 15 \times [(700 \times 10^3 - 10 \times 10^5 \times 0.0485) - 600 \times 10^3]$$

$$= 15\,[51500]$$

$$\boxed{Q_{1\text{-}2} = 772.5 \text{ kJ}}$$ *Ans.*

Problem 46: *A valve is fitted in a 60 mm diameter horizontal pipeline; the valve and pipeline are well insulated. The valve, which is partially open, throttles the steam flowing steadily along the pipeline from a pressure of 2×10^6 N/m^2 to a pressure of 200×10^3 N/m^2. The enthalpy of the steam approaching the valve is 2.77×10^6 J/kg and mass flow rate of steam is 0.03 kg/s.*

(i) Calculate the enthalpy of the steam downstream of the valve by neglecting ΔKE of steam.

(ii) Determine steam velocity upstream and downstream of the valve and the corresponding specific volumes of the steam are 0.098 m^3/kg and 0.9602 m^3/kg respectively.

Solution:

(i) $D = 60$ mm, $Q = 0$

$\Delta KE = 0$

The SFEE can be written as

$$\left[h_1+\frac{\cancelto{0}{V_1^2}}{2}+\cancelto{0}{z_1 g}\right]+\cancelto{0}{Q}=\left[h_2+\frac{\cancelto{0}{V_2^2}}{2}+\cancelto{0}{z_2 g}\right]+\cancelto{0}{W}$$

$$h_1 = h_2$$

$$h_1 = 2.77 \times 10^3 \text{ kJ/kg} \qquad \text{(given)}$$

$$\therefore \quad h_1 = h_2 = 2.77 \times 10^3 \text{ kJ/kg} \qquad \textit{Ans.}$$

(ii) When $v_1 = 0.098$ m^3/kg, $v_2 = 0.9602$ m^3/kg

From continuity equation, we can write

$$m = \frac{A_1V_1}{v_1} = \frac{A_2V_2}{v_2}$$

$$V_1 = \frac{mV_1}{A_1} = \frac{0.03 \times 0.098}{\frac{\pi}{4}(0.06)^2} = 1.0398 \text{ m/s} \qquad \textit{Ans.}$$

$$V_2 = \frac{mV_2}{A_2} = \frac{0.03 \times 0.9602}{\frac{\pi}{4}(0.06)^2} = 10.18 \text{ m/s} \qquad \textit{Ans.}$$

Problem 47: *A long, well-insulated pipeline consists of two pipes connected in series, the internal diameters of which are 90 mm and 30 mm respectively. A steady flow of steam enters the 90 mm diameter pipe at a pressure of 350×10^3 N/m^2, a specific volume of 0.684 m^3/kg and an enthalpy of 2.98×10^3 kJ/kg. At a point downstream in the 30 mm diameter pipe, the pressure is 300×10^3 N/m^2 the specific volume is 0.790 m^3/kg and enthalpy is 2.968×10^3 kJ/kg.*

Determine the velocity of the steam at the two points in the pipeline and the mass flow rate of the steam.

Solution:

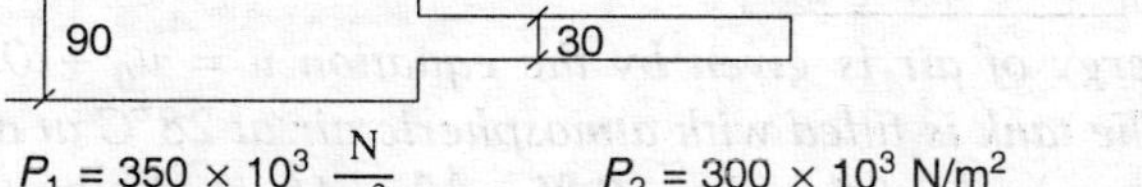

$P_1 = 350 \times 10^3 \dfrac{\text{N}}{\text{m}^2}$ $\qquad$ $P_2 = 300 \times 10^3$ N/m^2 $\qquad$ $m = ?$

$V_1 = 0.684$ m^3/kg $\qquad$ $V_2 = 0.790$ m^3/kg $\qquad$ $v_1 = ?$

$h_1 = 2.98 \times 10^3 \dfrac{\text{kJ}}{\text{kg}}$ $\qquad$ $h_2 = 2.968 \times 10^3$ kJ/kg $\qquad$ $v_2 = ?$

Fig. 3.50

The SFEE per unit mass can be written as

$$\left[h_1 + \frac{V_1^2}{2} + \overset{0}{\cancel{z_1 g}}\right] + \overset{0}{\cancel{Q}} = \left[h_2 + \frac{V_2^2}{2} + \overset{0}{\cancel{z_2 g}}\right] + \overset{0}{\cancel{W}}$$

$$h_1 + \frac{V_1^2}{2} = h_2 + \frac{V_2^2}{2}$$

$$\frac{V_2^2 - V_1^2}{2} = h_1 - h_2 = (2.98 - 2.968) \times 10^6$$

$$V_2^2 = (12000 \times 2) + V_1^2 \qquad (1)$$

from continuity eqn, we can write

$$m = \frac{A_1 V_1}{v_1} = \frac{A_2 V_2}{v_2}$$

$$V_2 = \left(\frac{A_1}{A_2}\right) V_1 \left(\frac{v_2}{v_1}\right) = \frac{\frac{\cancel{\pi} D_1^2}{\cancel{4}} V_1 (v_2)}{\frac{\cancel{\pi} D_2^2}{\cancel{4}} (v_1)}$$

$$V_2 = \frac{0.09^2 (0.79)}{0.03^2 (0.684)} V_1$$

$$\boxed{V_2 = 10.39\ V_1}$$

$\therefore$ $$V_2^2 = 107.9\ V_1^2$$

from (1) $$107.9\ V_1^2 = 24000 + V_1^2$$

$$106.9\ V_1^2 = 24000$$

$\therefore$ $$\boxed{V_1 = 14.98 \text{ m/s}}$$ *Ans.*

$$V_2 = 155.64 \text{ m/s}$$ *Ans.*

$$m = \frac{A_1 V_1}{v_1} = \frac{\frac{\pi D_1^2}{4} \times 14.98}{0.684}$$

$$\boxed{m = 0.139 \text{ kg/s}}$$ *Ans.*

Problem 48: *The internal energy of air is given by the equation $u = u_0 + 0.718\ t$. 2 kg of air is contained in a tank at 28°C. The tank is filled with atmospheric air at 28°C at a pressure of 1 bar by operating a valve. The specific volume of the air is 0.85 m³/kg. After 30 minutes, air has reached a temperature 45°C and mass of air was 2.5 kg. By neglecting ΔKE and work interaction in the pipe, evaluate heat transfer?*

Solution:
For tank filling process, we can write

$$(m_2 - m_1)\left[hp + \frac{Vp^2}{2}\right] + Q = m_2h_2 - m_1u_1 + W$$

$$Q = m_2u_2 - m_1u_1 - (m_2 - m_1)\,hp$$

Given: $u = u_0 + 0.718\,t$

$\therefore$ $u_1 = u_0 + 0.718\,(28)$

$u_1 = u_0 + 20.1$

Similarly $u_2 = u_0 + 0.718\,(45)$

$u_2 = u_0 + 32.31$

Now $hp = up + (PV)_p$

$= [u_0 + 0.718 \times 28] + 1 \times 10^5 \times 0.85$

$hp = u_0 + 85020.1$

$\therefore$ $Q = 2.5\,[u_0 + 32.31] - 2[u_0 + 20.1]$

$- (2.5 - 2)\,(u_0 + 85020.1)$

$= 2.540 + 80.775 - 240 - 40.2 - 0.540 - 42510.05$

$\boxed{Q = -42.46 \text{ kJ}}$ *Ans.*

Problem 49: *A steam turbine receives steam with a flow rate of 900 kg/min and experiences a heat loss of 840 kJ/min. The exit pipe is 3 m below the level of the inlet pipe. Find the power developed by the turbine if the pressure decreases from 62 bar to 9.86 kpa. Velocity increases from 30.5 m/s to 274.3 m/s and internal energy decreases by 938.8 kJ/kg. The specific volume changes from 0.058 m³/kg to 13.36 m³/kg.* *(VTU, Feb. 2002)*

Given data

$$\text{Steam flow rate} = \dot{m} = 900 \text{ kg/min}$$

$$= \frac{900}{60} = 15 \text{ kg/sec}$$

$$\text{Heat loss} = Q = -840 \text{ kJ/min} = -14 \text{ kJ/sec}$$

$$z_1 = 3 \text{ m}$$

$$z_2 = 0$$

$$P_1 = 62 \text{ bar} = 62 \times 10^2 \text{ kpa}$$

$$P_2 = 9.86 \text{ kpa}$$

$$V_1 = 30.5 \text{ m/s},\ V_2 = 274.3 \text{ m/s}$$

$$U_2 - U_1 = 938.8 \text{ kJ/kg}$$

$$V_1 = 0.058 \text{ m}^3\text{/kg}$$

$$V_2 = 13.36 \text{ m}^3\text{/kg}$$

Solution: The SFEE can be written as

$$\left[h_1 + \frac{V_1^2}{2} + z_1 g\right] + Q = \left[h_2 + \frac{V_2^2}{2} + z_2 g\right] + W$$

$$h_2 - h_1 = (U_2 + P_2V_2) - (U_1 + P_1V_1)$$
$$= (U_2 - U_1) + (P_2U_2 - P_1V_1)$$
$$= -938.5 + (9.86 \times 13.36) - (62 \times 10^2 \times 0.058)$$
$$h_2 - h_1 = -1166.37 \text{ kJ/kg}$$

$$\therefore \quad W = \dot{m}\left[(h_1 - h_2) + \frac{V_1^2 - V_2^2}{2} \times 10^{-3} + (z_1 - z_2) \times 10^{-3} \times g\right] + Q$$

$$W = 15\left[1166.37 + \frac{30.5^2 - 274.3^2}{2} \times 10^{-3} + 3 \times 10^{-3} \times 9.81\right] - 14$$

$$= 15[1166.37 - 37.155 + 0.02943] - 14$$

$$\boxed{W = 16924.66 \text{ kW}}$$ *Ans.*

Problem 50: *Air flows steadily through a rotary compressor. At entry the air is 20°C and 101 kpa. At exit the same air is at 200°C and 600 kpa. Assuming the flow to be adiabatic*

(i) Evaluate the work done per unit mass of air if velocities at inlet and exit are negligible.

(ii) What would be the increasing work input if the velocities at inlet and exit are 50 m/s and 110 m/s. *(VTU, Feb 2004)*

Solution: The SFEE is

$$\left[h_1 + \frac{V_1^2}{2} + z_1 g\right] + Q = \left[h_2 + \frac{V_2^2}{2} + z_2 g\right] + W$$

As $\Delta PE = 0$

$\Delta KE = 0$

$Q = 0$ (adiabatic process)

$\therefore \quad W = h_1 - h_2$

$= CP(T_1 - T_2)$

But $CP = \frac{\gamma}{\gamma - 1} \times R = \frac{1.4}{1.4 - 1} \times 287$

$CP = 1004.5$ J/K

$\therefore \quad W = 1004.5\ [20 - 200]$

$$\boxed{W = -180.81 \text{ kJ}}$$ *Ans.*

In the second case, it is given that

$$V_1 = 50 \text{ m/s and } V_2 = 110 \text{ m/s}$$

$$-W = (h_2 - h_1) + \left(\frac{V_2^2 - V_1^2}{2}\right) \times 10^{-3}$$

$$= 180.81 + \left(\frac{110^2 - 50^2}{2}\right) \times 10^{-3}$$

$$\boxed{W = 185.61 \text{ kJ/kg}}$$ *Ans.*

3.18 REVIEW QUESTIONS

1. State first law of thermodynamics for a non-cyclic process.
2. Define internal energy, enthalpy.
3. Prove that internal energy is a property of the system.
4. Derive an expression for displacement work for a polytropic process.
5. With P-V diagram, derive an expression for work in the following cases:
   (i) Constant volume process
   (ii) Constant pressure process
   (iii) Constant temperature process
   (iv) Isontropic process
6. Derive an expression of SFEE and state the assumptions made.
7. Apply the SFEE for the following systems:
   (i) Compressor
   (ii) Centrifugal pump
   (iii) Steam or gas turbine
   (iv) Internal combustion engine
   (v) Boiler
   (vi) Throttling process
8. Write down energy equation for:
   (i) filling of a tank
   (ii) emptying of tank

CHAPTER 4

Second Law of Thermodynamics

4.1 INTRODUCTION

First law of thermodynamics deals with conservation of energy. It says that energy is neither created nor destroyed but can be converted from one form to another. It states that, when a system executes a closed cycle, the cyclic integral of heat is equal to the cyclic integral of work. It places no restriction on the direction of flow of heat and work and it can be assumed that change of state can take place in both directions. In a system, a given amount of heat is transferred to do an equal amount of work and flow of heat and work can be reversed during the execution of a cycle to satisfy the first law. But in practical, this is not the case, especially in the inter-conversion of heat and work. For example,

1. Heat always flows from high temperature region to lower temperature side in the absence of other effects. From experience, we can say that the reverse process is impossible i.e original temperature of a body can not be restored without the aid of external work.
2. A battery will discharge through a resistor releasing certain amount of energy. But the battery can not be recharged by heating the resistor, i.e. reverse of this will not happen.
3. When a vehicle is running, it can be stopped by applying the brakes during which kinetic energy of automobile is converted in to heat. But we can not accelerate the vehicle by cooling the brake drum.
4. It is impossible to have a device which will operate continuously by receiving heat from a single reservoir and producing an equivalent amount of work.

Limitations of First Law

1. It deals only with conservation of energy and will not specify the direction of flow of heat and work.
2. Spontaneous processes in nature occur only in one direction. It does not deny the feasibility of a spontaneous process reversing itself.
3. It provides necessary, but not sufficient condition for a process to occur.

4.2 THERMAL RESERVOIR

A Thermal or Heat reservoir is that part of environment in which heat exchange takes the place with a system. The heat capacity of the reservoir is very high and the quantity of heat transferred to or from the reservoir will not affect its temperature. Hence the temperature of a reservoir remains constant. The heat source is one which is at high temperature and supplies heat, for example, furnace of a boiler, combustion chamber and nuclear reactor. Heat sink is one which is maintained at low temperature and extracts the heat. Example: atmospheric air, ocean, river, etc.

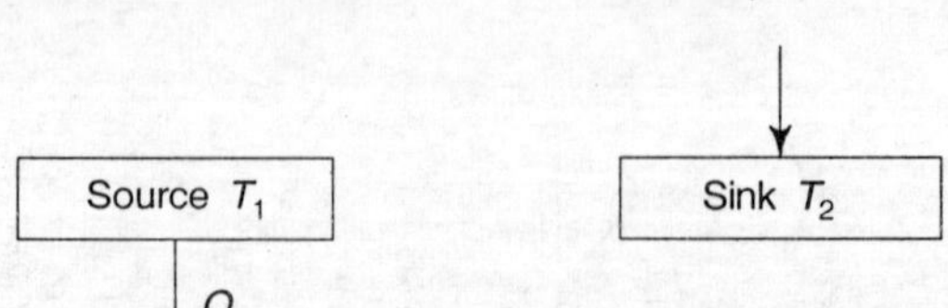

4.3 HEAT ENGINE REFRIGERATOR AND HEAT PUMP

Heat engine is a thermodynamic device used to convert heat energy into useful work when operating in a thermodynamic cycle. A thermal power plant is an example for heat engine.

A heat engine is precisely defined as "a device operating in cycle between high temperature source and low temperature sink and thus producing work".

The figure 4.1 shows block diagram of a heat engine that receives 'Q_1' quantity of heat from a source which is maintained at temperature 'T_1' and rejects 'Q_2' quantity of heat to the sink which is at temperature T_2. Some of the heat is utilized to obtain the work W which is equal to $(Q_1 - Q_2)$.

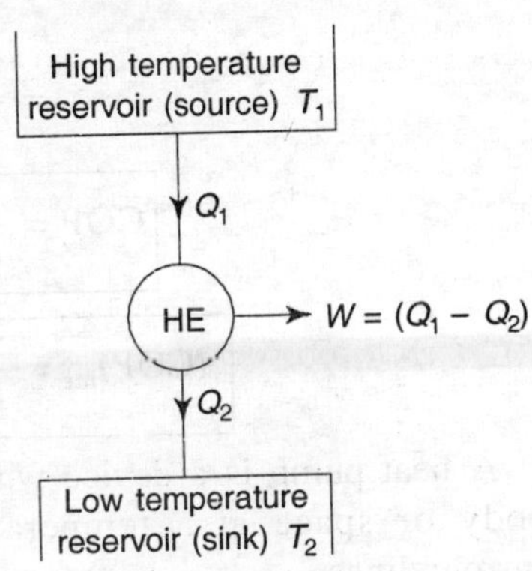

Fig. 4.1 Heat Engine

The heat engine is a cyclic/continuously operating device and the working substance flows through the engine.

Generally, the performance of the heat engine is expressed in terms of "efficiency of the engine". The term thermal efficiency explains useful utilization of heat received in a heat engine. It is, the ratio of net work output to the total heat supplied.

$$\therefore \quad \eta th = \frac{\text{Net work output}}{\text{Heat supplied}}$$

$$= \frac{Q_1 - Q_2}{Q_1}$$

$$\eta th = 1 - \frac{Q_2}{Q_1}; \text{ It is always less than unity.}$$

By increasing Q_1 or by decreasing Q_2, thermal efficiency of heat engine can be increased.

Refrigerator and Heat Pumps

Refrigerator and heat pumps are reversed heat engines, i.e., the direction of flow of heat and work are opposite to that of a heat engine, i.e., Work is supplied as input and heat is the output. These devices are used to remove heat from a body at lower temperature and then transfer this heat to another body which is at higher temperature.

Refrigerator is a device which is operating in a cycle and the purpose of which is to maintain a body at a temperature lower than the temperature of the surroundings. It removes heat from the cooled space and operates between temperature of surroundings (T_1) and temperature below that of surroundings (T_2). The refrigerators are used to preserve food items and medicines at low temperature.

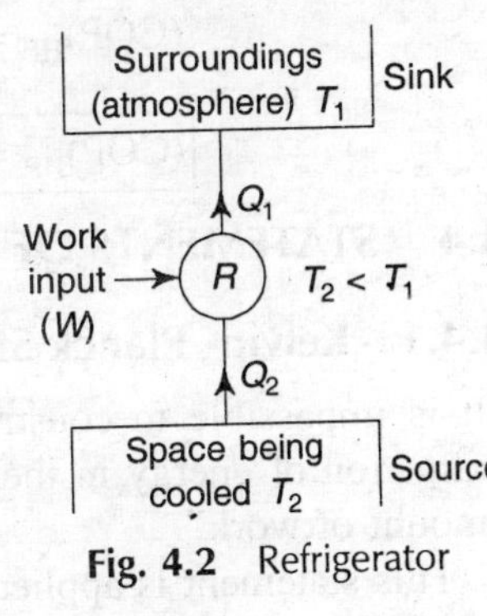

Fig. 4.2 Refrigerator

The refrigerator shown in figure operates between temperatures T_1 and T_2 ($T_2 < T_1$, i.e. sink is at higher temperature than source) extracts Q_2 quantity of heat, thereby cool the space. The desired effect of device is to cool the space by extracting Q_2 quantity of heat from it. W is the work input supplied to the device for heat transfer against reverse temperature gradient.

The performance of the refrigerator can be expressed in terms of "coefficient of performance" or COP.

$$\text{COP} = \frac{\text{Desired effect}}{\text{Work input}} = \frac{\text{Heat extracted at temperature } T_2}{\text{Work input}}$$

$$= \frac{Q_2}{W} \quad (\text{But } W = Q_1 - Q_2)$$

$$\therefore \quad \boxed{\text{COP} = \frac{Q_2}{Q_1 - Q_2}}$$

$$\boxed{(\text{COP})_{\text{ref}} = \frac{Q_2}{Q_1 - Q_2}}$$

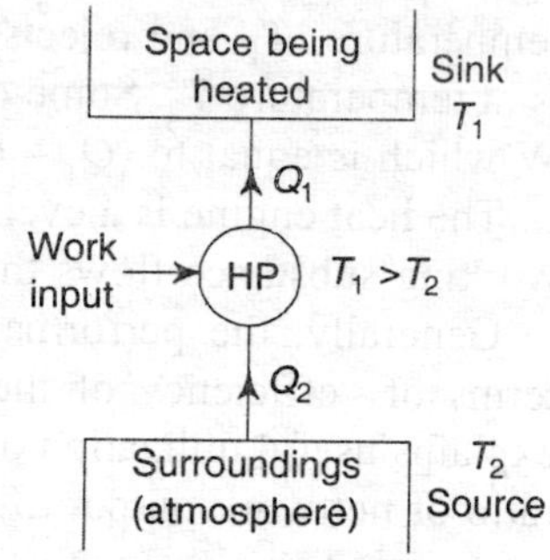

Fig. 4.3 Heat pump

A heat pump is a device which is operating in a cycle, maintains a body or space at a temperature higher than the temperature of surroundings.

The purpose of the device is to heat the medium which may already be warmer than its surroundings. The device operates between temperature of surroundings (T_2) and temperature above that of the surroundings (T_1). These devices are used in the winter seasons to keep the room warm.

The heat pump shown in figure, extracts 'Q_2' quantity of heat from atmosphere and adds 'Q_1' quantity of heat to the space. The sink is at higher temperature than the surroundings i.e $T_1 > T_2$. 'W' be the work in put supplied to the device for transfer of heat against a reverse temperature gradient. The desired effect is to heat the space by adding 'Q_1' quantity of heat

The performance of heat pump can be expressed as

$$(\text{COP})_{\text{HP}} = \frac{\text{Desired effect}}{\text{Work input}} = \frac{\text{Heat to high temperature sink}}{\text{Work input}}$$

$$(\text{COP})_{\text{HP}} = \frac{Q_1}{W} = \frac{Q_1}{Q_1 - Q_2} = 1 - \frac{Q_1}{Q_2}$$

$$(\text{COP})_{\text{HP}} = \frac{Q_1}{Q_1 - Q_2} = 1 + \frac{Q_2}{Q_1 - Q_2} = 1 + (\text{COP})_{\text{Reff}}$$

$$\boxed{(\text{COP})_{\text{HP}} = 1 + (\text{COP})_{\text{Ref}}}$$

4.4 STATEMENTS OF SECOND LAW OF THERMODYNAMICS

4.4.1 Kelvin- Planck Statement

"It is impossible to construct a device that will operate in a cycle and gives no effect other than extraction of energy in the form of heat from a single thermal reservoir and performs an equivalent amount of work".

This statement is applied to heat engines.

The statement says that no heat engine can be developed that extracts 'Q' quantity of heat from high temperature reservoir and converts that into an equivalent amount of work. i.e. $W = Q$. The

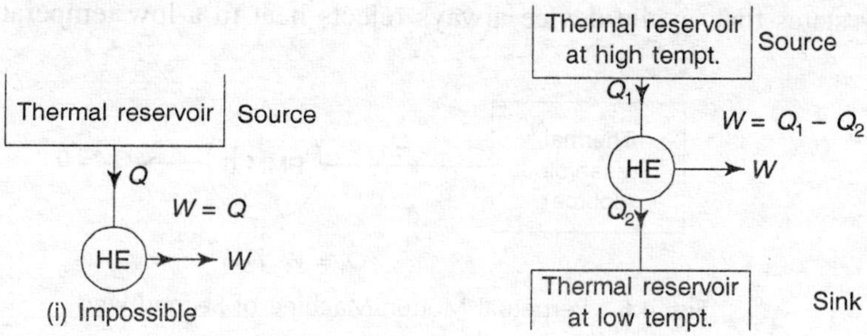

(i) Impossible (ii) Possible

Fig. 4.4 Representation of Kelvin-Planck statement

efficiency of such heat engine would be $\dfrac{O/P}{I/P} = \dfrac{W}{Q} = 100\%$. The device satisfies first law of thermodynamics (principle of conservation of energy), but violates the Kelvin statement of second law. Obviously, no heat engine exists with 100% thermal efficiency.

Thus a heat engine receives 'Q_1' quantity of heat from a high temperature thermal reservoir, a portion of which is used to do work 'W', before rejecting 'Q_2' quantity of heat to the low temperature thermal reservoir (atmosphere).

4.4.2 Clausius Statement

"It is impossible to construct a cyclic device which produces no effect other than the transfer of heat from a low temperature body to high temperature body".

The statement is related to refrigerator or heat pump and says that without any work input, it is not possible to operate a refrigerator. It means that heat cannot flow itself from low temperature region to high temperature region without the aid of work input.

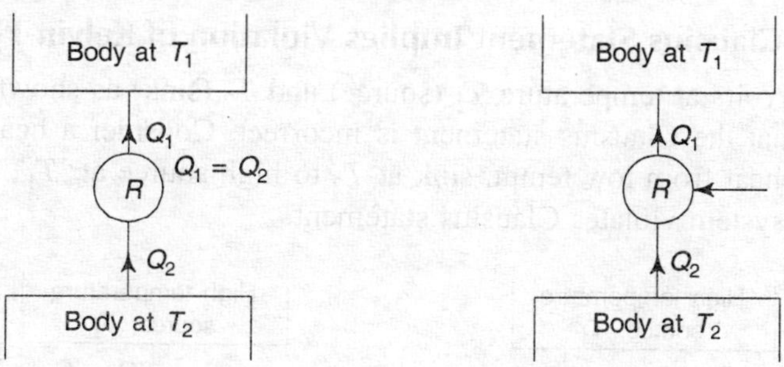

(i) Impossible (ii) Possible

Fig. 4.5 Representation of Clausius Statement

The schematic arrangement to prove the Clausius statement is shown in figure. The statement is concerned with heat pump and refrigerators.

Both of the above two statements are negative statements and are based on experimental observations. Both of these statements have no mathematical proof.

4.5 PERPETUAL MOTION MACHINE OF THE SECOND KIND (PMM II)

The second law puts a restriction on the thermal efficiency of a heat engine and is always less than

100%. The law states that, cyclic device always rejects heat to a low temperature sink to do useful work '*W*'.

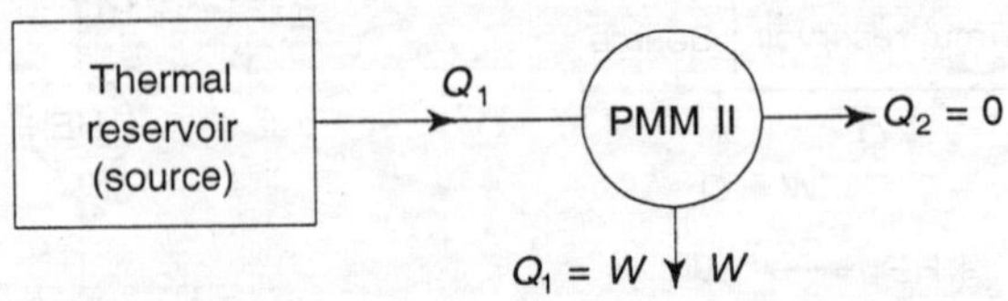

Fig. 4.6 Perpetual Motion Machine of Second kind.

The PMM II is a device operating in a cycle and produces work by excha. ging heat with a single heat reservoir. Such a device is impossible as per Kelvin Planck's statement, but it does not violate first law of thermodynamics.

When a heat engine is operating between two thermal reservoirs, the thermal efficiency is given by

$$\eta\text{th} = \frac{Q_1 - Q_2}{Q_1}\text{; when } Q_2 = 0$$

Then
$$\eta\text{th} = \frac{Q_1}{Q_1} = 100\%$$

This violates Kelvin- Planck statement of second law.

4.6 EQUIVALENCE OF KELVIN-PLANCK AND CLAUSIUS STATEMENTS

The kelvin-planck and clausius statements appear to be different all together, but are interlinked and are complementary to each other. A device satisfying one statement and violating the other cannot exist. Thus violation of one statement leads to violation of other and can not the proved like a mathematical theorem.

4.6.1 Violation of Clausius Statement Implies Violation of Kelvin-Planck Statement

Consider thermal reservoirs at temperature T_1 (source) and T_2 (sink) as shown in figure 4.7.

(1) Let us assume that the Clausius statement is incorrect. Consider a heat pump which transfers 'Q_2' quantity of heat from low tempt. sink at T_2 to high source at 'T_1', without any work input i.e. $W = 0$. This system violates Clausius statement.

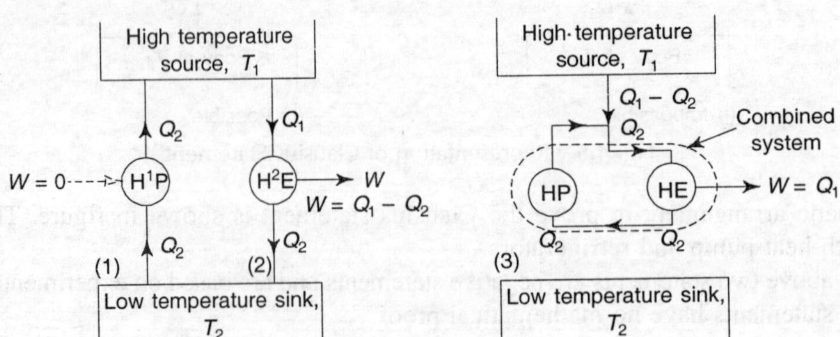

Fig. 4.7

(ii) Consider a heat engine which works as per Kelvin-Planck's statement. The heat engine receives 'Q_1' quantity of heat from a high temperature source and does useful work $W = Q_1 - Q_2$ before rejecting Q_2 quantity of heat to a low temperature sink. By applying first law to the device, we can write $W = Q_1 - Q_2$.

(iii) Consider combination of HP and HE. This system absorb '$Q_1 - Q_2$' quantity of heat from a high temperature source and delivers work $W = Q_1 - Q_2$, leaving the sink (T_2) unaffected. The resulting device works as a PMM II which is the violation of Kelvin-Planck's statement.

Thus violation of Clausius statement leads to violation of Kelvin Planck's statement.

4.6.2 Violation of Kelvin Planck Statement Implies Violation of Clausius Statement

Consider thermal reservoirs at temperature 'T_1' (source) and 'T_2' (sink) as shown in figure.

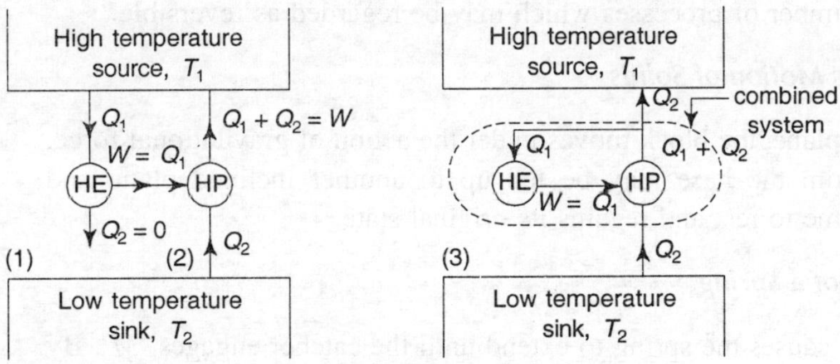

Fig. 4.8

(i) Let us assume that the Kelvin Planck statement is incorrect. Consider a cyclic heat engine absorbs ' Q_1' quantity of heat from source (T_1) and does an equivalent amount of work $W = Q_1$.

(ii) Consider a heat pump, which operates in a cycle, absorbs 'Q_2' from sink and delivers $W = Q_1 + Q_2$. Let 'W' be the work done on the device. The device operates in conformity with the Clausius statement.

(iii) Consider the comined system. In this case, work delivered by HE is used to drive HP. Then HP delivers $Q_1 + Q_2$ among which 'Q_1' is used to run HE and the combined system delivers Q_2 to source at T_1. But heat (Q_2) itself can not flow from 'T_2' and 'T_1' as ($T_1 > T_2$) without using any external work. Thus it violates Clausius statement.

Thus violation of Kelvin Planck statement leads to violation of Clausius statement.

4.7 REVERSIBLE AND IRREVERSIBLE PROCESSES

A Reversible or Ideal process is one in which both the systems and surroundings can be restored to their respective initial states by reversing the direction of process. These processes do not leave any permanent changes in both systems and surroundings. A reversible process must be a quasi-static process. These reversible processes gives maximum work output in the engines and reduces work required to run devices like refrigerator, heat pump, compressor. etc...

In many processes, when reversed, system and surroundings do not regain initial states without any other inputs and leave net change in both system and surroundings. These processes are called Irreversible or Natural processes.

The reasons for making the process irreversible are termed as irreversibilities and are usually grouped into internal and external irreversibilities. The first type is caused due to internal factors ie, dissipative effects inside the fluid, whereas the latter type is due to factors at the system surrounding interface. The factors that render processes irreversible are:

(i) Friction
(ii) Electrical resistance
(iii) In elastic solid deformation
(iv) Free expansion (unrestrained expansion)
(v) Transfer of energy in the form of heat through a finite temperature difference
(vi) Lack of equilibrium during the process
(vii) Mixing of two different substances, etc.

4.8 EXAMPLES OF REVERSIBLE PROCESSES

Consider a number of processes which may be regarded as reversible.

1. Frictionless Motion of Solids

On a smooth plane, the block moves under the action of gravitational force. The block from the base may be led up to another inclined plane and eventually come to rest and regains its original state.

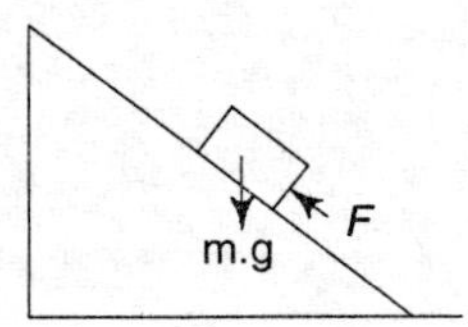

Fig. 4.9 (a) FrictionLess Relative Motion

2. Expansion of a Spring

The force 'F' causes the spring to extend until the catcher engages 'M'. By lowering the catcher and slowly relaxing the 'F' the process can be made reversible and initial state can be restored.

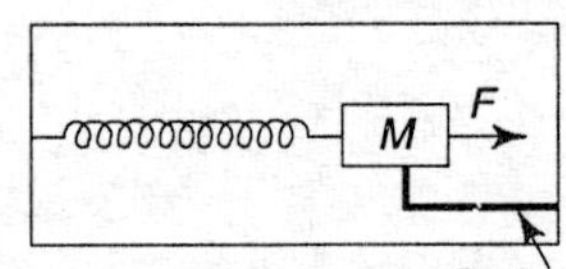

Fig. 4.9 (b)

3. Slow Adiabatic Expansion of Gas

The gas in the cylinder expands against a frictionless piston, with $q = 0$, which is moved by an external force. The process is reversible because an infinitesimal increase in 'F' is sufficient to bring back the system to it's initial state.

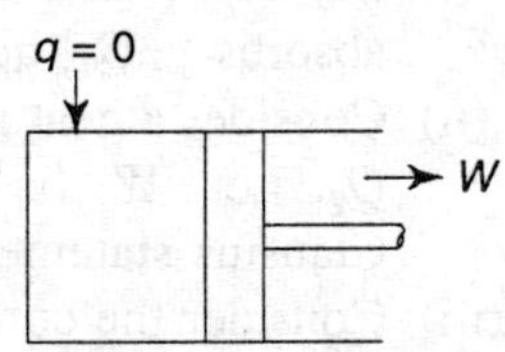

Fig. 4.9 (c)

4. Slow Isothermal Compression

The steam in the cylinder is compressed by pushing the frictionless piston inward. The steam condenses and transfers heat to the cylinder walls. This heat transfer necessiates infinitesimally higher temperature in the steam than in it's surroundings. The process can be made reversible, by reducing the force infinitesimally. The movement of the piston reverses results in drop in steam temperature infinite- similly. This also reverses direction of heat transfer and condensate reevaporates. The system and surroundings regain their initial states.

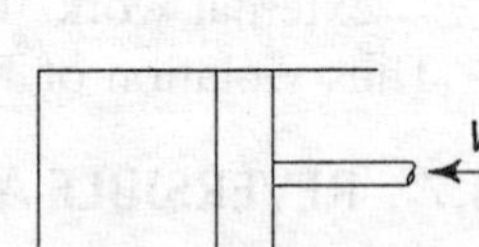

Fig. 4.9 (d)

4.9 EXAMPLES OF IRREVERSIBLE PROCESS

Friction is always present in the real systems. Friction cannot be reduced completely, but may be reduced by suitable lubrication. The presence of friction in a moving device causes irreversibility in the process as work done does not show equivalent rise in kinetic or potential energy of the system.

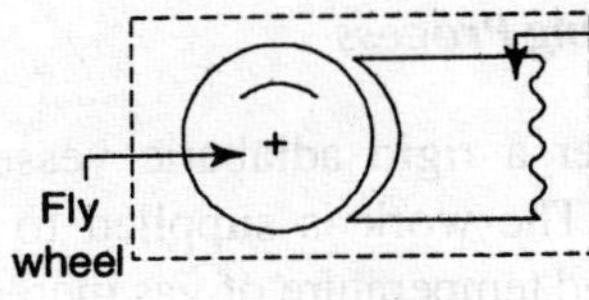

Fig. 4.10 (a)

1. Consider a flywheel and brake arrangement as shown in figure. The fly-wheel which was rotating at N rpm, was brought to rest by applying the brake. Neglecting work required to move the brake block and heat transfer due to brake application, we can write

$$Q_{1-2} = E_2 - E_1 + W_{1-2} \quad \text{1– before braking}$$
$$0 = E_2 - E_1 + W_{1-2} \quad \text{2 – After braking}$$
$$E_2 = E_1$$

The energy may exist in the forms of KE, PE and molecular internal energy, we write

$$U_2 + \frac{mV_2^2}{2} + m\,z_2 g = U_1 + \frac{mV_1^2}{2} + mz_1 g$$

PE = 0, and $V_2 = 0$, as wheel is brought to rest

$$\therefore \quad U_2 = U_1 + \frac{mV_1^2}{2}$$

∴ In the system, molecular internal energy increases by absorbing KE of the wheel. The reverse is not possible, i.e., by converting internal molecular energy into KE, one cannot cause the wheel to rotate.

2. Heat Transfer Through a Finite Temperature Difference

Whenever there is temperature difference, heat transfer take place. During heat transfer, if heat addition is accomplished in infinite number of steps, then after every step, the new state would be an equilibrium state. Thus every time infinitesimal temperature variation is caused by infinitesimal heat transfer and the process is said to be quasi-static, therefore reversible. Therefore heat transfer through finite temperature difference which practically occurs is carried out by irreversible state changes and the process is said to be irreversible.

3. Free Expansion Process [Unrestrained expansion]

Consider an adiabatic vessel separated by a partition wall as shown in figure. If the partition wall is suddenly removed, the air will fill the entire volume of the vessel instantaneously, and this process is not quasi-static and the air is not interacting with surroundings.

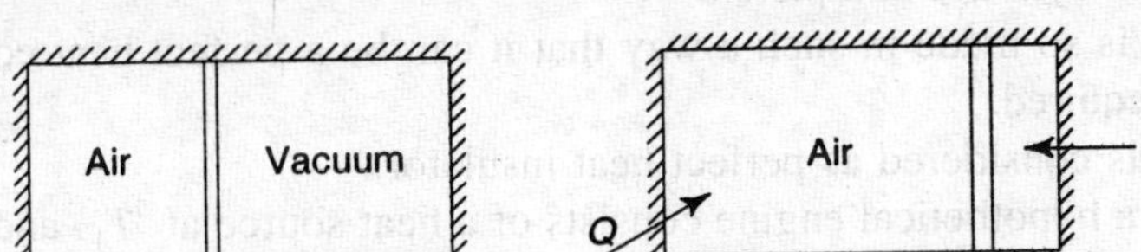

Fig. 4.10 (b)

To restore original state of the system, the air is to be compressed by an imaginary piston and heat transfer also occurs. These work and heat interactions causes a change in the surroundings and the surroundings are not restored to their initial states. This indicates that, free expansion is an irreversible process.

4. Stirring Process

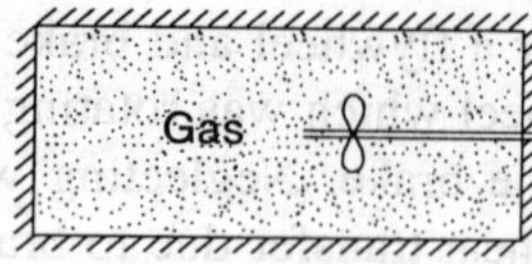

Fig. 4.10 (c)

Consider a rigid adiabatic vessel, containing gas as shown in figure. The work is supplied to the paddle wheel and hence it turns and temperature of gas increases. This stirring action changes state of the system and molecular internal energy of the system increases.

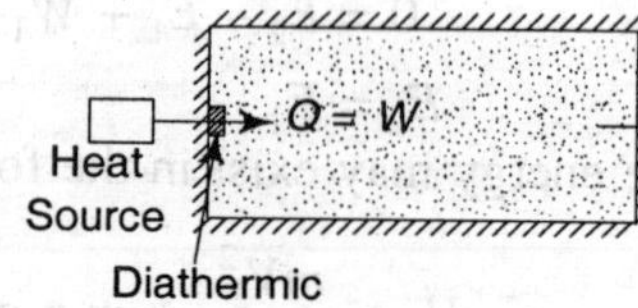

Fig. 4.10 (d)

Assume that, same amount of work is delivered by the system at the expense of its molecular internal energy. This decreases temperature of the system and a heat source is used to bring back the system to its initial state. These two processes constitute a cycle during which it delivers work and system exchanges heat with single reservoir. It becomes PMM-II. Hence the process of dissipation of stirring work to internal energy is irreversible.

4.10 CARNOT'S PRINCIPLE

It is well known that a reversible process is more efficient process, as it involves no loss. Carnot, in 1824 deduced a concept that has been called Carnot's principle.

"No engine can be more efficient than a reversible engine working between the same temperature limits".

The amount of heat which a heat engine transfers in to work is dependent on the temperature range and no reference is given to the maximum amount of work which can be produced for any given temperature range. The evaluation of maximum amount of work can be made by the use of Carnot cycle.

4.11 CARNOT ENGINE AND CARNOT CYCLE

A Carnot cycle is a hypothetical cycle describes a series of events occurring in a theoretical heat engine known as the Carnot engine.

The following assumptions are made in the analysis of Carnot cycle.

(1) A perfect gas is used as working fluid.
(2) Large source or sink are to be considered in the cycle hence the transfer of heat does not effect source or sink temperature.
(3) No friction between cylinder and piston.
(4) The cylinder head is so made in such a way that it can be a perfect heat conductor or a perfect heat insulator as required.
(5) The cylinder wall is considered as perfect heat insulator.

The Carnot engine is a hypothetical engine consists of a heat source at 'T_1' and sink at temperature 'T_2', both of these are assumed to remain at constant temperatures at all the times.

The sequence of operation for different processses constituting a Carnot cycle is

1-2; Iso Thermal Expansion

During this process, the heat source at temperature 'T_1' is put in communication with working fluid (gas) in the cylinder through a perfect conductor. As heat is supplied, gas expands isothermally from state '1' to '2', during which heat supplied is equal to workdone. ($\Delta u_{1\text{-}2} = 0$)

$$Q_{1\text{-}2} = Q_1 = W_{1\text{-}2} = P_1V_1 \log_e \frac{V_2}{V_1} = mRT_1 \log_e \frac{V_2}{V_1}.$$

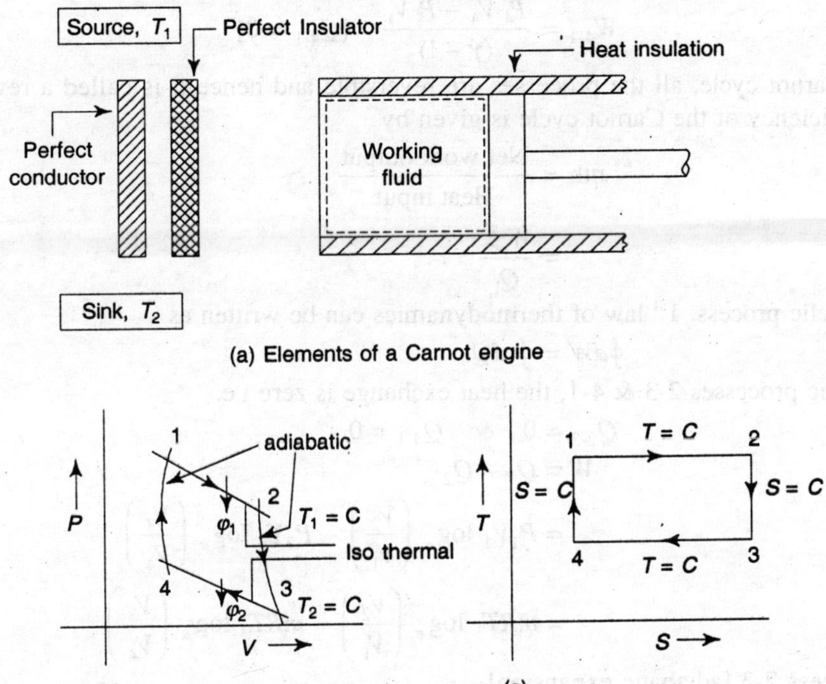

Fig. 4.11

2-3; Adiabatic Expansion

During this process, heat source is removed and insulator is put in communication with cylinder head. The gas in the cylinder is allowed to expand reversibly and adiabatically, till the temperature decreases to T_2. Work is done by the working fluid at the expense of internal energy and is given by

$$W_{2\text{-}3} = \frac{P_2V_2 - P_3V_3}{\gamma - 1} \qquad [Q_{2\text{-}3} = 0]$$

3-4; Isothermal Compression

During this process, the heat sink at temperature T_2 and is put in communication with gas in the cylinder through the conductor. The piston moves inward and gas in the cylinder is compressed isothermally (at constant temperature 'T_2'). The expansion continues up to state point '4', and it lies on the path of frictionless reversible adiabatic process drawn from state '1'. The system rejects heat to the sink and is equal to workdone on the system as $\Delta u = 0$. This woks is given by

$$Q_{3\text{-}4} = Q_2 = W_{3\text{-}4} = P_3V_3 \log_e (V_4/V_3) = mRT_2 \log_e (V_4/V_3)$$

4-1; Adiabatic compression

During this process, heat sink is removed and insulator is brought in contact with cylinder head. The compression of the gas will be done reversibly and adiabatically till the system returns to its original state 1. The internal energy of the system increases and temperature raises to 'T_1'. The work is done on the system and is given by

$$W_{4\text{-}1} = \frac{P_4 V_4 - P_1 V_1}{(\gamma - 1)} \quad [Q_{4\text{-}1} = 0]$$

In the Carnot cycle, all the processes are reversible and hence it is called a reversible cycle. The thermal efficiency of the Carnot cycle is given by

$$\eta th = \frac{\text{Net work output}}{\text{Heat input}} \ldots$$

$$= \frac{\dot{W}_{\text{net}}}{Q_1} \ldots \tag{A}$$

Far a cyclic process, 1st law of thermodynamics can be written as

$$\oint dW = \oint dQ$$

For adiabatic processes 2-3 & 4-1, the heat exchange is zero i.e.

$$Q_{2\text{-}3} = 0 \quad \& \quad Q_{4\text{-}1} = 0.$$

$$W = Q_1 - Q_2$$

$$= P_1 V_1 \log_e \left(\frac{V_2}{V_1}\right) - P_3 V_3 \log_e \left(\frac{V_3}{V_4}\right)$$

$$= mRT_1 \log_e \left(\frac{V_2}{V_1}\right) - mRT_3 \log_e \left(\frac{V_3}{V_4}\right)$$

for the process 2-3 [adiabatic expansion]

$$\left(\frac{T_2}{T_3}\right) = \left(\frac{V_3}{V_2}\right)^{\gamma - 1} \qquad T_2\text{: temperature at state } 2 = T_1$$

$$\therefore \left(\frac{T_1}{T_2}\right) = \left(\frac{V_3}{V_2}\right)^{\gamma - 1} \qquad T_3\text{: temperature at state } 3 = T_2$$

$$\therefore \left(\frac{V_2}{V_3}\right) = \left(\frac{T_2}{T_1}\right)^{\frac{1}{\gamma - 1}} \tag{1}$$

Similarly for the process 4-1 [adiabatic compression]

$$\left(\frac{T_4}{T_1}\right) = \left(\frac{V_1}{V_4}\right)^{\gamma - 1} \qquad T_4\text{: temperature at state } 4 = T_2$$

$$\left(\frac{T_2}{T_1}\right) = \left(\frac{V_1}{V_4}\right)^{\gamma - 1} \qquad T_1\text{: temperature at state } 1 = T_1$$

$$\left(\frac{V_1}{V_4}\right) = \left(\frac{T_2}{T_1}\right)^{\frac{1}{\gamma - 1}} \tag{2}$$

from (1) and (2), we can write,

$$\frac{V_2}{V_3} = \frac{V_1}{V_4}$$

$$\therefore \quad \frac{V_2}{V_1} = \frac{V_3}{V_4}$$

$$\therefore \quad W = mR \log_e \frac{V_2}{V_1} (T_1 - T_2)$$

$$\therefore \quad Q_1 = mRT_1 \log_e (V_2/V_1)$$

Substitute all these values in Eqn. (A), we get,

$$\eta\text{th} = \frac{mR \log_e V_2/V_1 \, (T_1 - T_2)}{mR \log_e (V_2/V_1) \, T_1}$$

$$\boxed{\eta\text{th} = \frac{T_1 - T_2}{T_1}} = \boxed{1 - \frac{T_2}{T_1}}$$

CONCLUSIONS

(1) The efficiency of Carnot engine depends only on source and sink temperatures and is independent of working fluid. The efficiency becomes maximum (100%) when sink temperature $T_2 = 0$, but this is impossible and it violates Kelvin Planck statement of second law.

(2) As the temperature difference between source and sink increases, efficiency also increases and is directly proportional to $(T_1 - T_2)$.

(3) The efficiency can be increased either by increasing source temperature or by decreasing sink temperature.

The Carnot engine is a hypothetical device and is not possible due to following reasons:

(1) All four processes are reversible. For this to happen there should not be any internal friction between particles of working fluid and no friction exists between the piston and cylinder walls.

(2) The heat absorption and rejection have to take place with infinitesimal temperature differences.

(3) The piston has to move very slowly to achieve Isothermal compression or expansion. In the mean time, piston movement must be very fast to achieve adiabatic compression or expansion. It is impossible to achieve different speeds of the piston during different processess.

4.12 REVERSED HEAT ENGINE (CARNOT HEAT PUMP)

The figure shows a Carnot heat pump (reversed Carnot heat engine). All the processes in a Carnot cycle are reversed individually and hence energy transfers during those processes gets reversed in direction, but magnitude remains the same. The sequence of operation is:

1-4; Reversible adiabatic expansion of the gas in the cylinder and temperature falls from T_1 to T_2.

4-3; Isothermal expansion during which the system absorbs 'Q_2' quantity of heat from the space being cooled at 'T_2'.

3-2; Reversible adiabatic compression, during which temperature of gas rises from T_2 to 'T_1'.

2-1; Isothermal compression during which, system rejects 'Q_1' quantity of heat, at higher temperature 'T_1'.

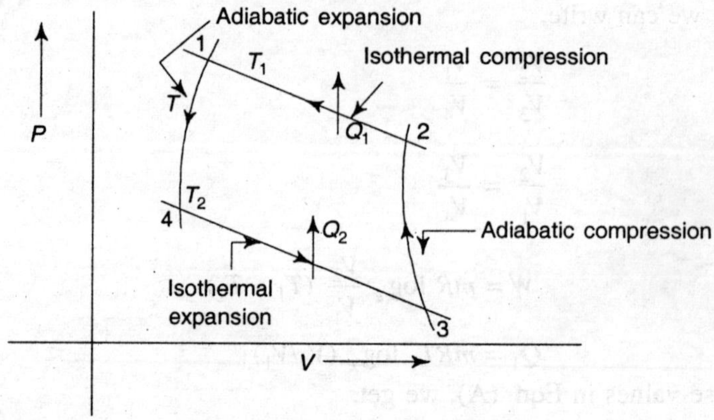

Fig. 4.12

Now $$\eta\text{th} = \frac{\text{Net work output}}{\text{Heat input}}$$

and $$W_{net} = Q_1 - Q_2 = mRT_1 \ln\left(\frac{V_2}{V_1}\right) - mRT_2 \ln\left(\frac{V_3}{V_4}\right)$$

$[\because \quad V_3/V_4 = V_2/V_1]$

$$Q_1 - Q_2 = mR \ln (V_2/V_1)(T_1 - T_2)$$

For a Heat pump,

$$\text{COP} = \frac{Q_1}{Q_1 - Q_2} = \frac{mR\, T_1 \ln (V_2/V_1)}{mR \ln (V_2/V_1)(T_1 - T_2)}$$

$$(\text{COP})_{HP} = \frac{T_1}{(T_1 - T_2)} = 1 - \frac{T_1}{T_2}$$

For a refrigerator,

$$\text{COP} = \frac{Q_2}{Q_1 - Q_2} = \frac{mR\, T_2 \ln (V_2/V_1)}{mR \ln (V_2/V_1)(T_1 - T_2)}$$

$$(\text{COP})_{ref} = \frac{T_2}{T_1 - T_2} = \frac{T_2}{T_1} - 1$$

4.13 CARNOT THEOREM

Carnot theorem states that "No heat engine operating in a cycle between two heat reservoirs each having a fixed temperature, can be more efficient than a reversible engine operating between same heat reservoirs."

Proof: Let us consider

E_{R1} = reversible engine

E_{IR} = Irreversible engine.

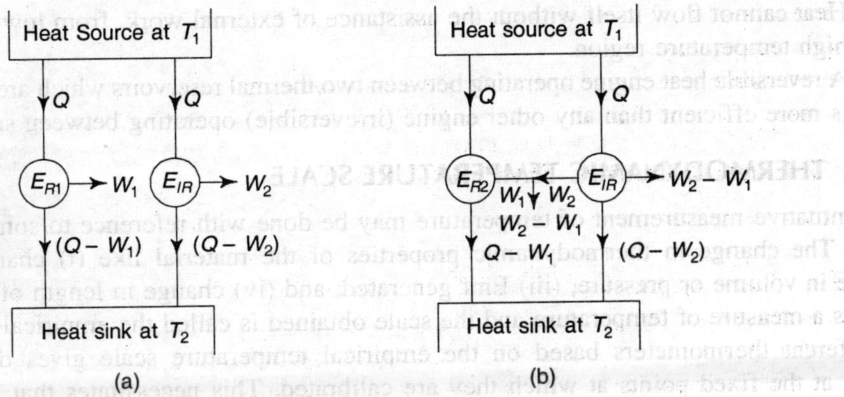

Fig. 4.13

Both E_{R1} and E_{IR} operate between same thermal reservoirs at T_1 and T_2, absorbs 'Q' quantity of heat and delivers W_1 and W_2 (work output) respectively. The engine 'E_{R1}' rejects $(Q - W_1)$ and $(Q - W_2)$ is the quantity of heat rejected by the engine E_{IR}.

Let us assume that engine 'E_{IR}' is more efficient than 'E_{R1}' i.e. $nE_{IR} > nE_{R1}$. Then $\frac{W_2}{Q} > \frac{W_1}{Q}$; also $W_2 > W_1$ and $(Q - W_1) > (Q - W_2)$ i.e. E_{IR} gives more output than E_{R1}.

Let the engine 'E_{R1}' be reversed and now it works as a refrigerator or heat pump (E_{R2}), for which magnitudes of heat and work interactions will remain the same, but in the opposite directions.

Then E_{IR} drives E_{R2}. Since $W_2 > W_1$, a portion of W_2 (equal to W_1) is used to drive 'E_{R2}'. The heat rejected from E_{IR} may be used to drive 'E_{R2}. Therefore source may be eliminated.

The net result is E_{IR} and E_{R2} constitute a heat engine operating in a cycle produces net work ($W_2 - W_1$), exchanging heat with a single reservoir at T_2. This violates the Kelvin Planck statement of second law. Therefore the assumption $\eta E_{IR} > \eta E_{R1}$ is wrong.

Therefore we can write

$$\eta_{\text{Rev engine}} \geq \eta_{\text{Irr. Engine}}$$

Corollaries of Second Law (Carnot's Theorem)

1. All reversible heat engines operating between the two thermal reservoirs, with fixed temperatures, have the same efficiency.

   Consider heat engines E_{R1} and E_{R2} operating between same thermal reservoirs and at temperatures, 'T_1' and 'T_2'. Both of these engines absorb 'Q' quantity of heat from the high temperature source.

   Suppose, if the efficiency of one engine is more than the other [i.e. $\eta E_{R1} > \eta E_{R2}$ or $\eta E_{R2} > \eta E_{R1}$] and engine with lower efficiency is reversed to operate as a heat pump or refrigerator, it violates Kelvin Planck statement of second law [as discussed earlier]. Therefore all reversible engines operating between same thermal reservoirs have same efficiencies.

2. The efficiency of any reversible heat engine operating between source and sink depends only on the temperatures T_1 and T_2, where 'T_1' is source temperature and 'T_2' is sink temperature. The efficiency is independent of nature of working fluid.

   If an attempt is made to prove that, efficiency also depends upon nature of working fluid, it violates Kelvin Planck statement of second law.

3. Heat cannot flow itself without the assistance of external work, from low temperature region to high temperature region.
4. A reversible heat engine operating between two thermal reservoirs which are at fixed temperatures, is more efficient than any other engine (irreversible) operating between same temperatures.

4.14 THERMODYNAMIC TEMPERATURE SCALE

A quantitative measurement of temperature may be done with reference to some condition or datum plane. The change in thermodynamic properties of the material like (i) change in resistance, (ii) change in volume or pressure, (iii) Emf generated, and (iv) change in length of mercury column, are used as a measure of temperature and the scale obtained is called the empirical scale.

Different thermometers based on the empirical temperature scale gives different temperatures except at the fixed points at which they are calibrated. This necessitates that the temperature scale which is independent of the nature of thermometric material is to be found: "A temperature scale which is independent of the property of thermometric substance is known as Thermodynamic temperature scale". We know that

$$\eta_{rev} = f(t_1, t_2)$$

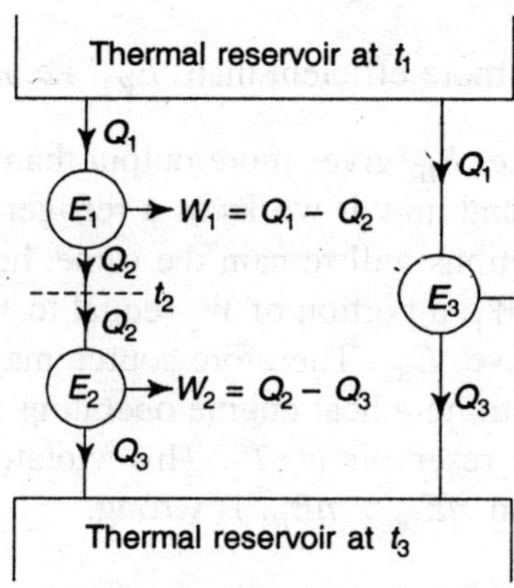

Fig. 4.14

Consider two reversible heat engines E_1 and E_2 operating between temperatures t_1 and t_2 and t_2 and t_3 as shown in figure. 'E_1' absorbs 'Q_1' quantity of heat from source at 't_1' & rejects 'Q_2' to sink at 't_2' by delivering work $W_1 = Q_1 - Q_2$. E_2 receives 'Q_2' from a source at 't_2' (which acts as sink for E_1) and before rejecting Q_3 to sink at 't_3', delivers $W_2 = Q_2 - Q_3$.

∴ For reversible heat engine E_1,

$$\eta_1 = \frac{Q_1 - Q_2}{Q_1} = 1 - \frac{Q_2}{Q_1}$$

$$1 - \frac{1}{\frac{Q_1}{Q_2}} = 1 - \frac{1}{f(t_1, t_2)}$$

i.e.

$$\frac{Q_1}{Q_2} = f(t_1, t_2)$$

Similarly, for E_2, we can write

$$\frac{Q_2}{Q_3} = f(t_2, t_3)$$

for engine E_3, $$\frac{Q_1}{Q_3} = f(t_1, t_3)$$

Now $$\frac{Q_1}{Q_2} = \frac{Q_1 / Q_3}{Q_2 / Q_3}$$

$$f(t_1, t_2) = \frac{f(t_1, t_3)}{f(t_2, t_3)}$$

The RHS of equation must be a function of 't_1' and 't_2' only. Therefore we can write

$$f(t_1, t_2) = \frac{\psi(t_1)}{\psi(t_2)}$$

Similarly $$f(t_2, t_3) = \frac{\psi(t_2)}{\psi(t_3)}$$

where 'ψ' is another function of 't'. Therefore $\psi(t) = T$ and hence we can write

$$\frac{Q_1}{Q_2} = f(t_1, t_2) = \frac{T_1}{T_2}$$

i.e. In a reversible heat engine, heat interactions 'Q_1' and 'Q_2'are proportional to absolute temperature 'T_1' and 'T_2'. This equation gives basis for a thermodyna-mic temperature scale suggested by Lord Kelvin. By fixing one temparature, the other temperature can be determined.. At present, the reference temperature is taken as triple point temperature of water which has a numerical value 273.16 K.

∴ For a Carnot engine, operating between thermal reservoirs at temperatures T and Ttp,

$$T = T_{tp} \times \frac{Q}{Q_{tp}} = 273.16 \times \frac{Q}{Q_{tp}}$$

If $Q = 0$, $T = 0$.

Thus the thermodynamic temperature scale gives an idea about absolute zero.

4.15 A SERIES OF REVERSIBLE HEAT ENGINES

Consider a number of reversible heat engines E_{R1}, E_{R2}, E_{R3} and E_{R4}, as shown in figure. In the arrangement shown in figure.

- E_{R1}, E_{R2} etc. operates between only two reservoirs.
- Each sink of Reversible heat engine forms a source of the next engine.
- Work done by each engine is same as temperature difference between source and sink is same.

$$W_1 = W_2 = W_3 = W_4$$

i.e. $$Q_1 - Q_2 = Q_2 - Q_3 = Q_3 - Q_4$$
$$= Q_4 - Q_5 \text{ etc.}$$

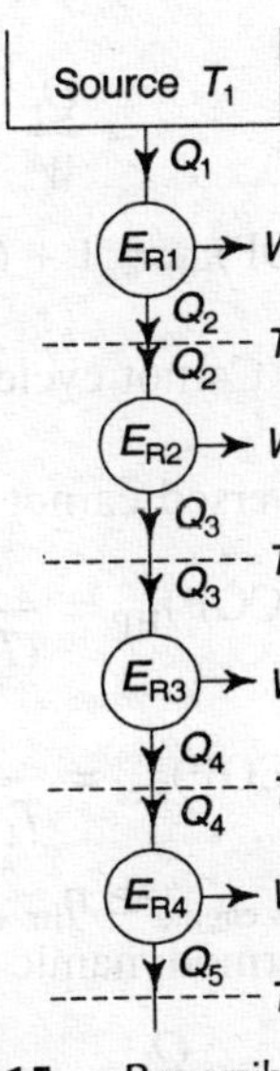

Fig. 4.15 Reversible Engines in Series

from Kelvin scales, we can write

$$\frac{Q_1}{Q_2} = \frac{T_1}{T_2}; \frac{Q_2}{Q_3} = \frac{T_2}{T_3}; \frac{Q_3}{Q_4} = \frac{T_3}{T_4} \ldots$$

Hence $$\frac{Q_1}{T_1} = \frac{Q_2}{T_2} = \frac{Q_3}{T_3} = \frac{Q_4}{T_4} = \ldots\ldots\ldots \quad (1)$$

Also $$W = \eta_1 Q_1 = \eta_2 Q_2 = \eta_3 Q_3 = \ldots\ldots\ldots$$

$$= \frac{T_1 - T_2}{T_1} Q_1 = \frac{T_2 - T_3}{T_2} Q_2 = \frac{T_3 - T_4}{T_3} Q_3 = \ldots\ldots\ldots \quad (2)$$

from (1) and (2), we can write

$$T_1 - T_2 = T_2 - T_3 = T_3 - T_4 = \ldots\ldots$$

4.16 FORMULAE USED IN SOLVING PROBLEMS

1. For heat engine; $\eta th = \frac{\text{Network output}}{\text{Heat supplied}}$

$$\eta th = \frac{Q_1 - Q_2}{Q_1} = 1 - \frac{Q_2}{Q_1}$$

2. $(COP)_{Ref} = \frac{\text{Desired effect}}{\text{Work input}} = \frac{\text{Heat extracted}}{\text{Work input}}$

$$\frac{Q_2}{W} = \frac{Q_2}{Q_1 - Q_2}$$

3. $(COP)_{HP} = \frac{\text{Desired effect}}{\text{Work input}} = \frac{\text{Heat to high temerature sink}}{\text{Work input}}$

$$= \frac{Q_1}{W} = \frac{Q_1}{Q_1 - Q_2}$$

4. $(COP)_{HP} = 1 + (COP)_{Ref}$
5. For Carnot cycle; $\eta th = \frac{T_1 - T_2}{T_1} = 1 - \frac{T_2}{T_1}$
6. Reversed carnot engine;

$$(COP)_{HP} = \frac{T_1}{(T_1 - T_2)} = 1 - \frac{T_1}{T_2}$$

$$(COP)_{Ref} = \frac{T_2}{T_1 - T_2} = \frac{T_2}{T_1} - 1$$

7. $\eta_{\text{Rev engine}} \geq \eta_{\text{Irr. engine}}$
8. Thermodynamic temperature scale

$$\frac{Q_1}{Q_2} = f(t_1, t_2) = \frac{T_1}{T_2}$$

4.17 NUMERICAL PROBLEMS

Problem 1: *A heat engine in a cycle, delivering work equal to 22 × 10³ Nm and absorbs heat of 95 kJ. Determine the efficiency of the engine and heat transfer from the working fluid.*

Solution:

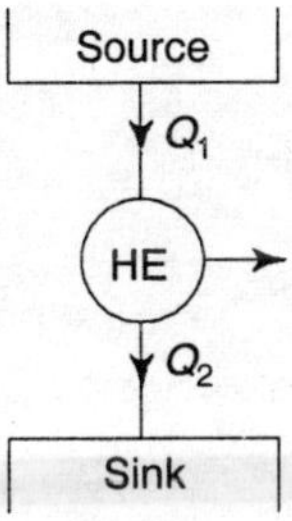

Fig. 4.16

$$W = 22 \times 10^3 \text{ Nm}$$
$$Q_1 = 95 \times 10^3 \text{ J}$$
$$\therefore \quad W = Q_1 - Q_2$$
$$22 \times 10^3 = 95 \times 10^3 - Q_2$$
$$\boxed{Q_2 = 73 \text{ kJ}} \qquad \textit{Ans.}$$
$$\eta th = \frac{W}{Q_1} = \frac{22 \times 10^3}{95 \times 10^3} = 23.15\% \qquad \textit{Ans.}$$

Problem 2: *A heat engine working at the rate of 105 kW has an efficiency of 22%. Determine the quantity of heat received and transferred from the working fluid.*

Solution: Referring the same figure in problem (1).

$$P = 105 \text{ kW}, \ \eta = 22\%$$
$$\eta = \frac{W}{Q_1} = \frac{105 \times 10^3}{Q_1}$$
$$Q_1 = \frac{105 \times 10^3}{0.22} = 477.27 \times 10^3 \text{ J/s.} \qquad \textit{Ans.}$$
$$W = Q_1 - Q_2$$
$$105 \times 10^3 = 477.27 \times 10^3 - Q_2$$
$$\therefore \quad Q_2 = 372.27 \times 10^3 \text{ J/s} \qquad \textit{Ans.}$$

Problem 3: *In an ideal closed cycle gas turbine, the efficiency is given by*

$\eta = 1 - \dfrac{1}{(r_p)^{\gamma-1}}$, where r_p = presure ratio, i.e. out let pressure/Inlet pressure of a compressor and γ = 1.4 for air,

(i) Evaluate η for r_p = 7, with air as working fluid.

(ii) The heat transfer from air in the cooler is 3.2 × 10⁶ J/s.

Determine heat transfer rate to the air in the heater and the power developed, assuming all other heat transfers to be negligible.

Solution: Given: $\eta = 1 - \dfrac{1}{(r_p)^{\gamma-1}}$, where $r_p = 7$, $\gamma = 1.4$.

$$Q_2 = 3.2 \times 10^6 \text{ J/s}$$
$$\therefore \quad \eta = 1 - \frac{1}{(7)^{1.4-1}} = 54.08\% \qquad \textit{Ans.}$$

$$\eta = \frac{W}{Q_1}$$

$$Q_{1\times\eta} = W = Q_1 - Q_2$$

$$Q_2 = Q_1 - Q_{1\times\eta}$$

$$Q_2 = Q_1(1-\eta)$$

$$\therefore \quad Q_1 = \frac{Q_2}{1-\eta} = \frac{3.2\times 10^6}{1-0.54} = 6.95\times 10^6 \text{ J/s}$$

$$Q_1 = 6.95\times 10^6 \text{ J/s}$$ *Ans.*

$$\therefore \quad W = Q_1 - Q_2 = 6.95\times 10^6 - 3.2\times 10^6$$

$$= 3.75\times 10^6 \text{ J/s}$$ *Ans.*

Problem 4: *A reversed heat engine absorbs 220 kJ of heat from a low tempt. region and has a mechanical input of 90 kJ. Determine the quantity of heat transferred to high temperature region. Also determine COP of the reversed engine when working as a refrigerator and as a heat pump.*

Solution:

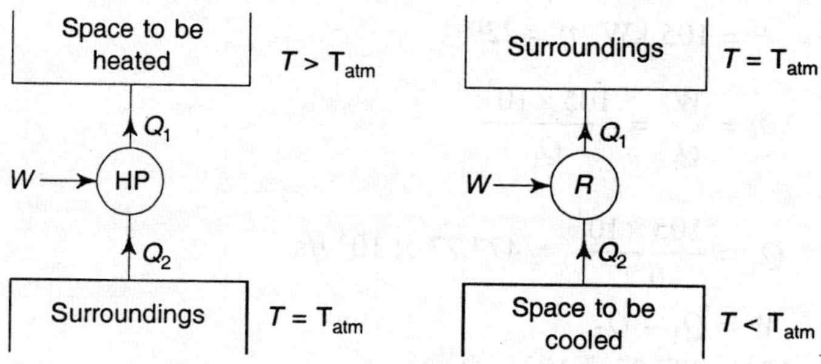

Fig. 4.17

$$W = Q_1 - Q_2$$

$\therefore$ Heat transferred to high tempt. region

$$Q_1 = W + Q_2 = 90 + 220 = 310 \text{ kJ}$$

$$(\text{COP})_{\text{Ref}} = \frac{Q_2}{W} = \frac{220}{90} = 2.44$$ *Ans.*

$$(\text{COP})_{\text{HP}} = \frac{Q_1}{W} = \frac{310}{90} = 3.44$$ *Ans.*

Problem 5: *An inventor claims to have developed a refrigerator which maintains the refrigerated space at –10°C. While operating in a room at 25°C and it has a COP of 8.5. How would you evaluate his claim is valid.* *[VTU, March 2000]*

Solution: $$(\text{COP})_{\text{ref}} = \frac{Q_2}{W} = \frac{T_2}{T_1 - T_2} = \frac{273-10}{298.-263} = 7.51$$

The inventor claims that, its refrigerator has a COP of 8.5. As 8.5 > 7.51, his claim is invalid.

Problem 6: *A reversible heat engine operates between two reservoirs at 820°C and 27°C. Engine drives a reversible refrigerator which operates between reservoirs at temperatures of 27°C and –15°C. The heat transfer to the engine is 2000 kJ and the network available for the combined cycle is 300 kJ. (i) How much heat is transferred to the refrigerant and also determine the total heat rejected to the reservoir at 27°C. (ii) If the efficiency of heat engine and COP of refrigerator are each 40% of their maximum values, determine heat transferred to refrigerant and total heat rejected as in (1).*

Solution:

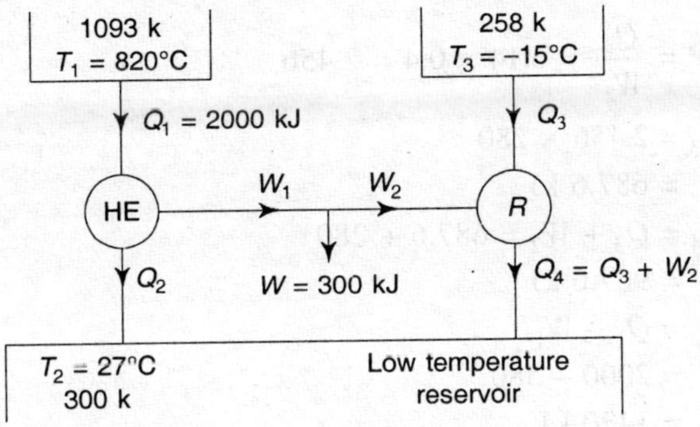

Fig. 4.18

(i) ∴ For heat engines, we can write

$$\eta_{max} = 1 - \frac{T_2}{T_1} = 1 - \frac{300}{1093} = 0.725 = \frac{W_1}{Q_1}$$

∴ $W_1 = 0.725 \times 2000 = 1450$ kJ

for a refrigerator cycle.

$$(COP)_{Ref} = \frac{T_3}{T_2 - T_3} = \frac{258}{300 - 258} = 6.14 = \frac{Q_3}{W_2}$$

∵ $W = W_1 - W_2 = 300$ kJ

$W_2 = 1450 - 300 = 1150$ kJ

∴ $Q_3 = 1150 \times 6.14 = 7061$ kJ

$Q_4 = Q_3 + W_2$

$= 7061 + 1150$

$= 8211$ kJ

$Q_2 = Q_1 - W_1$

$= 2000 - 1450$

$= 550$ kJ

∴ Heat transferred to the refrigerant

$Q_3 = 7061$ kJ *Ans.*

Total heat transferred to the low tempt. reservoir at 27°C.

$= Q_2 + Q_4$

$= 550 + 8211$

$= 8761$ kJ *Ans.*

(ii) $\therefore$ Efficiency of the actual heat engine cycle

$$\eta = 0.4 \cdot \eta_{max} \mid \eta_{max} = 72.5\%$$
$$\eta = 0.4 \times 0.725$$
$$\eta = 0.29$$
$$W_1 = 1450 \times 0.4 = 580 \text{ kJ}$$
$$W_2 = 580 - 300 = 280 \text{ kJ}$$

COP of the actual refrigerator cycle is

$$\text{COP} = \frac{Q_3}{W_2} = 6.14 \times 0.4 = 2.456$$

$$\therefore \quad Q_3 = 2.456 \times 280$$
$$= 687.6 \text{ kJ} \qquad \textit{Ans.}$$
$$Q_4 = Q_3 + W_2 = 687.6 + 280$$
$$= 967.6 \text{ kJ}$$
$$Q_2 = Q_1 - W_1$$
$$= 2000 - 580$$
$$Q_2 = 1420 \text{ kJ}$$

Heat transferred to the refrigerant

$$Q_3 = 687.6 \text{ kJ}$$

Total heat transferred to the low temperature reservoir at 27°C

$$= Q_2 + Q_4$$
$$= 1420 + 967.6$$
$$= 2387.6 \text{ kJ.} \qquad \textit{Ans.}$$

Problem 7: *In a reversed heat engine, the work done on the engine is 75 × 10³ Nm and heat transfer from low temperature region to engine is 225 × 10³ J. Determine the heat transfer to high temperature region and coefficient of performance as a refrigerator and as a heat pump.*

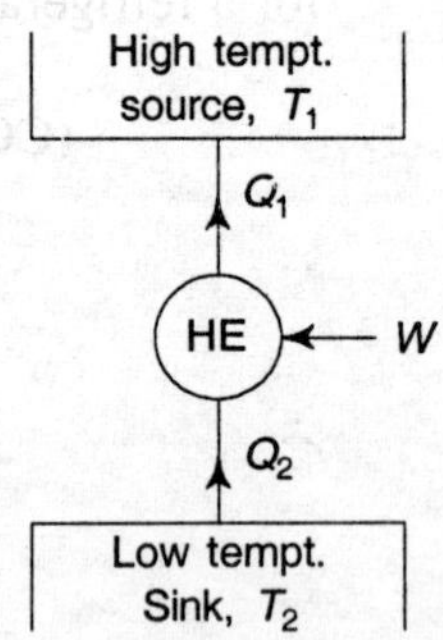

Fig. 4.19

Solution:

$$W = 75 \times 10^3 \text{ Nm}$$
$$Q_2 = 225 \times 10^3 \text{ J}$$
$$Q_1 = ? \quad (\text{COP})_R = ? \quad (\text{COP})_{HP} = ?$$
$$W = Q_1 - Q_2$$
$$Q_1 = W + Q_2$$
$$= 75 \times 10^3 + 225 \times 10^3 = 300 \text{ kJ} \qquad \textit{Ans.}$$

$$(\text{COP})_R = \frac{Q_2}{W} = \frac{225 \times 10^3}{75 \times 10^3} = 3.00 \qquad \textit{Ans.}$$

$$(\text{COP})_{HP} = \frac{Q_1}{W} = \frac{300 \times 10^3}{75 \times 10^3} = 4.00 \qquad \textit{Ans.}$$

Problem 8: *When 50 KW of power is supplied to a heat pump, its coefficient of performance is 5.*
(i) Determine the rate of heat transfers to and from the working fluid.
(ii) If the heat pump is used to heat the water in a heat exchanger while its temperature increases from 50°C to 75°C in flowing through the heat pump. Determine mass flow rate of heated water by neglecting water velocity.

Solution:

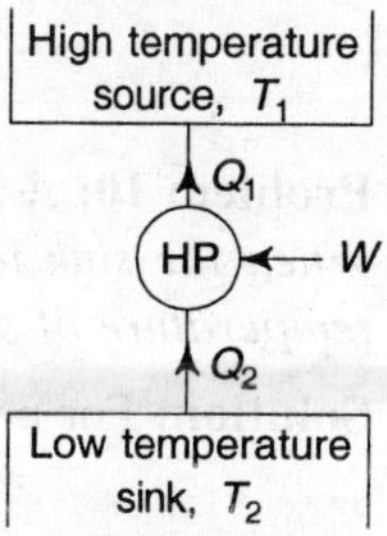

Fig. 4.20

$$P = 50 \text{ KW}$$
$$(\text{COP})_{\text{HP}} = 5$$
$$(\text{COP})_{\text{HP}} = \frac{Q_1}{W}$$
$$5 = \frac{Q_1}{50 \times 10^3}$$
$$\boxed{Q_1 = 250 \times 10^3 \text{ J/s}} \qquad \textit{Ans.}$$
$$W = Q_1 - Q_2$$
$$50 \times 10^3 = 250 \times 10^3 - Q_2$$
$$\boxed{Q_2 = 200 \times 10^3 \text{ J/s}} \qquad \textit{Ans.}$$
$$Q_1 = \dot{m}\, C_p\, (t_2 - t_1)$$
$$250 \times 10^3 = \dot{m} \times 4180 \times (75 - 50)$$
$$\dot{m} = 2.39 \text{ kg/s.} \qquad \textit{Ans.}$$

Problem 9: *A heat pump is driven by a heat engine. The heat transfer from the heat engine and heat pump are used to heat the water circulating through the radiators of a building. The heat engine efficiency is 30% and COP of heat pump is 3.8. Evaluate the ratio of heat transfer to the circulating water to the heat transfer to the heat engine.*

Solution:

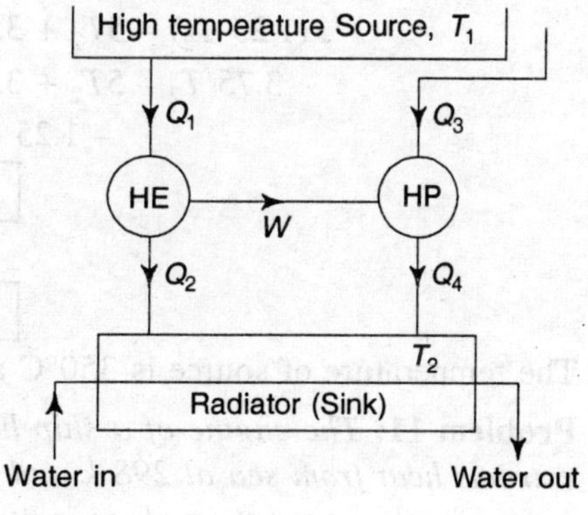

Fig. 4.21

We know that, $$\eta_{\text{HE}} = \frac{W}{Q_1}$$
$$(\text{COP})_{\text{HP}} = \frac{Q_4}{W}$$
$$W = Q_1\, \eta_{\text{HE}}, \qquad (1)$$
$$W = \frac{Q_4}{(\text{COP})_{\text{HP}}} \qquad (2)$$

Also $$Q_2 = Q_1 - W = Q_1 - Q_1\, \eta_{\text{HE}}$$
$$= (1 - \eta_{\text{HE}})\, Q_1$$
$$= (1 - 0.3)\, Q_1$$
$$= 0.7\, Q_1$$

from (1) and (2) we can write

$$Q_1 \eta_{\text{HE}} = \frac{Q_4}{(\text{COP})_{\text{HP}}}$$

$$Q_4 = Q_1\ \eta_{HE}\ (COP)_{HP}$$
$$= Q_1\ (0.3)(3.8)$$
$$= 1.14\ Q_1$$

$$\therefore \quad \boxed{\frac{Q_4}{Q_1} = 1.14}$$ **Ans.**

Problem 10: *A heat engine working on Carnot cycle converts one-fifth of the heat input in to work. When the sink temperature is reduced by 70°C, the heat engine efficiency gets doubled. Determine temperature of source and sink.*

Solution: For a Carnot heat engine

$$\eta_{HE} = \frac{T_1 - T_2}{T_1}$$

T_1 = source tempt.

T_2 = sink tempt.

Given:
$$\frac{T_1 - T_2}{T_1} = \frac{1}{5}$$

i.e.
$$5T_1 - 5T_2 = T_1$$
$$4T_1 = 5T_2$$
$$\boxed{T_1 = 1.25\ T_2} \qquad (1)$$

Also,
$$\frac{T_1 - (T_2 - 70)}{T_1} = 2 \times \frac{1}{5} = \frac{2}{5}$$
$$5T_1 - 5T_2 + 350 = 2T_1$$
$$3T_1 - 5T_2 + 350 = 0$$

from (1)
$$3(1.25\ T_2) - 5T_2 + 350 = 0$$
$$3.75\ T_2 - 5T_2 + 350 = 0$$
$$-1.25\ T_2 = -350$$
$$\boxed{T_2 = 280°C}$$ **Ans.**
$$T_1 = 1.25 \times 280$$
$$\boxed{T_1 = 350°C}$$ **Ans.**

The temperature of source is 350°C and that of sink is 280°C.

Problem 11: *The engine of a ship has a thermal efficiency 75% of that of carnot cycle. The engine extracts heat from sea at 298 k and rejects heat at 280 k to atmosphere. The work delivered by the engine is dissipated through an agitator to heat 480 kg of sea water to 350 k. What quantity of heat must be extracted from the sea water to provide the required heating effect? Take specific heat of water C_p = 4.186 kJ/kg k.*

Solution:
$$\eta_{\text{Carnot cycle}} = \frac{T_1 - T_2}{T_1} = \frac{298 - 280}{298} = \frac{18}{298}$$

For the given engine, $\eta_{HE} = 0.8 \times \text{Carnot eff.}$

$$= 0.75 \times \frac{18}{298} = 0.0453$$

Work required for heating water

$$= mC_p\,\Delta T = 480 \times 4.186 \times (350 - 298)$$

$$= 104482.56 \text{ kJ}$$

But $\eta = \dfrac{W}{Q_1}$

$$\therefore \quad Q_1 = \frac{W}{\eta} = \frac{104482.56}{0.0453}$$

$$\boxed{Q_1 = 2306458.27 \text{ kJ}}$$ ***Ans.***

Problem 12: *A Carnot engine operates between high temperature source at 'T_1' and low temperature sink at 'T_2' and it drives a refrigerator which extracts heat from a reservoir at temperature 'T_3'. The refrigerator rejects heat at temperature 'T_2'. Determine the minimum quantity of heat extracted from the source at 'T_1'. ($T_1 > T_2 > T_3$).*

If T_1 = 820 k, T_2 = 310 k, and T_3 = 265 k,

Determine minimum value of heat extracted from reservoir at 'T_1', if heat extracted from the reservoir at T_3 is equal to 100 kJ.

Solution:

We know that,

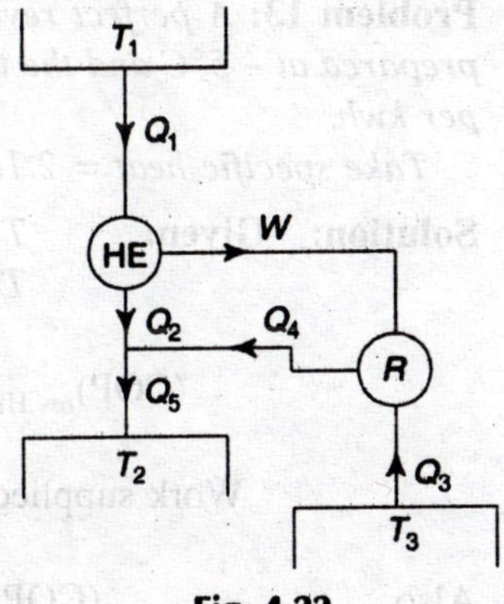

Fig. 4.22

$$\eta_{HE} = \frac{W}{Q_1} = \frac{T_1 - T_2}{T_1}$$

$$Q_1 = W\left(\frac{T_1}{T_1 - T_2}\right) \qquad (1)$$

$$(\text{COP})_R = \frac{Q_3}{W} = \frac{Q_3}{Q_4 - Q_3}$$

$$(\text{COP})_R = \frac{T_3}{T_2 - T_3}$$

$$\therefore \quad W = Q_3\,\frac{(T_2 - T_3)}{T_3} \qquad (2)$$

from equations (1) and (2), we can write

$$Q_1 = Q_3\,\frac{(T_2 - T_3)}{T_3} \cdot \left(\frac{T_1}{T_1 - T_2}\right) \qquad (3)$$

$$Q_1 = 100\,\frac{(310 - 265)}{265}\left(\frac{820}{820 - 310}\right)$$

$$\boxed{Q_1 = 1056 \text{ kJ}}$$ *Ans.*

Also, $W = Q_4 - Q_3$

$\therefore$ $$Q_4 = W + Q_3 = Q_3 + Q_3\left(\frac{T_2 - T_3}{T_3}\right)$$

$$\left[\text{from eqn. (2) } W = Q_3\left(\frac{T_2 - T_3}{T_3}\right)\right]$$

$$Q_4 = 100 + 100\left(\frac{310 - 265}{265}\right)$$

$$\boxed{Q_4 = 116.98 \text{ kJ}}$$

$\therefore$ $$W = Q_3\left(\frac{T_2 - T_3}{T_3}\right)$$

$$= 100\left(\frac{310 - 265}{265}\right)$$

$$\boxed{W = 16.98 \text{ kJ}}$$ *Ans.*

Problem 13: *A perfect reversed heat engine is used for making ice from water at 25°C. The ice is prepared at – 5°C and the temperature of freezing mixture is –10°C. Calculate quantity of ice formed per kwh.*

Take specific heat = 2.1kJ/Kg k and latent heat = 335 kJ/kg for ice.

Solution: Given: $T_1 = 25°C = 298$ k

$T_2 = -10°C = 263$ k

$$(\text{COP})_{\text{rev.HE}} = \frac{T_2}{T_1 - T_2} = \frac{263}{298 - 263} = 7.51$$

Work supplied = 1 kwh = 1 × 3600 kJ

Also $$(\text{COP}) = \frac{Q_2}{Q_1 - Q_2} = \frac{Q_2}{W}$$

$$Q_2 = (\text{COP})\ W = 7.51 \times 3600 = 27050 \text{ kJ/hr.}$$

Q_2 = Total heat extracted from ice

Removal of heat from 1 kg of water at 25°C to produce ice at – 5°C.

$$= mC_{pW} \cdot \Delta T + \text{mL.} + mC_{pi}\Delta T$$

$$= 1 \times 4.186\ (25 - 0) + 1 \times 335 + 1 \times 2.1 \times (0 - (-5))$$

$$= 450.15 \text{ kJ}$$

$\therefore$ Ice formed per kwh $= \dfrac{27050}{450.15} = 60.09$ kg *Ans.*

Problem 14: *An Inventor claims to have developed a heat engine with specifications as follows;*

Power developed = 76 kW

Fuel burnt = 4.8 kg/hr

Heating value of fuel = 73000 kJ/kg.

Temperature limits = 980 k and 393 k.

How would you evaluate his claim is valid.

Solution: For the given engine

$$\text{Work done} = 76 \text{ kW} = 76 \times 3600 \qquad \left| \; \frac{\text{kJ}}{S} \times 3600 \right.$$

$$= 273600 \frac{\text{kJ}}{\text{hr}} \qquad \left| \; \frac{\text{kJ}}{\text{hr}} \right.$$

$$\text{Heat supplied} = (\text{mass of fuel}) \times \text{calorific value of fuel}$$

$$= 4.8 \times 73000$$

$$= 350400 \frac{\text{kJ}}{\text{hr}} \qquad \left| \; \frac{\text{kg}}{\text{hr}} \times \frac{\text{kJ}}{\text{kg}} \right.$$

$$\eta_{th} = \frac{\text{Work done}}{\text{Heat supplied}} = \frac{273600}{350400} = 0.78 = 78\%$$

$$\eta_{rev} = \frac{T_1 - T_2}{T_1} = \frac{980 - 393}{980} = 59.89\%$$

Since the claimed engine shows efficiency (78%) greater than that of a reversible engine, the claim is wrong.

Problem 15: *A reversible heat engine takes 900 kJ of heat from a source at 700 k. The engine develops 350 kJ of net work and rejects heat to two low temperature reservoirs at 600 k ad 500 k. Determine engine thermal efficiency and heat rejected to each low temperature reservoir.*

Solution:

$$\eta_{th} = \frac{\text{Work done}}{\text{Heat supplied}}$$

$$= \frac{350}{900} = 38.8\%$$

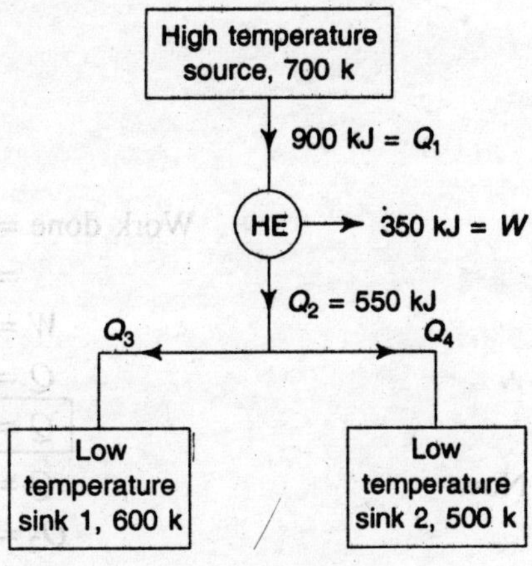

Fig. 4.23

also

$$W = Q_1 - Q_2$$

$$Q_2 = Q_1 - W$$

$$= 900 - 350$$

$$= 550 \text{ kJ}$$

The heat rejected to sink at 500 k

$$= Q_4 = Q_2 - Q_3$$

$$= 550 - Q_3$$

As the engine is reversible, according to Clausius theorem

$$\oint \frac{\gamma Q}{T} = 0 = \frac{900}{700} - \frac{Q_3}{600} - \frac{Q_4}{500} = 0$$

$$\frac{900}{700} - \frac{Q_3}{600} - \frac{550 - Q_3}{500} = 0$$

$$= 1.285 - 1.66 \times 10^{-3}\ Q_3 - 1.1 + 2 \times 10^{-3}\ Q_3 = 0$$

$$= 0.185 + 3.4 \times 10^{-4}\ Q_3 = 0$$

$$Q_3 = \frac{-0.185}{3.4 \times 10^{-4}}$$

$$\boxed{Q_3 = -\ 544.117\ \text{kJ}}$$ **Ans.**

$$Q_4 = 550 - 544.117$$

$$\boxed{Q_4 = 5.8\ \text{kJ}}$$ **Ans.**

Problem 16: *A reversible heat engine extracts heat from three reservoirs at 1000 k, 810 k and 595 k. The engine delivers 10 × 10³ J/S of net work and rejects 400 kJ/min of heat to a sink at 298 k. If the heat supplied to the reservoir at 1000 k is 55% of the heat supplied by the reservoir at 595 k. Determine quantity of heat absorbed by each reservoir.*

Solution:

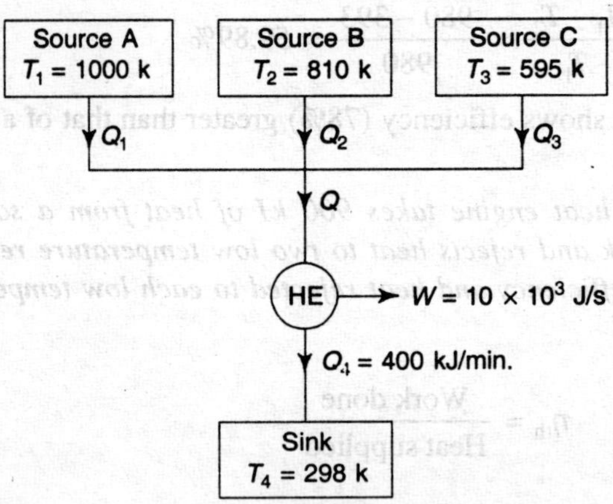

Fig. 4.24

Work done $= 10 \times 10^3$ J/s

$= 10 \times 10^3 \times 60 = 600 \times 10^3$ J/min

$W = Q - Q_4$

∴ $Q = W + Q_4 = 600 \times 10^3 + 400 \times 10^3$

$$\boxed{Q = 1000 \times 10^3\ \text{J/min}}$$

Now $Q = Q_1 + Q_2 + Q_3$ | **Given:** $Q_1 = 55\%\ Q_3$

$Q_2 = Q - Q_1 - Q_3$

$Q_2 = 1000 \times 10^3 - 0.55\ Q_3 - Q_3$

$Q_2 = (1000 \times 10^3 - 1.55\ Q_3)$ J/min

From, clausius theorem,

$$\oint \frac{\delta Q}{T} = 0$$

i.e $$\frac{Q_1}{T_1} + \frac{Q_2}{T_2} + \frac{Q_3}{T_3} - \frac{Q_4}{T_4} = 0$$

$$\frac{0.55Q_3}{1000} + \frac{(1000 - 1.55Q_3)}{810} + \frac{Q_3}{595} - \frac{400}{298} = 0$$

$$5.5 \times 10^{-4}\ Q_3 + 1.234 - 1.913 \times 10^{-3}\ Q_3 + 1.68 \times 10^{-3}\ Q_3$$
$$1.342 = 0$$
$$3.17 \times 10^{-4}\ Q_3 = 0.108$$
$$\boxed{Q_3 = 340.69 \text{ kJ/min}}$$ **Ans.**

$$Q_2 = 1000 - 1.55\ Q_3$$
$$= 1000 - 1.55\ (340.69)$$
$$\boxed{Q_2 = 471.93 \text{ kJ/min}}$$ **Ans.**

$$Q_1 = 0.55\ Q_3$$
$$= 0.55\ (340.69)$$
$$\boxed{Q_1 = 187.37 \text{ kJ/min}}$$ **Ans.**

Problem 17: *A reversible heat engine operates between heat reservoirs A, B and C. The engine receives equal quantities of heat from reservoirs A and B at Ta and Tb respectively and rejects heat to reservoir C at Tc. If efficiency of this engine is 'α' times greater than efficiency of another reversible engine operating between reservoirs A and C only, then show that*

$$\alpha = \frac{1}{2}\frac{Ta}{Tb}\left[\frac{Tb}{Ta} + \frac{Tb - Tc}{Ta - Tc}\right]$$

Solution:

(i) For a reversible engine

$$\oint \frac{\delta Q}{T} = 0$$

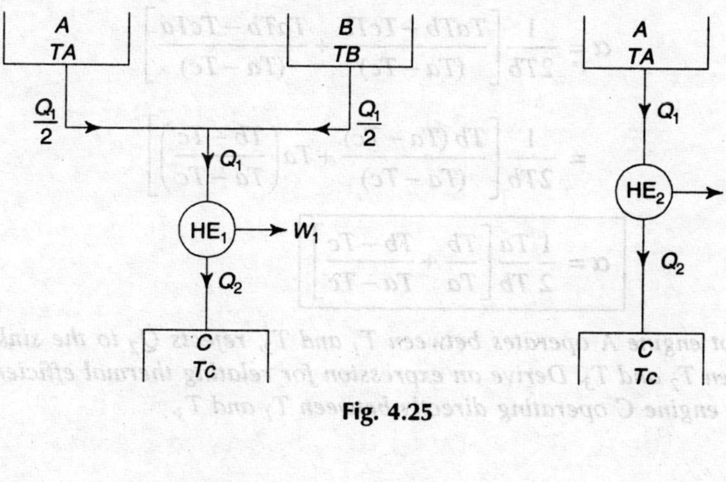

Fig. 4.25

$$\therefore \quad \frac{Q_1}{2Ta} + \frac{Q_1}{2Tb} = \frac{Q_2}{Tc}$$

$$W = Q_1 - Q_2$$

$$Q_2 = Q_1 - W_1$$

$$\therefore \quad \frac{Q_1}{2Ta} + \frac{Q_1}{2Tb} = \frac{Q_1 - W_1}{Tc}$$

$$Q_1 - W_1 = Tc\left[\frac{Q_1}{2Ta} + \frac{Q_1}{2Tb}\right]$$

$$\therefore \quad W_1 = Q_1 - \frac{Tc\,Q_1}{2Ta} - \frac{Tc\,Q_1}{2Tb}$$

$$W_1 = Q_1\left[1 - \frac{Tc}{2Ta} - \frac{Tc}{2Tb}\right]$$

$$= Q_1\left[\frac{2TaTb - TcTb - TcTa}{2Ta\,Tb}\right]$$

$$\eta_1 = \frac{WD}{Q_{sup}} = \frac{WD}{Q_1} = \boxed{\frac{2TaTb - TcTb - TcTa}{2Ta\,Tb}}$$

(ii) **For the reversible engine operating between Ta and Tc.**

$$\eta_2 = \frac{Ta - Tc}{Ta}, \text{ but } \eta_1 = \alpha\,\eta_2 \text{ (given)}$$

$$= \frac{2TaTb - TcTb - TcTa}{2Ta\,Tb} = \alpha\left(\frac{Ta - Tc}{Ta}\right)$$

$$\alpha = \frac{\not{Ta}\,[2TaTb - TcTb - TcTa]}{2\,\not{Ta}\,Tb\,(Ta - Tc)}$$

$$\alpha = \frac{1}{2Tb}\left[\frac{TaTb - TcTb}{(Ta - Tc)} + \frac{TaTb - TcTa}{(Ta - Tc)}\right]$$

$$= \frac{1}{2Tb}\left[\frac{Tb\,(Ta - Tc)}{(Ta - Tc)} + Ta\left(\frac{Tb - Tc}{Ta - Tc}\right)\right]$$

$$\boxed{\alpha = \frac{1}{2}\frac{Ta}{Tb}\left[\frac{Tb}{Ta} + \frac{Tb - Tc}{Ta - Tc}\right]} \qquad \textit{Ans.}$$

Problem 18: *A Carnot engine A operates between T_1 and T_2, rejects Q_2 to the sink which is at T_2, and B operates between T_2 and T_3. Derive an expression for relating thermal efficiency of engines A and B, so that Carnot engine C operating directly between T_1 and T_3.*

Solution:

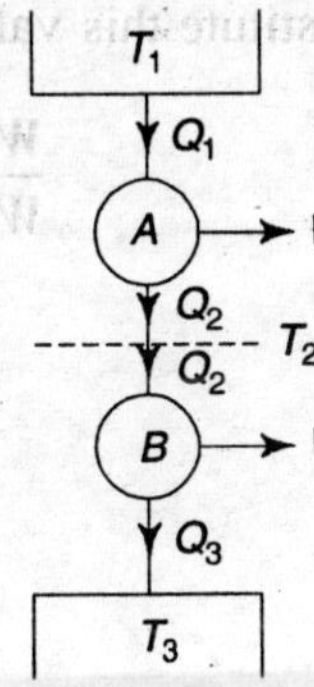

Fig. 4.26

$$\eta_A = \frac{T_1 - T_2}{T_1} = 1 - \frac{T_2}{T_1}$$

$$1 - \eta_A = \frac{T_2}{T_1}$$

Similarly $$1 - \eta_B = \frac{T_3}{T_2}$$

$\therefore$ $$(1 - \eta_A)(1 - \eta_B) = \frac{T_2}{T_1} \times \frac{T_3}{T_2} = \frac{T_3}{T_1}$$

far engine C operating between T_1 and T_3

$$\eta_C = \frac{T_1 - T_3}{T_1} = 1 - T_3/T_1$$

$$1 - \eta_C = T_3/T_1$$

$\therefore$ $$(1 - \eta_A)(1 - \eta_B) = (1 - \eta_C)$$

$\therefore$ $$(1 - \eta_C) = 1 - \eta_B - \eta_A + \eta_A \eta_B$$

$$\boxed{\eta_C = \eta_A + \eta_B - \eta_A \eta_B}$$ ***Ans.***

Problem 19: *Series combinations of three Carnot engines A, B & C operates between temperatures of 1000 k and 300 k. Calculate the intermediate temperatures, if the amount of work produced by the engines is in the productivity of 5:4:3.*

Solution:

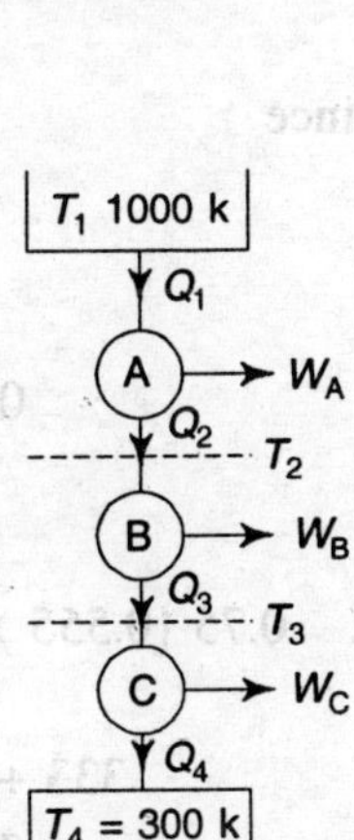

Fig. 4.27

We know that, $$\eta_A = \frac{Wa}{Q_1} = \frac{T_1 - T_2}{T_1} = \frac{1000 - T_2}{1000} \quad (1)$$

$$\eta_B = \frac{Wb}{Q_2} = \frac{T_2 - T_3}{T_2} \quad (2)$$

$$\eta_C = \frac{Wc}{Q_3} = \frac{T_3 - T_4}{T_3} = \frac{T_3 - 300}{T_3} \quad (3)$$

and $$W_a : W_b : W_c = 5:4:3$$

Also, $$Q_1 = W_a + Q_2$$

Substitute this value in Eqn. (1),

$$\frac{Wa}{Wa + Q_2} = \frac{1000 - T_2}{1000}$$

$$(wa + Q_2)(1000 - T_2) = 1000\ Wa$$

$$Wa\ 1000 + Q_2\ 1000 - T_2\ Wa - Q_2 T_2 = 1000\ Wa$$

$$Q_2\ (1000 - T_2) = T_2 Wa$$

$$Q_2 = Wa\left(\frac{T_2}{1000 - T_2}\right)$$

Substitute this value in eqn (2),

$$\frac{Wb}{Wa}\left(\frac{1000-T_2}{T_2}\right)=\frac{T_2-T_3}{T_2}; \quad \text{also given} \quad \frac{Wb}{Wa}=\frac{4}{5}$$

$$\frac{4}{5}\left(\frac{1000-T_2}{\cancel{T_2}}\right)=\left(\frac{T_2-T_3}{\cancel{T_2}}\right)$$

$$800-\frac{4}{5}T_2=T_2-T_3$$

$$800+T_3=T_2\,(1.8)$$

$$\therefore \quad T_2=0.555\,(800+T_3)$$

Similarly $\quad Q_2=Q_3+W_6$

Substitute this value in Eqn (2), we write

$$\left(\frac{Wb}{Q_3+Wb}\right)=\frac{T_2-T_3}{T_2}$$

$$\therefore \quad Q_3=W_6\left(\frac{T_3}{T_2-T_3}\right)$$

Substitute this value in Eqn (3), we get

$$\frac{Wc}{Wb}\,\frac{(T_2-T_3)}{T_3}=\frac{T_3-300}{T_3};$$

Since $\quad Wc: Wb = 3:4$

$$\frac{3}{4}\frac{(T_2-T_3)}{\cancel{T_3}}=\frac{T_3-300}{\cancel{T_3}}$$

$$0.75\,T_2-0.75\,T_3=T_3-300$$

$$0.75\,T_2+300=T_3\,(1+0.75) \quad \because \quad T_2=0.555\,(800+T_3)$$

$$0.75\,T_2+300=T_3\,(1.75)$$

$$\therefore \quad 0.75\,\{0.555\times 800+0.555\,T_3\}+300=T_3\,(1.75)$$

$$333+0.416\,T_3+300=T_3\,(1.75)$$

$$T_3\,(1.75-0.416)=333+300$$

$$\boxed{T_3=474.51\text{ k}} \qquad \textit{Ans.}$$

$$T_2=0.555\,(800+T_3)$$

$$=0.555\,(800+474.51)$$

$$\boxed{T_2=707.35\text{ k}} \qquad \textit{Ans.}$$

Problem 20: *A source at temperature T_1K supplies heat to a reversible heat engine which rejects heat to a low temperature sink at 780 k. The sink at 780 k acts as a source for second reversible engine which rejects heat to a cold reservoir at 280 k. Determine T_1, (i) for equal thermal efficiencies of the two engines (ii) when two engines deliver the same amount of work.*

Solution:

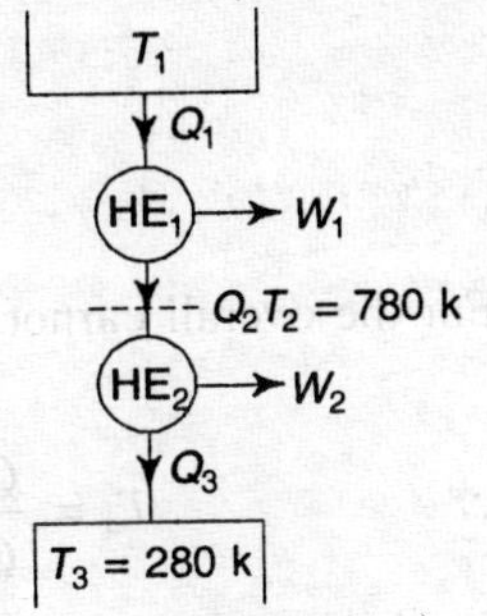

Fig. 4.28

For $\quad HE_1, \eta_1 = \dfrac{T_1 - T_2}{T_1} = 1 - \dfrac{T_2}{T_1}$

Similarly, for $\quad HE_2, \eta_2 = \dfrac{T_2 - T_3}{T_2} = 1 - \dfrac{T_3}{T_2}$

(i) when $\eta_1 = \eta_2$

$$1 - \frac{T_2}{T_1} = 1 - \frac{T_3}{T_2}$$

$$T_2 = \sqrt{T_1 T_3}$$

$$\therefore \quad T_1 = \frac{T_2^2}{T_3} = \frac{780^2}{280} = 2172.85 \text{ k}$$

$$\boxed{T_1 = 2172.85 \text{ k}} \qquad \textit{Ans.}$$

(ii) When $W_1 = W_2$,

For $\quad HE_1, W_1 = Q_1 - Q_2; W_2 = Q_2 - Q_3$

$\therefore \quad Q_1 - Q_2 = Q_2 - Q_3$

$\div$ by Q_2

$$\therefore \quad \frac{Q_1}{Q_2} + \frac{Q_3}{Q_2} = 2$$

Also, the engines are reversible, therefore

$$\frac{Q_1}{Q_2} = \frac{T_1}{T_2}; \quad \frac{Q_3}{Q_2} = \frac{T_3}{T_2}$$

$$\therefore \quad \frac{T_1}{T_2} + \frac{T_3}{T_2} = 2$$

$$T_2 = \frac{T_1 + T_3}{2}$$

$$\therefore \quad T_1 = 2T_2 - T_3$$

$$= 2 \times 780 - 280$$

$$T_1 = 1280 \text{ k}$$

$$\boxed{T_1 = 1280 \text{ k}} \qquad \textit{Ans.}$$

Problem 21: *In a series arrangement of three carnot engines, first engine receives 400 kJ of heat from a source at 360 k and delivers 180 kJ of work: The second and third heat engines delivers 120 kJ and 50 kJ of work respectively. Determine exhaust temperatures of second and third carnot engine.*

Solution:

The total work output of three heat engines = 180 + 120 + 50 = 350 kJ

From the principle of energy conservation.

$$Q_4 = Q_1 - W_{total}$$
$$= 400 - 350$$
$$Q_4 = 50 \text{ kJ}$$

For the overall carnot engine $\dfrac{T_1}{T_4} = \dfrac{Q_1}{Q_4}$

$\therefore$ $$T_4 = \frac{Q_4}{Q_1} \cdot T_1 = \frac{50}{400} \times 360 = 45 \text{ k}$$ *Ans.*

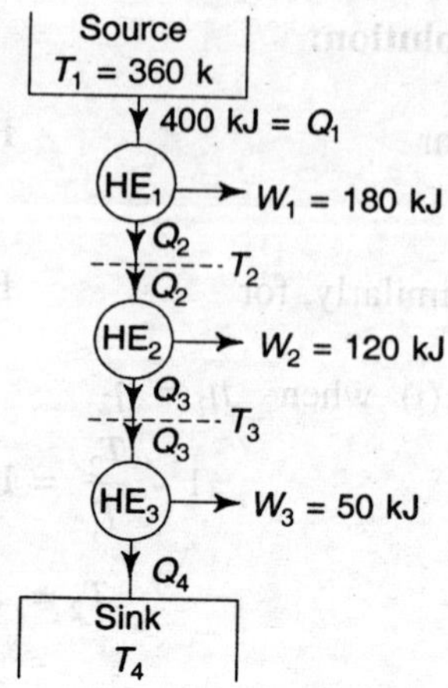

Fig. 4.29

for the heat engines HE_1 & HE_2 in series

$$Q_3 = Q_1 - W_1 - W_2$$
$$= 400 - 180 - 120 = 100 \text{ kJ}$$
$$\frac{T_1}{T_3} = \frac{Q_1}{Q_3}$$

$\therefore$ $$T_3 = \frac{Q_3}{Q_1} \cdot T_1 = \frac{100}{400} \times 360$$

$$\boxed{T_3 = 90 \text{ k}}$$ *Ans.*

Problem 22: *The thermal efficiency of three real heat engines which are in series is same. A source at 1175 k supplies 2250 kJ of heat to the first engine and the third heat engine 280 kJ of heat to a sink at 145 k. Determine work output from each engine.*

Solution:

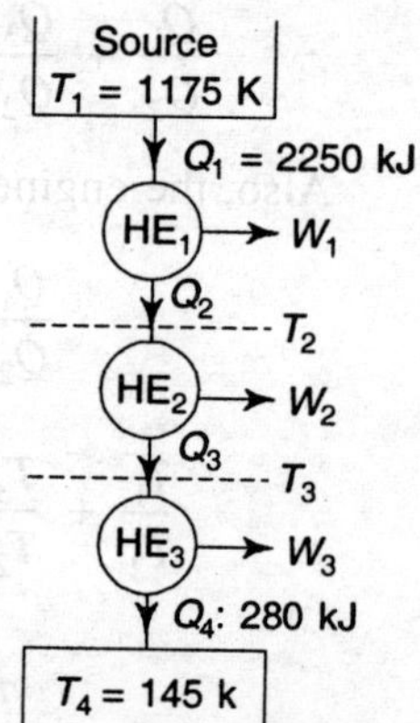

Fig. 4.30

For HE_1, $\eta_1 = \dfrac{Q_1 - Q_2}{Q_1} = 1 - \dfrac{Q_2}{Q_1}$

$\therefore$ $$\frac{Q_2}{Q_1} = 1 - \eta_1$$

Similarly HE_2, $$\frac{Q_3}{Q_2} = 1 - \eta_2$$

for HE_3, $$\frac{Q_4}{Q_3} = 1 - \eta_3$$

$\therefore$ $$\frac{\cancel{Q_2}}{Q_1} \times \frac{\cancel{Q_3}}{\cancel{Q_2}} \times \frac{Q_4}{\cancel{Q_3}} = \frac{Q_4}{Q_1}$$

$$\frac{Q_4}{Q_1} = (1 - \eta_1)(1 - \eta_2)(1 - \eta_3)$$

$\because$ $(\eta_1 = \eta_2 = \eta_3$, given)

$$\left(\frac{Q_4}{Q_1}\right) = (1 - \eta)^3$$

$$\therefore \qquad (1-\eta) = \left(\frac{280}{2250}\right)^{1/3} = 0.499$$

$$\eta = 0.5$$

We know that, $\quad \eta = \dfrac{W_1}{Q_1};\ W_1 = \eta Q_1 = 0.5 \times 2250 = 1125 \text{ kJ}$ **Ans.**

$$W_1 = Q_1 - Q_2$$

$$1125 = 2250 - Q_2$$

$$\boxed{Q_2 = 1125 \text{ kJ}}$$

$$\eta = \frac{W_2}{Q_2} = W_2 = 0.5 \times 1125 = 562.5 \text{ kJ}$$ **Ans.**

$$W_2 = Q_2 - Q_3$$

$$562.5 = 1125 - Q_3;\ Q_3 = 562.5 \text{ kJ}$$

$$\therefore \qquad \eta = \frac{W_3}{Q_3} = W_3 = 0.5 \times 562.5 = 281.25 \text{ kJ}$$ **Ans.**

$\therefore$ Total workdone $= W = W_1 + W_2 + W_3 = 1968.75$ kJ **Ans.**

Problem 23: *A reversible heat engine operates between two reservoirs at temperature 700°C and 50°C. The engine drives a reversible refrigerator which operates between reservoirs at tempts. of 50°C and –25°C. The heat transfer to the engine is 2500 kJ and net work output of the combined engine refrigerator plant is 400 kJ.*

(i) Determine heat transfer to the refrigerant & net heat transfer to the reservoir at 50°C.

(ii) Determine heat transfer to the refrigerant & net heat transfer to the reservoir at 50°C given that the efficiency of the heat engine and the COP of the refrigerator are each 45 percent of their maximum possible values. *(VTU, Aug, 2001)*

Solution:

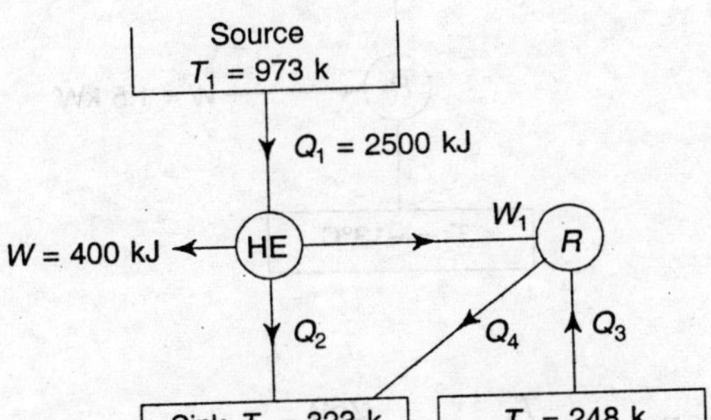

Fig. 4.31

(i) for the reversible heat engine

$$\eta_{HE} = \frac{T_1 - T_2}{T_1} = \frac{973 - 323}{973} = 66.8\%$$

also $$\eta_{HE} = \frac{W}{Q} = \frac{W_1 + 400}{Q_1}$$

$\therefore$ $\boxed{W_1 = 1270 \text{ kJ}}$

$\therefore$ $Q_2 = Q_1 - W_{net}$

$2500 - 400 - 1270 = 830$ kJ

Then COP of the reversible refrigerator can be calculated as

$$(COP)_R = \frac{Q_3}{.W_1} = \frac{Q_3}{Q_4 - Q_3} = \frac{T_3}{T_4 - T_3} = \frac{248}{323 - 248}$$

$\boxed{(COP)_R = 3.306}$ **Ans.**

also, $Q_3 = W_1 \times 3.306$

$Q_3 = 1270 \times 3.306$

$\boxed{Q_3 = 4198.62 \text{ kJ}}$ **Ans.**

$\therefore$ $Q_4 = W_1 + Q_3 = 1270 + 4198.62$

$= 5468.62$ kJ

(ii) Let $Q_3 = 4198.62$ kJ be the heat transferred to the refrigerator.

$\therefore$ Net heat transfer to the refrigerator at 323 k $= Q_4 - Q_3$

$= 5468.62 - 4198.62 = 1270$ kJ **Ans.**

Problem 24: *A refrigerator working between 27°C and –13°C is driven by an electric motor rated at 1.5 kW. What quantity of Ice at 0°C can be obtained from water at 20°C in 10 hours. Take CP for Ice = 2.1 kJ/kgk and CP for water = 4.2 kJ/kgk.* *(VTU March 2001)*

Solution:

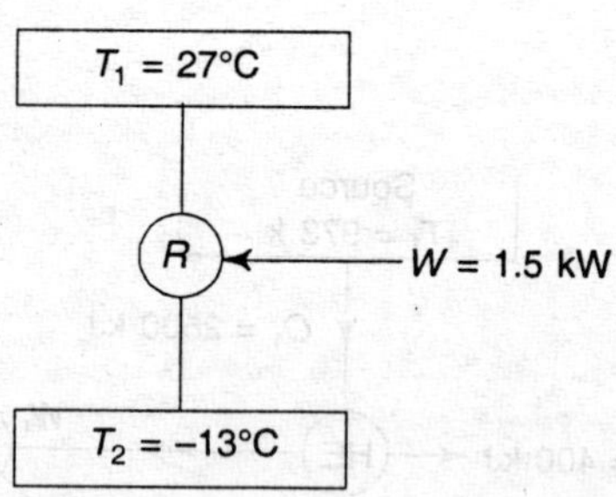

We know that

$$(COP)_R = \frac{T_2}{T_1 - T_2} = \frac{260}{300 - 260}$$

COP = 6.5

Also, $$COP = \frac{Q_2}{W}$$

$\therefore$ $Q_2 = 6.5 \times 1.5 = 9.75 \text{ kW} = 9.75 \text{ kJ/s}$

Heat removed from water at 20°C to convert into Ice at 0°C is equal to

$$[mcp\ (T_2 - T_1)]\ \text{water} + m \times \text{latent heat} = 1 \times 4.2\ (20 - 0) + 1 \times 335$$

$$= 419 \text{ kJ/kg of Ice}$$

$$\therefore \quad \text{Mass of Ice Produced} = \frac{9.75}{419} = 0.0232 \text{ kg/s.}$$

$\therefore$ Mass of Ice Produced in 10 hours $= 0.0232 \times 3600 \times 10 = \boxed{837.7 \text{ kg}}$ *Ans.*

Problem 25: *A heat engine with an efficiency of 40% drives a refrigerator with a COP of 4. Determine the ratio of total heat rejected to the atmosphere to the heat absorbed by the refrigerator.* *(VTU, Feb. 2002)*

Solution:

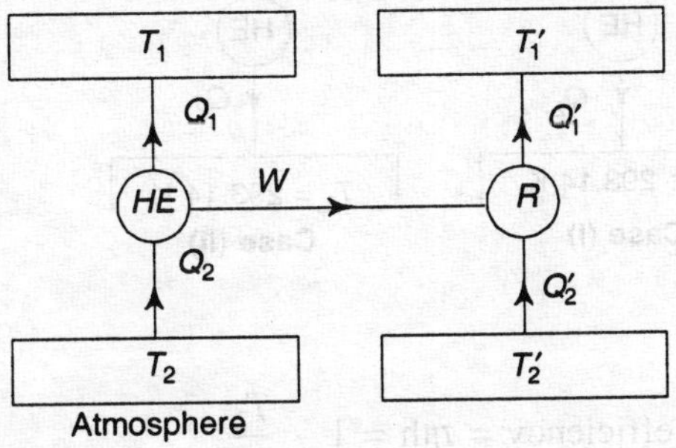

WKT,

$$\eta = \frac{W}{Q_1}$$

$$\therefore \quad Q_1 = \frac{W}{\eta} = \frac{W}{0.4} = 1.25 \text{ W}$$

$$\text{COP} = \frac{Q_2'}{W},\ Q_2' = 4 \text{ W}$$

also, $\quad W = Q_1 - Q_2$

$$\therefore \quad Q_2 = Q_1 - W = 1.25\ W - \text{W}$$

$$= 0.25 \text{ W}$$

$$Q_1' - Q_2' = W$$

$$Q_1' = Q_2' + W = 4 \text{ W} + W = 5 \text{ W}$$

$$\therefore \quad \frac{\text{Total heat rejected to atmosphere}}{\text{heat absorbed by the refrigerator}}$$

$$= \frac{Q_2 + Q_1'}{Q_2'} = \frac{0.25\text{ W} + 5\text{ W}}{4\text{ W}}$$

$$\boxed{= 1.3125}$$ *Ans.*

Problem 26: *A reversible heat engine operates with two environments. In the first, it draws 12000 kW from a source at 400°C and in the second it draws 25000 kW from a source at 100°C. In both the operations the engine rejects heat to a thermal sink at 20°C. Determine the operation in which the engine delivers more power.* *(VTU, Feb. 2004)*

Solution:

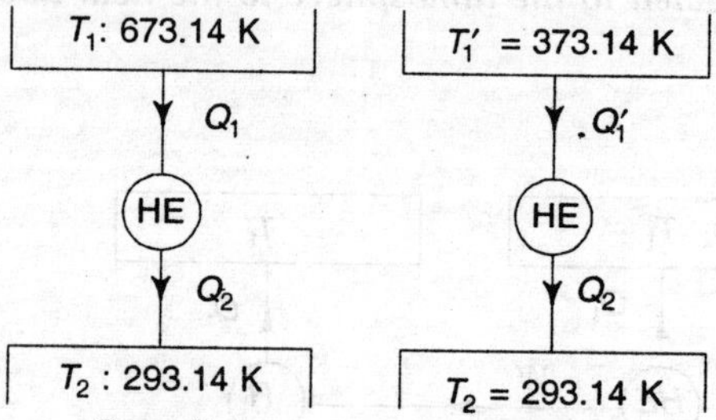

Case (i) Case (ii)

In the first case

$$\text{Thermal efficiency} = \eta\text{th} = 1 - \frac{T_2}{T_1}$$

$$= 1 - \frac{293.14}{673.14}$$

$$= 0.5646$$

$$WD = \eta \times Q_H = 0.5646 \times 12000 = 6775.2$$

$$\boxed{WD = 6775.2\text{ kW}}$$ *Ans.*

In the second case

$$\text{Thermal efficiency} = \eta\text{th} = 1 - \frac{T_2}{T_1'} = 1 - \frac{293.14}{373.14}$$

$$\eta\text{th} = 0.2144$$

$$WD = 0.2144 \times 25000$$

$$\boxed{WD = 5360\text{ kW}}$$ *Ans.*

It means, in the first case engine delivers more power. *Ans.*

4.18 REVIEW QUESTIONS

1. Explain what are the limitations of first law of thermodynamics.
2. Define the following terms.
   (i) Thermal reservoir, (ii) Heat engine
   (iii) Heat pump (iv) Refrigerator
3. Differentiate between heat engine and reversed heat engine.
4. Write the Kelvin-Planck and Clausius statements of second law, and prove that they are equivalent.
5. Define irreversibility and mention the factors which render a process irreversible.
6. Define COP?
7. What do you mean by PMM-II? Show that PMM-II is impossible.
8. Explain Carnot's principle.
9. What is Carnot cycle? What are the assumptions made in the Carnot cycle.
10. Define Carnot heat pump.
11. What are the corollaries of second law.
12. Define thermodynamic temperature scale & prove that $\dfrac{Q_1}{Q_2} = \dfrac{T_1}{T_2}$.

CHAPTER 5

Entropy

5.1 INTRODUCTION

Initially, the first law of thermodynamics was considered in terms of cycles and then it was applied to thermodynamic processes to deal with the definition of a property called, the internal energy. The definition of internal energy allowed us to use the first law quantitatively for processes. Similarly, the second law of thermodynamics was considered for cycles, when applied to thermodynamic processes leads to another property called entropy. The entropy enables us to treat the second law quantitatively for processes. The first law is considered to be the law of internal energy and second law is considered to be the law of entropy. As a result, we can claim that, thermodynamics is the science of three E's, namely Energy, Equilibrium and Entropy.

It is necessary to know the degree to which heat is transferred, during a process, which can affect any work transfer and end state. It can also be shown that the heat is transferred reversibly. A graph which performs a useful function could be developed, such that area underneath a process plotted on the graph gave the quantity of heat transferred reversibly. The concept of this graph is similar to the concept of pressure-volume diagram, the area under which gives work done during a reversible process.

In this new concept, absolute temperature '*T*' is to be considered on ordinate and a new property '*S*' (entropy) is to be considered on the absissa. The absolute temperature is closely related with the energy level of a substance *i.e.* internal energy and enthalpy. Hence area under this temperature-entropy diagram gives amount of heat transferred reversibly during a process.

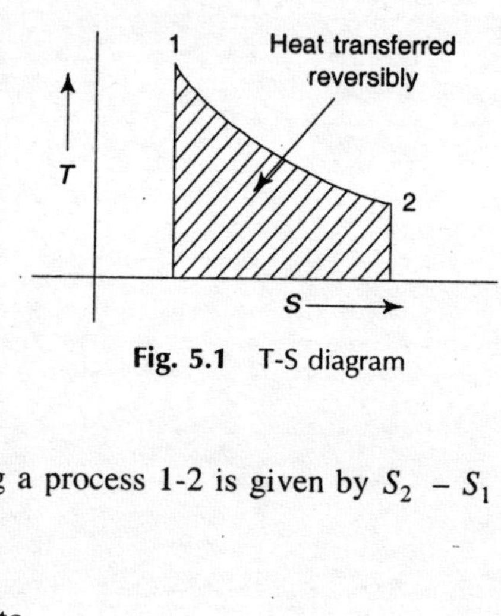

Fig. 5.1 T-S diagram

5.2 DEFINITION OF ENTROPY (S)

It is a property of a system, such that its increase, $S_2 - S_1$ during a process 1-2 is given by $S_2 - S_1 = \int_1^2 \frac{dQ_R}{T}$

Consider a system, which undergoes a change of state from 1 to 2 as shown in fig. 5.2. Consider a point on the graph where the coordinates are *T* and *S*. Let the state change from this point such that there is a small change in *S*, δs, then,

Heat transferred reversibly = Area swept out by this small change

$$= T\,\delta s$$

$\therefore$ Heat transferred reversibly from 1 to 2 = Total area under the curve 1-2

$$= \sum_{S=S_1}^{S=S_2} T\,\delta s$$

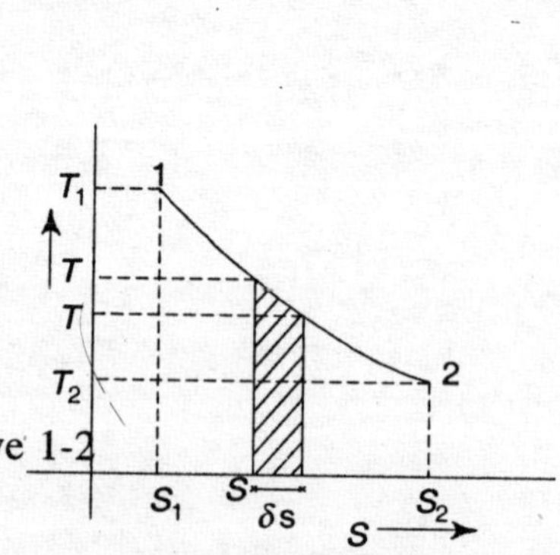

Fig .5.2 *T-S* diagram

In the limit as $\delta s \to 0$,

Heat transferred reversibly from 1 to 2 = $\int_{S_1}^{S_2} T \cdot ds = Q_{\text{rev}}$

After differentiation

$$dQ_{\text{rev}} = T \cdot ds$$

or

$$dS = \frac{dQ_{rev}}{T}$$

From the above equation it is clear that, Q_{rev} becomes positive if heat is received by the substance which increases its entropy.

Conversely, Q_{rev} becomes negative, if heat is rejected by the substance which decreases its entropy.

Thus, positive and negative changes of entropy show whether heat has been received or rejected during a process.

5.3 CLAUSIUS THEOREM

Consider a system which changes its state in a reversible path *i-f*. Let *i-a* and *b-f* are reversible adiabatic processes *a-b*; reversible isotherm and is drawn, such that (*i-a-b-f*) area = (*i-f*) area.

for the process *i-f*, from Ist law of thermodynamic,

$$Q_{i\text{-}f} = u_f - u_i + W_{i\text{-}f}$$

for the process *i-a-b-f*,

$$Q_{i\text{-}a\text{-}b\text{-}f} = u_f - u_i + W_{iabf}$$

$\because$ (*i-a-b-f*) area = (*i-f*) area

$$(W_{i\text{-}a\text{-}b\text{-}f}) = W_{i\text{-}f}$$

$\therefore$ We can write $Q_{i\text{-}f} = Q_{i\text{-}a\text{-}b\text{-}f}$

$$Q_{i\text{-}f} = Q_{ia} + Q_{ab} + Q_{bf}$$

$$Q_{ia} = 0,\ Q_{bf} = 0,\ (\because \text{ rev. adiabatic})$$

$\therefore$ $Q_{if} = Q_{a\text{-}b}$

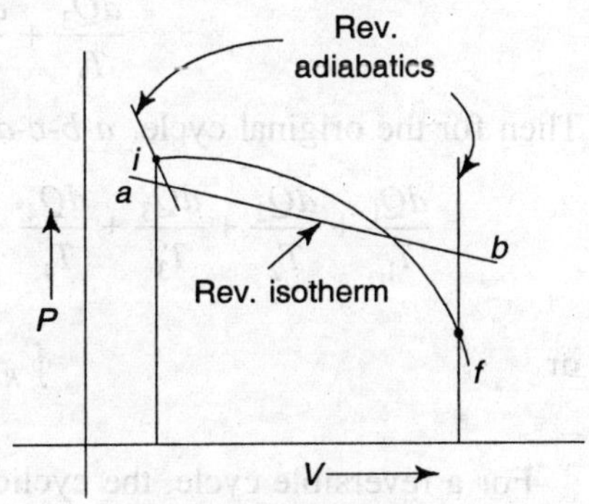

Fig. 5.3 Two rev. adiabatics and a rev. isotherm replaces a reversible path

It means, heat transferred in the process *i-f* (a reversible path) and isothermal process a-b, is equal.

Thus, any reversible path may be replaced by a reversible zig zag path, by providing the same end states which consists of two reversible adiabatic and a rev. isotherm.

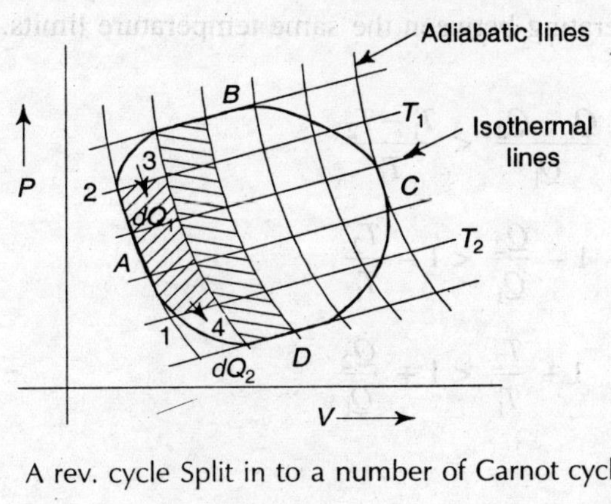

Fig. 5.4 A rev. cycle Split in to a number of Carnot cycles

Consider a reversible cycle *A-B-C-D* as shown in fig. 5.4. This closed cycle can be divided into a number of rev. cycles by drawing a number of rev. adiabatic and rev. isothermal curves. These elemental cycles are represented by 1-2-3-4, 5-6-7-8, etc.

For the elemental cycle 1-2-3-4,

dQ_1 – heat absorbed reversibly at T_1,

dQ_2 – heat rejected reversibly at T_2

$$\therefore \qquad \frac{dQ_1}{T_1} = \frac{dQ_2}{T_2}$$

Since heat supplied is to be considered as positive and heat rejected is to be considered as negative, we can write

$$\frac{dQ_1}{T_1} + \frac{dQ_2}{T_2} = 0$$

Secondly for elemental area 5-6-7-8

$$\frac{dQ_3}{T_3} + \frac{dQ_4}{T_4} = 0$$

Then for the original cycle, *a-b-c-d*, we can write

$$\frac{dQ_1}{T_1} + \frac{dQ_2}{T_2} + \frac{dQ_3}{T_3} + \frac{dQ_4}{T_4} + \ldots = 0$$

or

$$\oint_R \frac{dQ}{T} = 0$$

For a reversible cycle, the cyclic integral of $\frac{dQ}{T}$ is equal to zero. This is known as **'CLAUSIUS THEOREM'**. The equation is valid only for a reversible cycle for which thermal efficiency is given by $\eta t = 1 - \frac{Q_2}{Q_1}$ or $\eta t = 1 - \frac{T_2}{T_1}$.

5.4 CLAUSIUS INEQUALITY

The inequality of Clausius is a corollary or consequence of the second law of thermodynamics. We know that, according to Carnot theorem, the thermal efficiency of irreversible engine is less than that of reversible engine, operating between the same temperature limits.

Thus $\eta_{irv} < \eta_{rev}$

$$\frac{Q_1 - Q_2}{Q_1} < \frac{T_1 - T_2}{T_1}$$

$$1 - \frac{Q_2}{Q_1} < 1 - \frac{T_2}{T_1}$$

$$1 + \frac{T_2}{T_1} < 1 + \frac{Q_2}{Q_1}$$

$$\frac{Q_2}{Q_1} > \frac{T_2}{T_1}$$

or $$\frac{Q_2}{T_2} > \frac{Q_1}{T_1} \quad \text{or} \quad \frac{Q_1}{T_1} < \frac{Q_2}{T_2}$$

$$\therefore \quad \left(\frac{Q_1}{T_1}\right) - \left(\frac{Q_2}{T_2}\right) < 0$$

in the differential form for the irreversible engine, we can write,

$$\oint\left(\frac{dQ}{T}\right) < 0$$

In the general form, we can write

$$\oint\frac{dQ}{T} \leq 0$$

This equation is known as the inequality of Clausius which gives the criterion of the reversibility of a cycle.

If $\oint\frac{dQ}{T} = 0$, the cycle is reversible

$\oint\frac{dQ}{T} < 0$, the cycle is irreversible and possible.

$\oint\frac{dQ}{T} > 0$, this violates second law of thermodynamics,

hence it is impossible.

The differential $\frac{dQ}{T}$ is a perfect differential and the term $\int\frac{dQ}{T}$ is a property of the system. This property is entropy and is given by

$$S_2 - S_1 = \int_1^2 \frac{dQ}{T}$$

5.5 ENTROPY; AS A PROPERTY OF THE SYSTEM

Consider a system which undergoes a change of state from 1 to 2 in a reversible path '*A*'. The system may follow reversible paths *B* or *C* to reach initial state 1.

The paths *A* and *B* together constitute a reversible cycle. Also the paths *A* and *C* executes a reversible cycle. Then according to clausius theorem, we can write

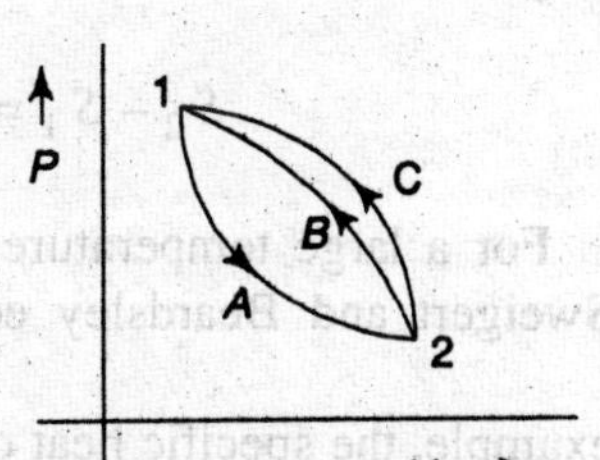

Fig. 5.5 Entropy is a point function

$$\int_{1A}^{2A}\frac{dQ}{T} + \int_{2B}^{1B}\frac{dQ}{T} = 0 \qquad (1)$$

also $$\int_{1A}^{2A} \frac{dQ}{T} + \int_{2C}^{1C} \frac{dQ}{T} = 0 \qquad (2)$$

By comparing equations (1) and (2), we can write,

$$\int_{2B}^{1B} \frac{dQ}{T} = \int_{2C}^{1C} \frac{dQ}{T}$$

It means that, the quantity $\int \frac{dQ}{T}$ is independent of the path followed by the process. It depends only on end states, hence it is a point function. It is an exact differential and is a property of the system.

5.6 ENTROPY CHANGE IN REVERSIBLE PROCESS

Consider a system that executes reversible process. The entropy change between the two state points can be written as

$$S_2 - S_1 = \int_1^2 \frac{dQ}{T}$$

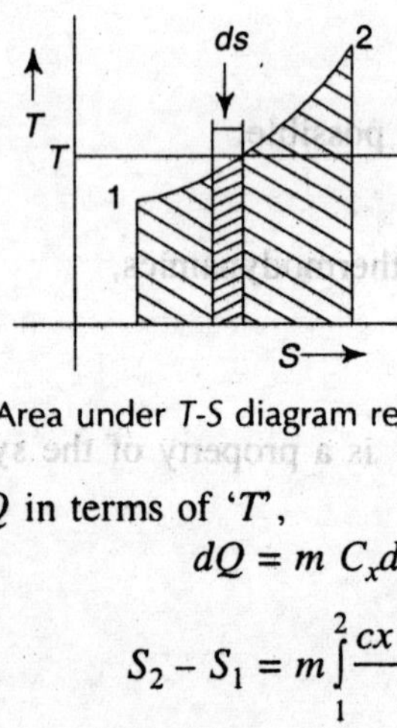

Fig. 5.6a Area under T-S diagram represent heat

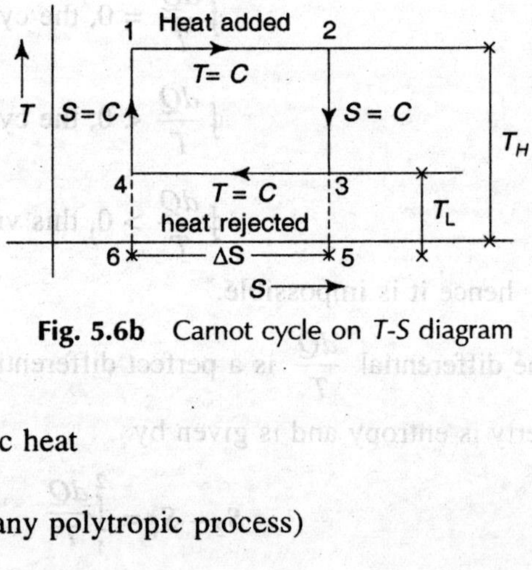

Fig. 5.6b Carnot cycle on T-S diagram

Express dQ in terms of 'T',

$$dQ = m\, C_x dT, \text{ where } C_x = \text{specific heat}$$

$$\therefore \quad S_2 - S_1 = m \int_1^2 \frac{cx \cdot dT}{T} \qquad \text{(Valid for any polytropic process)}$$

Neglect variation of specific heat with temperature,

$$S_2 - S_1 = m \cdot cx \cdot \log_e \left(\frac{T_2}{T_1}\right)$$

For a large temperature range, the changes of specific heat with temperature is significant. The Sweigert and Beardsley equations gives the relation between specific heat and temperature. For example, the specific heat of oxygen can be found by using he relation $C_p = 11.515 - \frac{172}{\sqrt{T}} + \frac{1530}{T}$ for a temperature range of 540 – 5000 R. Similarly for Nitrogen $C_p = 9.47 - \frac{3.47 \times 10^3}{T} + \frac{1.16 \times 10^6}{T^2}$ for a temperature range of 540 – 9000 *R*. Equations for the determination of specific heat of other gases like CO, H_2, H_2O, CO_2, etc. are also available.

For an isothermal process, specific heat is infinity.

$$\therefore \quad S_2 - S_1 = \int_1^2 \frac{dQ}{T} = \left(\frac{Q}{T}\right)_T$$

$$\therefore \quad dS = \frac{dQ}{T} \text{ or } dQ = T \cdot dS$$

dQ = differential amount of heat added which is a product of absolute temperature at which heat is added and a differential change in entropy. On this basis, the *T-S* diagram may be used to show the magnitude of heat transfer in reversible process.

The figure 5.6b shows a Carnot cycle, in which 1-2 represents heat added in the cycle. Therefore area 1-2-5-6 represents heat supplied and area 3-4-6-5 shows heat rejected. The difference between these two, say area corresponds to 1-2-3-4, gives cycle work.

The thermal efficiency of carnot cycle is given by $\eta th = \dfrac{\text{area 1-2-3-4}}{\text{area 1-2-5-6}}$

$$= \frac{(T_H - T_L)(\cancel{S_2 - S_1})}{T_H\,(\cancel{S_2 - S_1})} = \frac{T_H - T_L}{T_H}$$

Further, in a reversible adiabatic process, entropy remains constant, thereby termed as an isentropic process. Every reversible adiabatics must be an isentropic process, but the converse is not true (All isentropies are not necessarily reversible adiabatics)

5.7 ENTROPY CHANGES IN IRREVERSIBLE PROCESS

Consider a system that executes a process 1-2 as shown in figure. The system may be brought back to original state 1, by a reversible cycle or irreversible process (through the path *C*).

∴ For reversible cycle (*A-B*) we can write

$$\oint \frac{dQ}{T} = \int_{1A}^{2A} \frac{dQ}{T} + \int_{2B}^{1B} \frac{dQ}{T} = 0 \qquad (1)$$

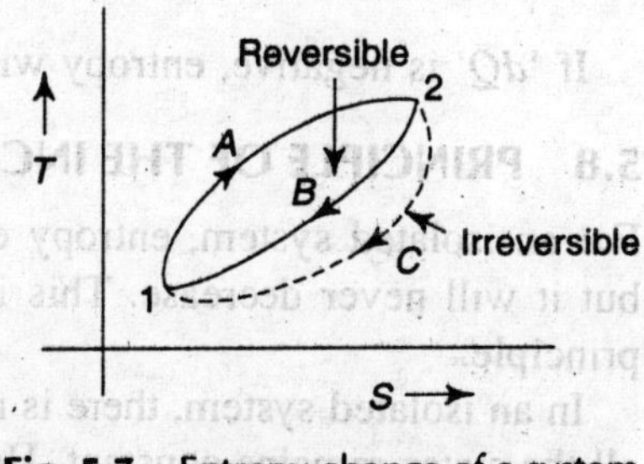

Fig. 5.7 Entropy change of a system during irreversible process

For irreversible cycle, the Clausius in- equality may be applied.

$$\therefore \quad \oint \frac{dQ}{T} = \int_{1A}^{2A} \frac{dQ}{T} + \int_{2C}^{1C} \frac{dQ}{T} < 0 \qquad (2)$$

from (1)

$$\int_{1A}^{2A} \frac{dQ}{T} = -\int_{2B}^{1B} \frac{dQ}{T}$$

Substitute in Eqn (2), we get

$$-\int_{2B}^{1B} \frac{dQ}{T} + \int_{2C}^{1C} \frac{dQ}{T} < 0 \quad \boxed{\text{or}}$$

$$\int_{2B}^{1B} \frac{dQ}{T} > \int_{2C}^{1C} \frac{dQ}{T} \tag{3}$$

As path '*B*' is a reversible path and entropy is a property.

$$\int_{2B}^{1B} \frac{dQ}{T} = \int_{2B}^{1B} dS = \int_{2C}^{1C} dS \tag{4}$$

[substitute (4) in Egn. (3),]

$$\therefore \quad \int_{2C}^{1C} ds > \int_{2C}^{1C} \frac{dQ}{T}.$$

for any irreversible process, we can write

$$dS > \frac{dQ}{T}$$

for a reversible process

$$dS = \frac{dQ_{rev}}{T}$$

In general, we can write

$$dS \geq \frac{dQ}{T}$$

or

$$S_2 - S_1 \geq \int_1^2 \frac{dQ}{T}$$

If 'dQ' is negative, entropy will tend to decrease due to heat transfer.

5.8 PRINCIPLE OF THE INCREASE OF ENTROPY

For an isolated system, entropy either increases or remains constant when the process is reversible, but it will never decrease. This is known as principle of increase of entropy or simply the entropy principle.

In an isolated system, there is no heat or work interaction with the surroundings and total energy of all the states remains constant. Hence $dQ = 0$, in an isolated system.

$\therefore \quad (dS)_{iso} \geq 0$

for a reversible process $(dS)_{iso} = 0$

$\therefore \quad S = \text{constant}$

i.e. for a reversible process, entropy remains constant.

For an irreversible process

$(dS)_{iso} > 0$ i.e. entropy increases.

An isolated system may be obtained by considering system and its surroundings within a single envelope, i.e. it becomes an universe. The isolated system includes every thing which is affected by the process.

Consider a process during which system at temperature T_{sys} receives 'δQ' quantity of heat from the surroundings at temperature (T_{surr}) ($T_{surr} > T_{sys}$). Let δW be the work done during this process.

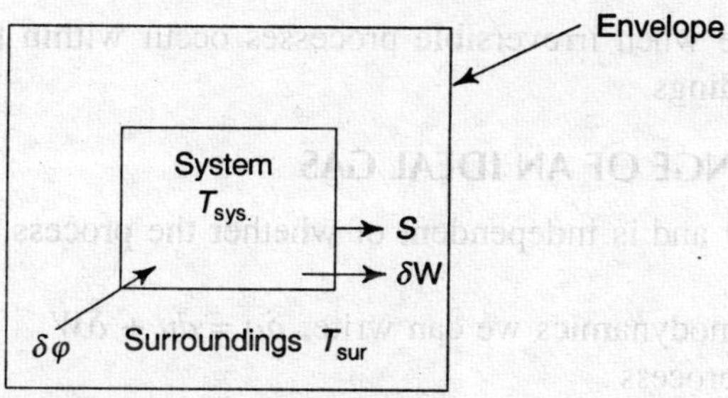

Fig 5.8 Isolated System

∴ For the system, we can write

$$dS_{sys} \geq \frac{\delta Q}{T_{sys}}.$$

For the surroundings 'δQ' is negative

$$\therefore \qquad dS_{surr} = \frac{\delta Q}{T_{surr}}$$

Therefore, the total change of entropy

$$\text{i.e.} \qquad dS_{sys} + dS_{surr} \geq \frac{\delta Q}{T_{sys}} - \frac{\delta Q}{T_{surr}}$$

$$\geq \delta Q \left(\frac{1}{T_{sys}} - \frac{1}{T_{surr}} \right)$$

since $T_{surr} > T_{sys}$, $\left(\frac{1}{T_{sys}} - \frac{1}{T_{surr}} \right)$ is positive

∴ We can write,

$$\boxed{dS_{sys} + dS_{surr} \geq 0}$$

Also, if $T_{sys} > T_{surr}$, then system rejects heat to the surroundings, then both terms 'δQ' and '$\left(\frac{1}{T_{sys}} - \frac{1}{T_{surr}} \right)$' becomes negative, yielding the same result.

Thus, for a system in the given surroundings, for all possible processes, we can write

$$dS_{sys} + dS_{surr} \geq 0 \quad \boxed{\text{or}}$$

$$(dS)_{iso} \geq 0$$

5.9 TOTAL ENTROPY CHANGES

In a system, the addition of heat increases its entropy and removal of heat decreases its entropy. When heat is removed from the system, it must be added to the surroundings. That means, the heat addition to the surroundings increases its entropy which must be exactly equal to decrease in entropy of the system. The entropy remains constant, when reversible adiabatic processes occur within the system.

The entropy will increase when irreversible processes occur within the system, when the system is isolated from its surroundings.

5.10 ENTROPY CHANGE OF AN IDEAL GAS

The entropy is a property and is independent of whether the process is performed in open system or closed system.

From first law of thermodynamics we can write, $\delta q = du + \delta W$

Further, for a reversible process

$$\delta W = PdV$$

$$\therefore \quad (\delta q)_{rev} = du + PdV$$

But, from the definition of entropy, we can write, $T \cdot dS = (\delta q)_{rev}$.

$$\therefore \quad T \cdot dS = du + PdV \qquad (1)$$

$$\therefore \quad TdS = dh - P\cancel{dV} - VdP + P\cancel{dV}$$

$$= dh - VdP \qquad (2)$$

$$h = u + PV$$

$$dh = du + PdV + VdP.$$

The equations (1) and (2) are used to determine entropy change for various processes.

(1) Constant Volume Process

For a perfect gas, we can write,

$$\delta q = du = C_v\, dT = T \cdot dS$$

$$\therefore \quad dS = \frac{C_v\, dT}{T}$$

after integration,

$$S_2 - S_1 = C_v \log_e \frac{T_2}{T_1} = C_v \log_e \frac{P_2}{P_1}$$

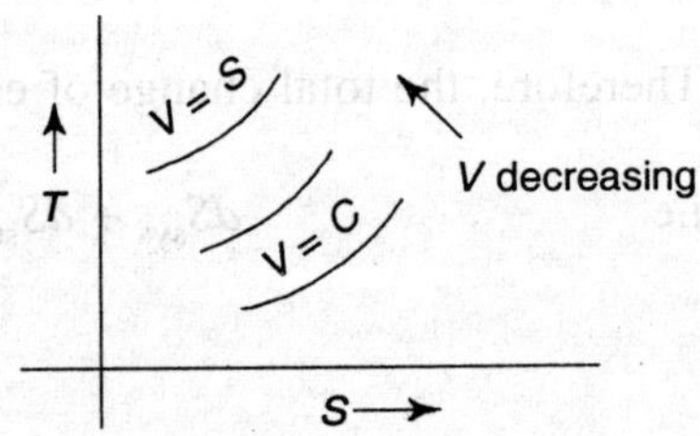

Fig. 5.9 Constant volume process

(2) Constant Pressure Process

We know that,

$$\delta q = dh = C_p\, dT = TdS$$

$$dS = C_p \frac{dT}{T}$$

after integration,

$$S_2 - S_1 = C_p \cdot \ln\left(\frac{T_2}{T_1}\right) = C_p \cdot \ln\left(\frac{V_2}{V_1}\right)$$

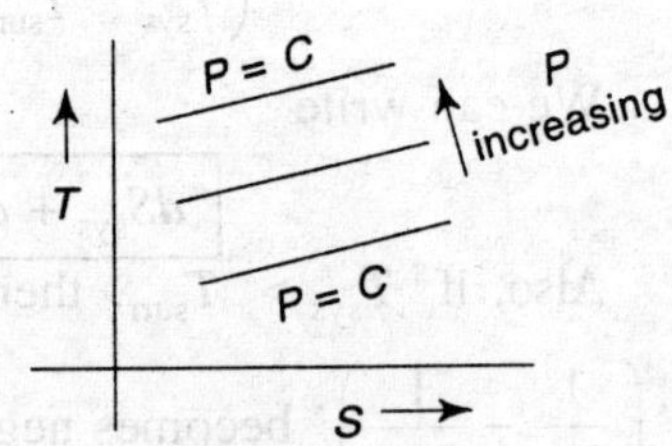

Fig. 5.10

(3) Isothermal Process

We know that, for an ideal gas,

$$\delta q = du + \delta W$$

$$\delta q = du + PdV$$

$$\delta q = PdV \ (du = 0) \text{ (as temperature remains constant)}$$

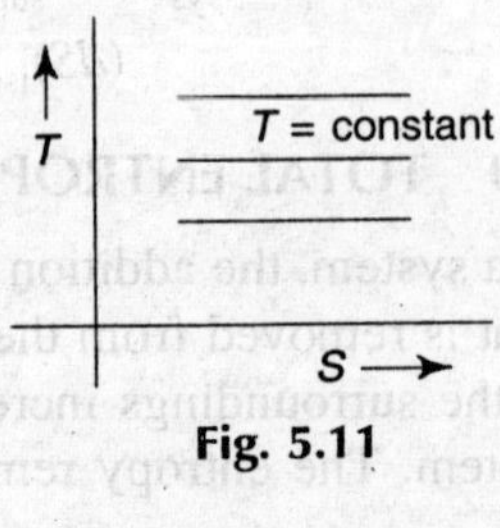

Fig. 5.11

$$\delta q = RT \cdot \frac{dV}{V} \quad (PV = RT)$$

$$\therefore \quad TdS = RT\frac{dV}{V} \quad (\delta q = TdS)$$

$$dS = R\frac{dV}{V}$$

After integration

$$S_2 - S_1 = R \log_e \frac{V_2}{V_1}$$
$$= R \log_e P_2/P_1$$

(4) Reversible Adiabatic Process for a Perfect Gas

$$\delta q = 0,\ dS = 0,$$

$$S_2 - S_1 = 0 = C_p \ln\left(\frac{V_2}{V_1}\right) + C_v \ln\left(\frac{P_2}{P_1}\right)$$

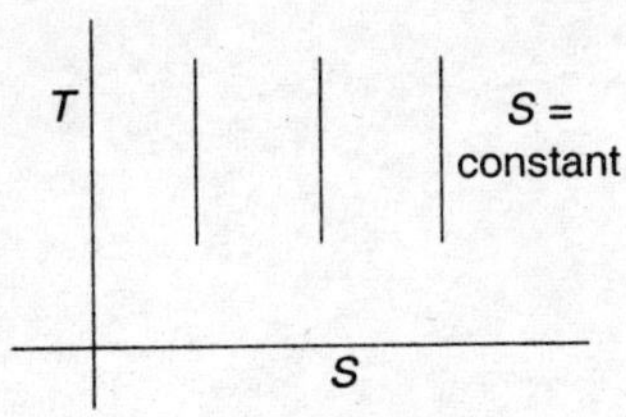

Fig. 5.12

i.e. $$\left(\frac{P_2}{P_1}\right)\left(\frac{V_2}{V_1}\right)^{C_p/C_v} = \left(\frac{P_2}{P_1}\right)\left(\frac{V_2}{V_1}\right)^{\gamma} = \text{constant}$$

$$P_2V_2^{\gamma} = P_1V_1^{\gamma} = C$$

or $$PV^{\gamma} = C$$

(5) Reversible Polytropic Process for a Perfect Gas

$$\delta q = du + PdV \quad (\delta w = PdV)$$

$$T \cdot dS = C_v dT + PdV$$

$$dS = C_v\frac{dT}{T} + P\frac{dV}{T}$$

$$= C_v\frac{dT}{T} + R\frac{dV}{T}$$

$$du = C_v\,dT$$
$$\delta q = TdS$$
$$PV = R.T$$
$$\frac{P}{T} = \frac{R}{V}$$

after integration,

$$\boxed{S_2 - S_1 = C_v \ln\left(\frac{T_2}{T_1}\right) + R\ln\left(\frac{V_2}{V_1}\right)}$$

This equation gives entropy change in terms of temperature and volume.

The same equation can be written as

$$S_2 - S_1 = C_v \log_e\left(\frac{P_2V_2}{P_1V_1}\right) + R\log_e\left(\frac{V_2}{V_1}\right)$$

$$= C_v \log_e \left(\frac{P_2}{P_1}\right) + C_v \log_e \left(\frac{V_2}{V_1}\right) + R \log_e \left(\frac{V_2}{V_1}\right)$$

$$= C_v \log_e \left(\frac{P_2}{P_1}\right) + (R + C_v) \log_e \left(\frac{V_2}{V_1}\right)$$

$$= C_v \log e \left(\frac{P_2}{P_1}\right) + C_p \log e \left(\frac{V_2}{V_1}\right)$$

$$(\because R = C_p - C_v)$$

This equation gives entropy change in terms of pressure and volume.

We know that,
$$S_2 - S_1 = C_v \ln \frac{T_2}{T_1} + R \ln \frac{V_2}{V_1} \qquad \left(\begin{array}{c} PV = T \\ V = \dfrac{T}{P} \end{array}\right)$$

$$= C_v \ln \frac{T_2}{T_1} + R \ln \left(\frac{T_2 P_1}{T_1 P_2}\right)$$

$$= C_v \ln \frac{T_2}{T_1} + R \ln \left(\frac{T_2}{T_1}\right) + R \ln \left(\frac{P_1}{P_2}\right)$$

$$= (C_v + R) \ln \left(\frac{T_2}{T_1}\right) + R \ln \left(\frac{P_1}{P_2}\right)$$

$$= C_p \ln \left(\frac{T_2}{T_1}\right) + R \ln \left(\frac{P_1}{P_2}\right)$$

This is the expression for entropy in terms of temperature and pressure.

5.11 ENTROPY CHANGE OF A PURE SUBSTANCE

The figure shows temperature entropy diagram for water. The entropy may be determined from specific entropies based on unit mass, which is usually tabulated in steam tables or may be obtained from chart.

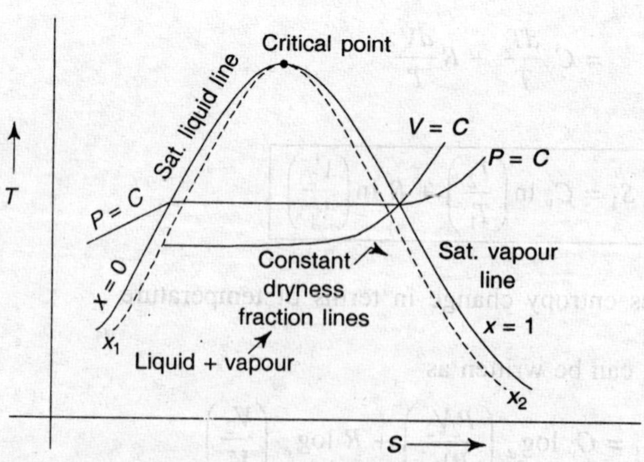

Fig. 5.13 *T-S* diagram for water

In the saturation region, entropy may be determined by knowing the dryness fraction of the steam. The relations being similar to those for specific volume and enthalpy and can be written as

$$S = Sf + x\,Sfg = Sf + x\,(Sg - Sf)$$
$$= Sf + x\,Sg - x\,Sf = (1 - x)\,Sf + xSg$$

In general, entropy of steam at different conditions can be written as follows:

(i) *for the wet steam*

$$S_{wet} = Sf + x\,Sfg; \quad Sf + x\,Sg$$

(ii) *for the dry saturated steam*

$$S_{dry} = Sf + Sg$$

(iii) *for the superheated steam*

$$S_{sup} = Sg + C_{pS} \ln\left(\frac{T_{sup}}{T_{sat}}\right)$$

Where C_{pS} = Specific heat of superheated steam at constant pressure,
T_{sup} = Temperature of superheated steam, and
T_{sat} = Saturation temperature.

5.12 AVAILABLE AND UNAVAILABLE ENERGY

The energy sources may be divided into high grade and low grade energy. Potential energy, kinetic energy and electrical energy can be completely transformed into useful shaft work without any loss and hence referred as high grade energy. On the other side, thermal energy cannot be fully converted into shaft work, i.e. a portion of energy is converted into useful work and hence referred as low grade energy.

High grade energy	**Low grade energy**
(i) Mechanical work	(i) Heat or thermal energy
(ii) Electrical energy	(ii) Heat from nuclear fusion or fission
(iii) Water, wind and tidal power	(iii) Heat derived from combustion of fuels, coal, wood, air, etc…
(iv) Kinetic energy of a jet	
(v) Animal and manual power	

The high grade energy is produced from low grade energy and the complete transformation of low grade energy into high grade energy is impossible. (by virtue of second law of thermodynamics)

A portion of thermal energy which is available for conversion is referred as available energy. The portion of thermal energy which is not usable and must be rejected is known as "Unavailable energy".

For a given amount of heat supply, the work output will be high when the heat rejection is at lower temperature which corresponds to atmospheric temperature or a large mass of water (ocean). The system delivers maximum work when its state is in equilibrium with that of the surroundings. This maximum work is referred as "availability". It is a composite property which depends on both the systems and surroundings. Under this equilibrium condition of system with the surroundings, all intensive and extensive properties will be common and no energy difference will exist to promote further work. Such an equilibrium state of a system is known as "dead state". When a system is in the dead state, it implies that the system is stable and uniform, in composition.

The change in state of system from the dead state 'O' is a measure of the available work. Available work of the system reduces as its state reaches the dead state.

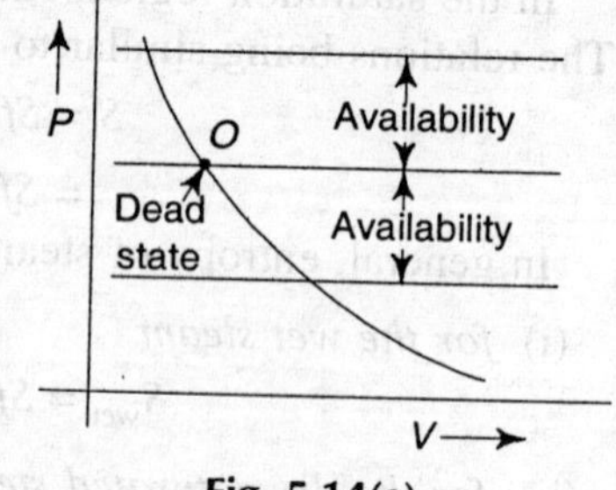

Fig. 5.14(a)

When the system temperature reduces to surrounding temperature, its capacity to do work by using temperature difference ceases and its available energy becomes zero. Similarly when the system pressure decreases to pressure of surroundings, the capacity of system to do work by using pressure difference or potential difference ceases and its available energy becomes zero.

The following two cases will be considered for available and unavailable portions of heat energy.

5.12.1 Heat Withdrawn from an Infinite Reservoir (at Constant Temperature)

Consider a reversible heat engine that operates between temperature T and T_0. Let Q be the quantity of heat supplied and W max is the available work.

$$\eta = \frac{T - T_0}{T} = \frac{W_{max}}{Q}$$

$$\therefore \quad W_{max} = Q\left[\frac{T - T_0}{T}\right]$$

$$= Q\left[1 - \frac{T_0}{T}\right]$$

$$= Q - \frac{Q}{T}T_0$$

$$= Q - T_0\, dS$$

$$[TdS = Q]$$

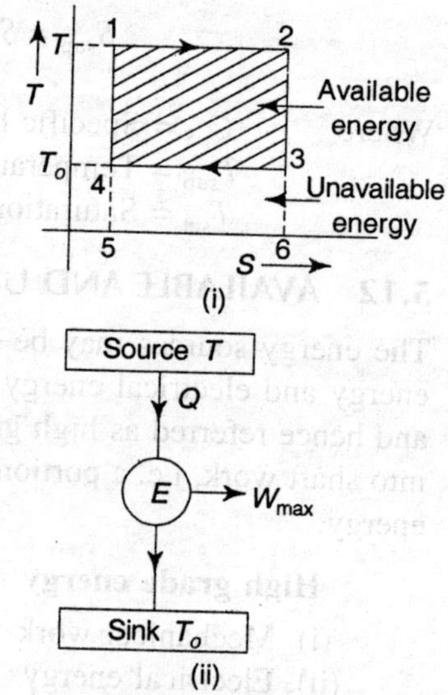

Fig. 5.14 (b)

Where $T_0\, dS$ – Unavailable energy equal to product of lowest temperature of heat rejection and change in entropy of the system.

5.12.2 Heat Withdrawn from a Finite Source (i.e. at Varying Temperature)

Consider a finite reservoir, in which the source temperature 'T_1' changes as the heat is withdrawn continuously from it. Therefore heat is supplied at variable temperature and rejected to the atmosphere at constant temperature T_0.

The maximum amount of work or available energy is equal to

$$W_{max} = \int\left(1 - \frac{T_0}{T}\right)\delta Q$$

$$= \int \delta Q - \int T_0 \frac{\delta Q}{T}$$

$$= \int \delta Q - T_0 \int \frac{\delta Q}{T}$$

$$Q - T_0\, dS$$

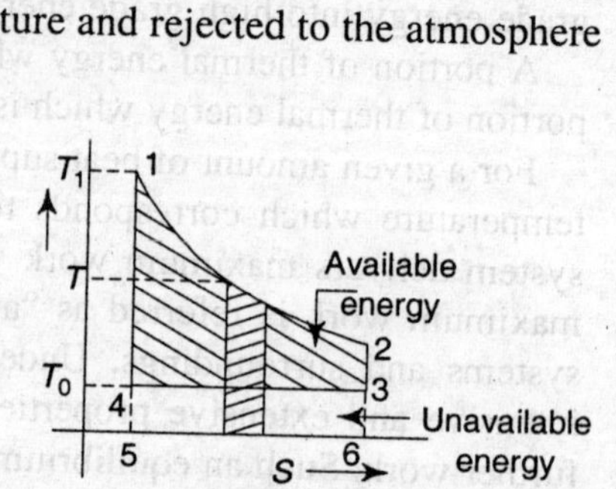

Fig. 5.14(c)

The result in this case is same as in case (i) except that the heat supply is calculated by integration, as the temperature of the source is varying during the process of heat supply.

5.13 APPLICATIONS OF ENTROPY PRINCIPLE

The principle of increase of entropy is the quantitative statement of the second law of thermodynamics. All irreversible processes are accompanied with entropy increase of the universe. The irreversibility of the process will be higher when the increase of entropy of the universe is high. Some of the applications of entropy principle are discussed as follows.

(1) Heat transfer through a finite temperature difference

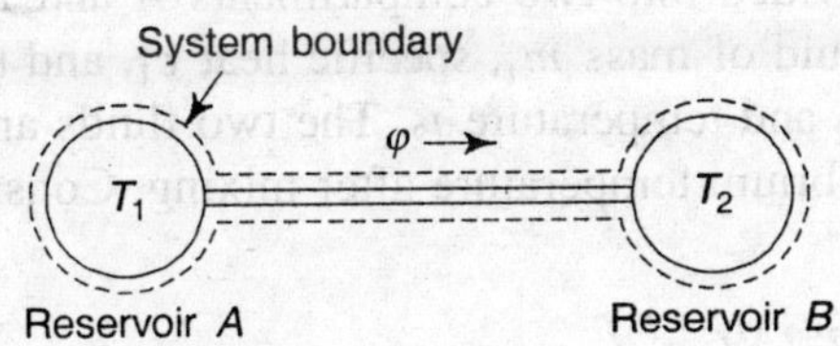

Fig. 5.15

Consider reservoir '*A*' at temperature 'T_1' and reservoir '*B*' at temperature 'T_2'. Let '*Q*' be the rate of heat transfer from *A* to *B*. In the reservoir *A*, we can write

$$\Delta S_A = -\frac{Q}{T_1} \qquad \text{('A' rejects 'Q' quantity of heat)}$$

In the reservoir *B*,

$$\Delta S_B = +\frac{Q}{T_2} \qquad \text{('B' receives 'Q' quantity of heat)}$$

There is no change in entropy in the line connecting these two reservoirs (steady state).

Consider reservoirs *A* and *B* and the connecting line as an isolated system and also we know that entropy is an additive property, we can write

$$S = S_A + S_B$$

$$\Delta S_{\text{univ}} = \Delta S_A + \Delta S_B$$

$$\Delta S_{\text{univ}} = -\frac{Q}{T_1} + \frac{Q}{T_2}$$

$$= Q\left[-\frac{1}{T_1} + \frac{1}{T_2}\right]$$

$$\Delta S_{\text{univ}} = Q\left[\frac{T_1 - T_2}{T_1 T_2}\right]$$

As $T_1 > T_2$, ΔS_{univ} is positive, hence the process is irreversible and possible.

If $T_1 = T_2$, $\Delta S_{\text{univ}} = 0$, the process is reversible.

If $T_1 < T_2$ Δs_{univ} is negative and the process is impossible.

2. *Mixing of two fluids*

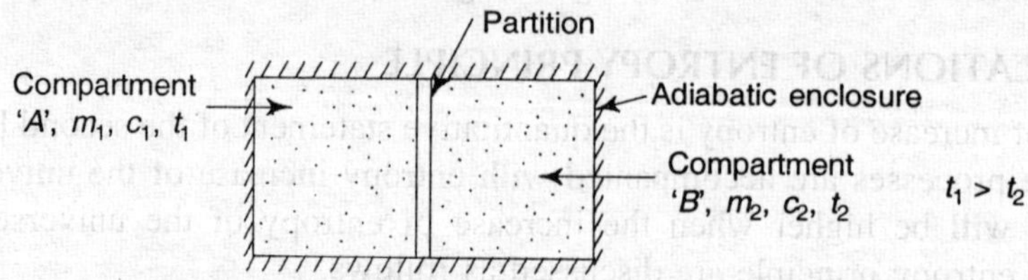

Fig. 5.16 Mixing of two fluids

Consider a container which is divided into two compartments A and B as shown in figure. Suppose the compartment 'A' is having fluid of mass m_1, specific heat c_1, and temperature t_1, and B having a fluid of mass m_2, specific heat c_2 and temperature t_2. The two fluids are mixed together by removing the partition and 't_f' be the equilibrium temperature after mixing. Consider the system as isolated one and $t_1 > t_f > t_2$, we can write,

$$m_1c_1\,(t_1 - t_f) = m_2c_2\,(t_f - t_2)$$

$$m_1c_1t_1 - m_1c_1t_f = m_2c_2t_f - m_2c_2t_2$$

$$m_1c_1t_1 + m_2c_2t_2 = m_2c_2t_f + m_1c_1t_f$$

$$\therefore \qquad t_f = \frac{m_1c_1t_1 + m_2c_2t_2}{(m_2c_2 + m_1c_1)}$$

Entropy change in the compartment 'A'.

$$\Delta S_A = \int_{t_1}^{t_f} \frac{dQ_{\text{rev}}}{T} = \int_{t_1}^{t_f} \frac{m_1c_1\,dT}{T}$$

$$= m_1c_1 \ln \frac{T_f}{T_1}$$

$$\Delta S_A = m_1c_1 \log_e \frac{T_f + 273}{T_1 + 273} \qquad \text{(It is negative as } T_1 > T_f\text{)}$$

Similarly entropy change in the compartment 'B'.

$$\Delta S_B = m_2c_2 \ln \frac{T_f + 273}{T_2 + 273} \qquad \text{(It is positive as } T_2 < T_f\text{)}$$

$$\therefore \qquad \Delta S_{\text{univ}} = \Delta S_A + \Delta S_B$$

$$= m_1c_1 \ln \frac{T_f}{T_1} + m_2c_2 \ln \frac{T_f}{T_2}$$

ΔS_{univ} is positive and mixing process is irreversible.

If $m_1 = m_2 = m$ and

$c_1 = c_2 = c$

then $$\Delta S_{\text{univ}} = mc \ln \frac{t_f{}^2}{t_1t_2}$$

and $$t_f = \frac{m_1c_1t_1 + m_2c_2t_2}{m_1c_1 + m_2c_2} = \frac{\cancel{m}c\,(t_1 + t_2)}{2\cancel{m}c}$$

$$\therefore \quad t_f = \frac{T_1 + T_2}{2}$$

$$\therefore \quad \Delta S_{\text{univ}} = mc \ln \frac{\left(\frac{T_1 + T_2}{2}\right)^2}{T_1T_2}$$

$$\Delta S_{\text{univ}} = 2mc \ln \frac{(T_1 + T_2)/2}{\sqrt{T_1T_2}}$$

It is also a positive value.

5.14 MAXIMUM WORK ATTAINABLE FROM TWO FINITE BODIES AT TWO DIFFERENT, BUT FIXED TEMPERATURES

Consider two identical finite bodies 'A' and 'B' with constant heat capacity and are maintained at fixed temperatures 'T_A' and 'T_B' respectively.

Let us consider T_A is greater than T_B. When 'A' and 'B' are brought in thermal contact, without producing any work, the final temperature obtained would be maximum,

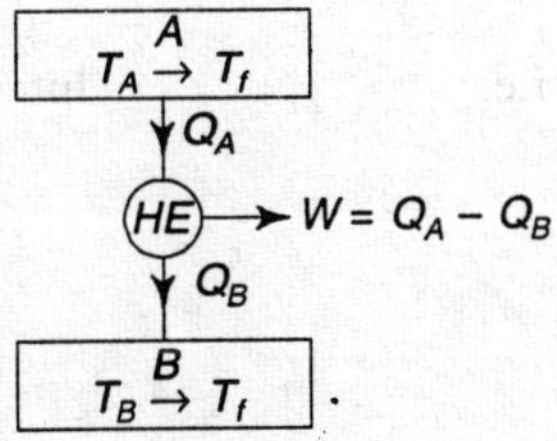

Fig. 5.17

i.e. $$T_f = \frac{T_A + T_B}{2}$$

Consider a heat engine, operating between 'A' and 'B', receives 'Q_A' quantity of heat, a part of which is converted in to work 'W' and rejects 'Q_B' to the body 'B'. When the value of 'W' is large, the corresponding value of 'T_f' is minimum. The heat engine decreases temperature of body 'A' and increases temperature of body 'B' and when both 'A' and 'B' attains the final temperature 'T_f', the heat engine will stop operating. Assume that the bodies 'A' and 'B' remains at constant pressure and there is no change of phase.

Total heat rejected from body 'A',

$$Q_A = C_p\,(T_A - T_f)$$

where C_p = heat capacity of A and B at constant pressure.

Total heat received by body 'B'.

$$Q_B = C_p\,(T_f - T_B)$$

The heat engine delivers 'W' quantity of work.

$$W = Q_A - Q_B$$
$$= C_p\,(T_A - T_f) - \{C_p(T_f - T_B)\}$$
$$= C_p\,(T_A + T_f - 2T_f)$$

If C_p, T_A and T_B are fixed, 'W' depends on 'T_f'. 'W' will be maximum when 'T_f' is minimum. Therefore, for body 'A', entropy change can be written as

$$\Delta S_A = \int_{T_1}^{T_f} C_p \frac{dT}{T} = C_p \cdot \ln \frac{T_f}{T_A}$$ and for body 'B', entropy change can be written as

$$\Delta S_B = \int_{T_2}^{T_f} C_p \frac{dT}{T} = C_p \cdot \ln \frac{T_f}{T_B}$$

For the working fluid in the heat engine, entropy will not change

i.e. $$\oint dS = 0$$

According to entropy principle,

$$\Delta S_{\text{univ}} \geq 0 \qquad [\Delta S_{\text{univ}} = \Delta S_A + \Delta S_B)$$

$$C_p \cdot \ln \frac{T_f}{T_A} + C_p \ln \frac{T_f}{T_B} \geq 0$$

$$C_p \cdot \ln \frac{T_{f^2}}{T_A T_B} \geq 0$$

For minimum value of 'T_f'.

$$C_p \ln \frac{T_{f^2}}{T_A T_B} = 0$$

i.e. $$\ln \frac{T_{f^2}}{T_A T_B} = 0 = 1$$

$$T_{f^2} = T_A T_B$$

$$\boxed{T_f = \sqrt{T_A T_B}}$$

∴ For maximum work, $T_f = \sqrt{T_A T_B}$.

But we know that, $W = C_p (T_A + T_B - 2T_f)$

∴ $$W_{\max} = C_p (T_A + T_B - 2\sqrt{T_A T_B})$$

$$W_{\max} = C_p \left(\sqrt{T_A} - \sqrt{T_B}\right)^2$$

The final temperatures of 'A' and 'B', initially were at T_A and T_B ranges from $\left(\frac{T_A + T_B}{2}\right)$ [No work delivered] to $\sqrt{T_A T_B}$ [work will be maximum].

5.15 MAXIMUM WORK ATTAINABLE FROM A FINITE BODY AND A THERMAL ENERGY RESERVOIR [TER]

Consider a finite body 'A' with thermal capacity 'C_p' and temperature 'T_A', and a TER at temperature T_R. Assume that the body temperature is higher than temperature of reservoir i.e., $T_A > T_R$.

Consider a heat engine, extracts 'Q_A' from body 'A' and delivers 'W' amount of work. The heat engine rejects $(Q_A - W)$ amount of heat to the 'TER'. The temperature of body 'A' decreases, but 'TER' would be at the same temperature 'T_R'. The engine will not work when T_A approaches T_R.

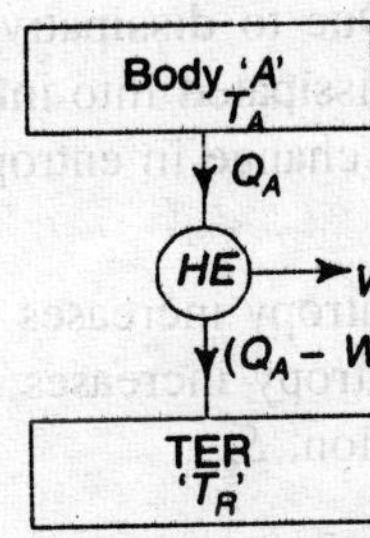

Fig. 5.18 Max. work attainable when TER is considered on one side of *HE*

We can write, for body '*A*',

$$\text{Entropy change, } \Delta S_A = \int_{T_A}^{T_R} C_p \frac{dT}{T} = C_p \ln \frac{T_R}{T_A}$$

For working fluid in the heat engine,

Entropy change, $\Delta S_{HE} = \oint dS = 0.$

for TER, entropy change

$$\Delta S_{TER} = \frac{Q_A - W}{T_R}$$

Consider the whole as an Isolated System, then,

$$\Delta S_{univ} = C_p \ln \frac{T_R}{T_A} + \frac{Q_A - W}{T_R}$$

Applying entropy principle,

$$\Delta S_{univ} \geq 0$$

$$C_p \ln \frac{T_R}{T_A} + \frac{Q_A - W}{T_R} \geq 0$$

i.e.

$$C_p \ln \frac{T_R}{T_A} \geq \frac{W - Q_A}{T_R}$$

$$\therefore \quad \frac{W - Q_A}{T_R} \leq C_p \cdot \ln \frac{T_R}{T_A}$$

$$\therefore \quad W \leq Q_A + T_R C_p \cdot \ln \frac{T_R}{T_A}$$

$$\therefore \quad W_{max} = Q_A + T_R C_p \ln \frac{T_R}{T_A}$$

or

$$W_{max} = Q_A + T_R C_p \ln \left(\frac{T_R}{T_A}\right)$$

$$Q_A = C_p (T_A - T_R)$$

$$\therefore \quad W_{max} = C_p (T_A - T_R) + T_R C_p \ln \left(\frac{T_R}{T_A}\right)$$

$$\boxed{W_{max} = C_p\left[(T_A - T_R) + T_R \ln\left(\frac{T_R}{T_A}\right)\right]}$$

5.16 ENTROPY GENERATION IN A CLOSED SYSTEM

In a closed system, the entropy increases by

(i) Heat interaction in which there is entropy transfer.

(ii) Due to dissipative effects or internal irreversibilities during which work or kinetic energy is dissipated into internal energy increase.

The change in entropy 'dS' of the system at temperature 'T', can be written as

$$dS = d_{es} + d_{is}$$

d_{es} =Entropy increases due to external heat interaction

d_{is} =Entropy increases due to internal irreversibility. It is also called entropy production or entropy generation, S_{gen}

$$dS \geq \frac{dQ}{T}$$

$$d_{is} \geq 0$$

During the expansion of steam in a turbine, entropy decreases as working fluid loses heat to the surroundings $\left[-\int \frac{dQ}{T}\right]$. This entropy decrease of the system is equal to entropy increase caused by internal irreversibilities, such as friction etc. $\left[\int d_{is}\right]$. It means $\int dS = 0$. i.e In the turbine entropy before and after expansion will remain the same. Therefore an isentropic process need not be adiabatic or reversible.

But a reversible isentropic process must be adiabatic. The entropy increase may also be caused by friction and other factors, hence an adiabatic process need not be isentropic. Also a reversible and adiabatic process must be isentropic.

In a closed system, we can write

$$dQ_R = dU_R + PdV$$

If the process is irreversible

$$dQ_I = dU_I + dW$$

But $\quad dU_R = dU_I \qquad$ [U is a property]

$\therefore \quad dQ_R - PdV = dQ_I - dW$

$\therefore \quad \left(\frac{dQ}{T}\right)_R = \left(\frac{dQ}{T}\right)_I + \frac{PdV - dW}{T} \qquad$ [÷ by T]

where [$PdV - dW$] → lost work, it is the work lost due to irreversibility.

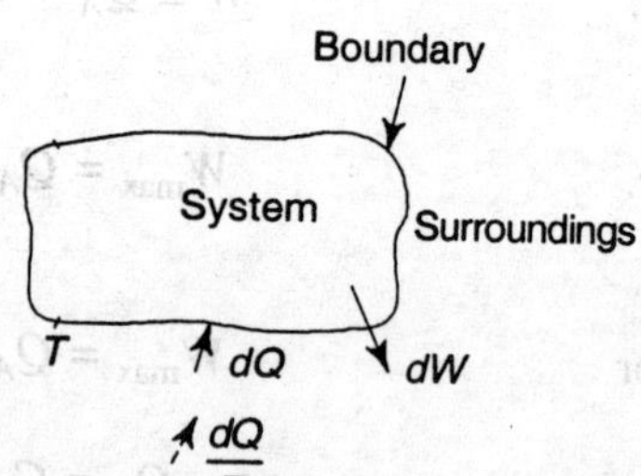

Fig. 5.19

dQ = Heat transfer

dW = Work transfer

$\frac{dQ}{T}$ = Entropy transfer

T = Boundary temperature

Consider a system that executes a process 1-2 during which entropy generation take place internally. According to first law

$$\int_1^2 dQ = E_2 - E_1 + \int_1^2 dW$$

$$Q_{1\text{-}2} = E_2 - E_1 + W_{1\text{-}2}.$$

According to second law

$$S_2 - S_1 \geq \int_1^2 \frac{dQ}{T}$$

$$\therefore \qquad S_2 - S_1 - \int_1^2 \frac{dQ}{T} = S_{gen}$$

(Entropy change) – (Entropy transfer) = (Entropy production)

$$\therefore \qquad S_{gen} \geq 0$$

The second law states that, in general, any thermodynamic process is accompanied by entropy generation.

5.17 FORMULAE USED IN SOLVING PROBLEMS

1. $$dS = \frac{dQ_{rev}}{T}$$

2. Clausius theorem;

$$\oint_R \frac{dQ}{T} = 0 \text{ i.e. } \frac{dQ_1}{T_1} + \frac{dQ_2}{T_2} + \frac{dQ_3}{T_3} + \ldots = 0.$$

3. Clausius inequality;

$$\oint \frac{dQ}{T} \leq 0.$$

For reversible cycle $\oint \frac{dQ}{T} = 0.$

If cycle is irreversible and possible $\oint \frac{dQ}{T} < 0.$

If cycle is impossible $\oint \frac{dQ}{T} > 0$

4. $$S_2 - S_1 = \int_1^2 \frac{dQ}{T}$$

5. $$dS \geq \frac{dQ}{T} \text{ or}$$

$$S_2 - S_1 \geq \frac{dQ}{T}$$

6. Principle of increase of entropy

$(dS)_{iso} > 0$ (for irreversible process)

$(dS)_{iso} = 0$, i.e. S = constant (for reversible process)

i.e. $dS_{sys} + dS_{surr} \geq 0.$

or $(dS)_{iso} \geq 0$

7. $$T \cdot dS = du + PdV$$

$$TdS = du - VdP$$

8. For constant volume process

$$S_2 - S_1 = C_v \ln \frac{T_2}{T_1} = C_v \ln \frac{P_2}{P_1}$$

9. For constant pressure process

$$S_2 - S_1 = C_p \ln \left(\frac{T_2}{T_1}\right) = C_p \ln \left(\frac{V_2}{V_1}\right)$$

10. For Iso thermal process

$$S_2 - S_1 = R \ln \left(\frac{V_2}{V_1}\right) = R \ln \left(\frac{P_2}{P_1}\right)$$

11. For Rev. adiabatic process

$$S_2 - S_1 = 0 \text{ i.e. } C_p \ln \left(\frac{V_2}{V_1}\right) + C_v \ln \left(\frac{P_2}{P_1}\right) = 0$$

12. For Rev. polytropic process

$$S_2 - S_1 = C_v \ln \left(\frac{T_2}{T_1}\right) + R \ln \left(\frac{V_2}{V_1}\right)$$

or

$$S_2 - S_1 = C_v \ln \left(\frac{P_2}{P_1}\right) + C_p \ln \left(\frac{V_2}{V_1}\right)$$

or

$$S_2 - S_1 = C_p \ln \left(\frac{T_2}{T_1}\right) + R \ln \left(\frac{P_1}{P_2}\right)$$

13. For wet steam

$$S_{wet} = S_f + x\, S_{fg};\ S_f + xS_g$$

14. For dry saturated steam

$$S_{dry} = S_f + S_g$$

15. For super heated steam

$$S_{sup} = S_g + C_{pS} \ln \left(\frac{T_{sup}}{T_{sat}}\right)$$

16. $$S_f = C_{pl} \log_e \frac{Tf}{273.16}$$

17. $$S_{wet} = C_{pl} \ln \frac{T_f}{273.16} + x\, \frac{hf_g}{T_f}$$

18. $$S_{sup} = C_{pl} \ln \frac{T_f}{273.16} + \frac{hf_g}{T_f} + C_{pV} \ln \frac{T_{sup}}{T_f}$$

5.18 NUMERICAL EXAMPLES

Problem 1: *Calculate specific entropy of water at 100°C.*

Solution: The specific entropy of water (liquid) can be written as

$$S_f = C_{pl} \log_e \frac{T_f}{273.16} \qquad C_p l = 4.187 \text{ kJ/kgK}$$

$$\therefore \quad S_f = 4.187 \ln \frac{373}{273.16} \qquad T_f = 100 + 273; \text{ k}$$

$$S_f = 1.304 \text{ kJ/kgK.} \quad \boxed{\text{or}}$$ ***Ans.***

From the steam table, the value of 'S_f' corresponding to t_s = 100 °C, can be taken as 1.3069 kJ/kgK.

Problem 2: *Determine the value of the specific entropy of wet steam at a pressure of 2×10^3 kN/m² (2 MPa) and 0.8 dry. Also compare the value obtained by calculation with value obtained from steam table.*

Solution: We know that, for the wet steam

$$S_{wet} = C_{pL} \ln \frac{T_f}{273.16} + x \frac{ht_g}{Tf}$$

$$hf_g = 188.87 \text{ kJ/kgK}$$

at P = 20 bar from steam table

$$P = 2 \times 10^3 \text{ kN/m}^2$$

$$= 2 \times 10^3 \times 10^3 \times 10^{-5}$$

$$= 20 \text{ bar}$$

$$t_s = 212.37 \text{ at 20 bar}$$

$$t_s = 212.37 + 273$$

$$= 485.37 \text{ k}$$

$$= 4.187 \ln \frac{485.37}{273.16} + \left(0.8 \times \frac{1888.7}{485.37}\right)$$

$$\boxed{S_{wet} = 5.52 \text{kJ/kgK}}$$ ***Ans.***

Also, $\quad S_{wet} = S_f + x\, Sf_g$

At 20 bar from steam table

$$S_f = 2.4468 \text{ kJ/kgK}$$

$$Sf_g = 3.8909 \text{ kJ/kgK}$$

$$S_{wet} = 2.4468 + 0.8\ (3.8909)$$

$$\boxed{S_{wet} = 5.559 \text{ kJ/kgK}}$$ ***Ans.***

This is more accurate value.

Problem 3: *Determine the value of specific entropy of steam at 1.5 MPa, with a temperature of 300°C.*

Solution: We know that the entropy of super heated steam is given by

$$S_{sup} = C_{pl} \ln \frac{T_f}{273.16} + \frac{hf_g}{T_f} + C_{pV} \ln \frac{T_{sup}}{T_f}.$$

At $P = 1.5 \times 10^6 \times 10^{-5}$

$= 15$ bar

$T_f = 198.28°C$

$= 198.28 + 273$

$= 471.28$ k

$hf_g = 1945.3$ kJ/kgK

$T_{sup} = 300 + 273 = 573$ k

$$= 4.187 \ln \frac{471.28}{273.16} + \frac{1945.3}{471.28} + 2.0934 \ln \frac{573}{471.28}$$

$$\boxed{S_{sup} = 6.820 \text{ kJ/kgK}}$$ ***Ans.***

or From super heat heated steam table at 15 bar and $T_{sup} = 300$ °C,

$$\boxed{S_{sup} = 6.9207 \text{ kJ/kgK}}$$ ***Ans.***

which is the accurate value.

Problem 4: *Dry saturated steam at 600 kN/m² flows into a turbine, where it expands, adiabatically to a pressure of 30 kN/m². Determine the maximum shaft work per unit mass of steam by neglecting kinetic energies at inlet and exit of the turbine.*

Solution: Let P_1, h_1 and S_1 be the pressure, enthalpy and entropy at inlet of the turbine and P_2, h_2 and S_2 are the corresponding values at the exit of the turbine.

∴ $S_1 = S_2$ (due to adiabatic expansion)

At $P_1 = 600$ kN/m²

$= 600 \times 10^3 \times 10^{-5}$ bar

$= 6$ bar

from stream table

$hg = h_1 = 2755.5$ kJ/kg

$S_1 = 6.7575 = Sg$

(The inlet steam is dry and saturated)

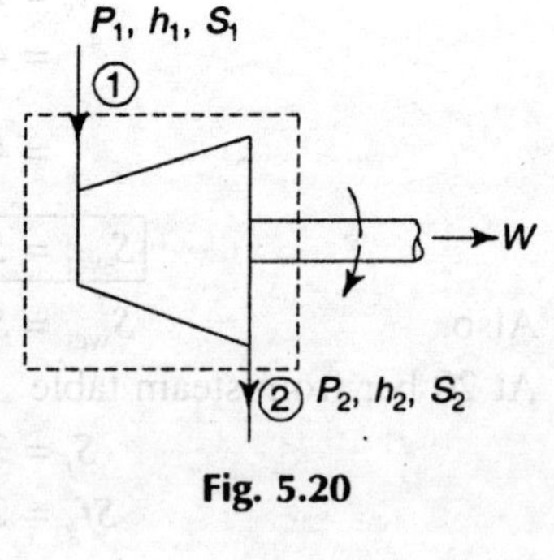

Fig. 5.20

30×10^3 N/m²

$= 30 \times 10^3 \times 10^{-5}$ bar

$= 0.3$ bar

∴ $S_2 = Sf_2 + x_2 Sf_{g2}$

At $P_2 = 0.3$ bar

$Sf_2 = 0.9441$ kJ/kgK

$Sf_{g2} = 6.8254$ kJ/kgK

∴ $6.7575 = 0.9441 + x_2\,(6.8254)$

∴ $\boxed{x_2 = 0.851}$

also $h_2 = hf_2 + x_2\, hfg_2$

At $P_2 = 0.3$ bar

$hf_2 = 289.3$ kJ/kg

$hf_{g2} = 2336.1$ kJ/kg

$= 289.3 + 0.851\,(2336.1)$

$h_2 = 2277.32$ kJ/kg

∴ Maximum shaft work per unit mass of steam

$$= W_{shaft} = h_1 - h_2$$
$$= 2755.5 - 2277.32$$
$$W = 478.18 \text{ kJ/kg}$$ ***Ans.***

Problem 5: *A fluid system at a temperature of 60°C and a pressure of 90. × 10³ N/m² executes a reversible process during which the temperature of the system remains constant. The heat transfer to the fluid is 120 kJ. Determine increase in entropy.*

If the mass of system is 2.5 kg, determine increase in specific entropy of the system.

Solution: Given: $T = 60 + 273 = 333$ k

$$dQ = 120 \text{ kJ}$$

We know that, from the principle of increase of entropy

$$S_2 - S_1 = \int \frac{dQ_{rev}}{T} = \frac{120}{333} = \boxed{0.360 \text{ kJ/K}}$$ ***Ans.***

If $m = 2.5$ kg

$$\therefore \quad \frac{dS}{m} = \frac{0.360}{2.5} = 0.144 \text{ kJ/kg K}$$

$$\boxed{= 0.144 \text{ kJ/kgK}}$$ ***Ans.***

Problem 6:

(a) A system has a mixture of 0.1 kg of saturated water and 0.7 kg of saturated steam in equilibrium at 300 × 10³ N/m². Determine dryness fraction, specific entropy and temperature of the mixture.

(b) Steam at a temperature of 250°C has a specific entropy of 6.980 kJ/kgK. Evaluate pressure, specific volume and specific internal energy of the steam.

Solution:

(a) By definition of dryness fraction, we can write

$$x = \frac{mg}{mf + mg} = \frac{0.7}{0.1 + 0.7} = 0.875$$

$$\boxed{x = 0.875}$$ ***Ans.***

At $300 \times 10^3 \times 10^{-5}$ bar = 3 bar

$$\boxed{t_s = 135.54 \text{ °C.}}$$ ***Ans.***

(Corresponding to 3 bar from steam table.)

Specific entropy

$$S_{wet} = Sf + xSf_g$$
$$S_f = 1.6717 \text{ kJ/kgK.}$$
$$Sf_g = 5.3192 \text{ kJ/kgK}$$
$$S = 1.6717 + 0.875\ (5.3192)$$
$$\boxed{S = 6{,}326 \text{ kJ/kgK}}$$ ***Ans.***

(b) From Superheated steam table corresponding to

$$T_{sup} = 250\text{°C and } S_{sup} = 6.980 \text{ kJ/kgK}$$
$$P = 9 \text{ bar and specific volume } V = 0.25963 \text{ m}^3\text{/kg}$$

and $h = 2946.8$ kJ/kg.

$\therefore \quad u = h - PV$

$= 2946.8 \times 10^3 - 9 \times 10^5 \,(0.25963)$

$= 2713.13$ kJ/kg.

Problem 7: *Determine increase in entropy of the universe (system and surroundings) when one kg of saturated water vapour at 100°C is condensed to saturated liquid at 100°C during a constant pressure process. The system rejects heat to the atmospheric air which is at 25°C.*

Solution: For the system, we can write

$$\Delta S_{system} = -m \cdot Sf_g$$

At $t_s = 100°C$;

from steam tables

$$S = 6.0485 \text{ kJ/kgK}$$

$$hf_g = 2256.9 \text{ kJ/kg.}$$

$$= -1 \times 6.0485 = -6.0485 \text{ kJ/kgK}$$

(entropy decreases)

for the surroundings

$$Q_{\text{to surroundings}} = m\,hfg = 1 \times 2256.9$$

$$= 2256.9 \text{ kJ/kg}$$

$$\therefore \quad \Delta S_{surr} = \frac{Q}{T} = \frac{2256.9}{25 + 273} = 7.573 \text{ kJ/K}$$

$\therefore$ ΔS for the universe

$$= \Delta S_{system} + \Delta S_{surr}$$

$$= -6.0485 + 7.573$$

$$\boxed{\Delta S_{unit} = 1.5245 \text{ kJ/K}}$$ *Ans.*

Problem 8: *A cylinder fitted with leak proof piston contains 1 kg of saturated water at 9 bar. The water undergoes a reversible (fully resisted) constant pressure process, as the piston moves slowly outwards so that water is converted into dry saturated steam and the process is accompanied by a heat transfer of 2050 kJ. The initial and final volume in the cylinder is 1115 cm³ and 0.25 m³. By neglecting friction, determine increase in entropy and internal energy in the cylinder.*

Solution: Given:

$m = 1$ kg $\qquad V_1 = 1115 \text{ cm}^3$

$P = 9$ bar $\qquad V_2 = 0.25 \text{ m}^3$

$Ts = 175.36°C$

$Q_{1\text{-}2} = 2050$ kJ

Increase in entropy

$$S_2 - S_1 = \int \frac{dQ_R}{T} = \frac{2050}{(175.36 + 273)}$$

$$S_2 - S_1 = 4.5722 \text{ kJ/K}$$

$$dS = \frac{S_2 - S_1}{m} = \frac{4.5722}{1} = \boxed{4.5722 \text{ kJ/kgK}}$$ *Ans*

Also, $\quad dQ = U + PdV$

$\therefore$ $U = dQ - PdV$

$= TdS - PdV$

$$U = (175.36 + 273)\ (4.5722) - \left[\frac{9 \times 10^5}{1000}(0.25 - 1.115 \times 10^{-3})\right]$$

$U = 1825.99$ kJ

$$u = \frac{U}{m} = \frac{1825.99}{1} = \boxed{1825.99 \text{ kJ/kg}}$$ **Ans.**

Problem 9: *Steam with quality 0.9 dry, at temperature 220°C undergoes a reversible constant temperature process to a pressure of 3 bar.*

(i) *Determine increase in entropy of the steam and heat transfer per unit mass of steam during the process.*

(ii) *Determine increase in internal energy and enthalpy of the steam.*

(iii) *If the steam is contained in a cylinder Piston machine, using the results of (i) and (ii) determine work done per unit mass of the steam.*

Solution:

(i) Given, At state 1,

Dryness fraction, $x = 0.9$.

$T = 220°C$

$S_1 = Sf + x.\ Sf_g = 2.5178 + 0.9\ (3.7639)$

$S_1 = 5.9053$ kJ/kgK

At $T = 220°C$
$S_f = 2.5178$ kJ/kgK
$Sf_g = 3.7639$ kJ/kgK

At state 2,

$P_2 = 3$ bar, $t = 220°C$

$t_{sat} = 135.54°C$

$\therefore$ The steam is being superheated. From the superheated steam table, at

$P_2 = 3$ bar and $t = 220°C$,

$S_2 = 7.4147$ (by interpolation approximately)

$\therefore$ Increase in entropy

$= S_2 - S_1 = 7.4147 - 5.9053$

$= \boxed{S_2 - S_1 = 1.5094 \text{ kJ/kgK}}$ **Ans.**

also we know that, $S_2 - S_1 = \dfrac{dQ}{T} = ds$

$dQ = T \cdot (dS)$

$= 493\ (1.5094)$

$\boxed{dQ = 744.13 \text{ kJ/kg}}$ **Ans.**

$T = 220 + 273$
$T = 493$ k

(ii) At state 1,

$h_1 = hf + x\ hf_g$

$h_1 = 943.7 + 0.9\ (1856.2)$

$h_1 = 2614.28$ kJ/kg

$x = 0.9$, $T = 220°C$.
$hf = 943.7$ kJ/kg
$hf_g = 1856.2$ kJ/kg

From superheated steam table, corresponding to

$P = 3$ bar and $T_{sup} = 220°C$.

$h_2 = 2916.7$ kJ/kg (by inter polation appx.)

Increase in enthalpy

$$h_2 - h_1 = 2916.7 - 2614.28$$

$$\boxed{h_2 - h_1 = 302.42 \text{ kJ/kg}}$$ ***Ans.***

Also we know that, $h = u + PV$

$$V_1 = V_f + xV_{fg}$$

$$V_1 = 0.00119 + 0.9\ (0.084848)$$

$$V_1 = 0.07755 \text{ m}^3\text{/kg}$$

and $P_1 = 23.198$ bar

At state 1,

$$V_f = 0.00119 \text{ m}^3\text{/kg}$$

$$V_{fg} = V_g - V_f = 0.084848$$

Now, from superheated steam tables

$V_2 = 0.7564$ m³/kg at 3 bar and 220°C by interpolation

$$u_1 = h_1 - P_1V_1$$

$$= 2614.28 - \frac{23.198 \times 10^5}{1000}\ (0.07755)$$

$$\boxed{u_1 = 2434.37 \text{ kJ/kg}}$$

$$u_2 = h_2 - P_2V_2 = 2916.7 - \frac{3 \times 10^5}{1000}\ (0.7564)$$

$$\boxed{u_2 = 2689.78 \text{ kJ/kg}}$$

∴ Increase in internal energy

$$= u_2 - u_1 = 2689.78 - 2434.37$$

$$\boxed{u_2 - u_1 = 255.41 \text{ kJ/kg}}$$ ***Ans.***

(iii) By applying first law,

$$dQ - dW = \Delta\left(h + \frac{V^2 \nearrow 0}{2g} + \frac{gz \nearrow 0}{gc}\right)$$

$$\therefore\ TdS - dW = \Delta h$$

$$dW = TdS - \Delta h = 493\ (1.5094) - 302.42$$

$$\boxed{dW = 441.71 \text{ kJ/kg}}$$ ***Ans.***

Problem 10: *A lump of steel of mass 9 kg at 1000 k is immersed in 85 kg of oil at 300 K. Determine change in entropy for universe. Assume specific heats of steel and oil as 0.5 kJ/kgK and 3.5 kJ/kgK respectively.*

Solution: By the principle of energy conservation

Heat lost by steel = heat gained by oil.

$$(mC_p\Delta T)_{steel} = (mC_p\Delta T)_{oil}$$

$$9 \times 0.5 \times (1000 - T_f) = 85 \times 3.5 \times (T_f - 300)$$

where T_f = equilibrium temperature

$$4500 - 4.5\ T_f = 297.5\ T_f - 89250$$

$$93750 = 302\ T_f$$

$$\boxed{T_f = 310.43 \text{ k}}$$

For steel, $(dS)_{steel} = mC_p \ln \dfrac{T_f}{T_1}$

$$= 9 \times 0.5 \ln \frac{310.43}{1000} = -5.264 \text{ kJ/K}$$

For oil, $(dS)_{oil} = mC_p \ln \dfrac{T_f}{T_1}$

$$= 85 \times 3.5 \ln \frac{310.43}{300} = 10.16 \text{ kJ/K}$$

∴ For the universe

$$(dS)_{univ} = (dS)_{steel} + (dS)_{oil} = -5.264 + 10.16$$

∴ $\boxed{\text{Entropy increase} = 4.896 \text{ kJ/K}}$ *Ans.*

Problem 11: *A system consists of 1 kg of ice (at °C) immersed in 4.5 kg of water (water at 28°C) and is open to atmosphere. Assuming adiabatic enclosure for the system, determine temperature of ice and water mixture and change in entropy of the system? Take specific heat of water = 4.187 kJ/kgK and latent heat of ice = 335 kJ/kg.*

Solution: Let T_m – Temperature of mixture
from the principle of energy conservation

heat lost by water = heat gained by ice

$$(mC_p\Delta T)_{water} = \text{latent heat} + C_p\,(\Delta T)$$

$$4.5 \times 4.187 \times (28 - T_m) = 335 + 4.187\,(T_m - 0)$$

$$527.56 - 18.84\,T_m = 335 + 4.187\,T_m$$

$$192.56 = 23.027\,T_m$$

$$\boxed{T_m = 8.362°\text{C}}$$

For water, change in entropy

$(dS)_{water}$ = entropy change due to cooling of water from 28°C to 8.362°C

$$= mC_p \cdot \ln \frac{T_m}{T_1}$$

$$= 4.5 \times 4.187 \times \ln \frac{8.362 + 273}{28 + 273} = -1.271 \text{ kJ/K}$$

for ice, entropy change

$(dS)_{ice}$ =entropy chang 'ue to ice melting at constant tempt. + due to heating of ice-water from 0°C to 8.362°C

$$= \frac{L}{T_1} + m\,C_p \cdot \ln \frac{T_m}{T_1}$$

$$= \frac{335}{0 + 273} + 1 \times 4.187 \times \ln \frac{8.362 + 273}{0 + 273}$$

$$= 1.353 \text{ kJ/K}$$

∴ Increase in entropy for the system

$$(dS)_{\text{system}} = (dS)_{\text{water}} + (dS)_{\text{ice}}$$
$$= -1.271 + 1.353$$
$$\boxed{(dS)_{\text{system}} = 0.082 \text{ kJ/K}}$$ **Ans.**

Problem 12: *A 30 kg steel ball at 427°C is dropped in 150 kg oil at 27°C, the specific heat of steel and oil are 0.5 kJ/kg and 2.5 kJ/kgK respectively. Estimate the entropy change of steel, oil and that of system containing oil and steel.* *(VTU, March 2001)*

Solution: Let T_f–final equilibrium temperature from the principle of energy conservation

Heat lost by steel ball = Heat gained by oil

$$(mC_p \Delta T)_{\text{steel ball}} = (mC_p \Delta T)_{\text{oil}}$$
$$30 \times 0.5\,(700 - T_f) = 150 \times 2.5 \times (T_f - 300)$$
$$10{,}500 - 15\,T_f = 375\,T_f - 112500$$
$$T_f = 315.38 \text{ K}$$

For steel ball, change in entropy

$$(dS)_{\text{steel ball}} = mC_p \ln\left(\frac{T_f}{T_1}\right)$$
$$= 30 \times 0.5 \ln \frac{315.38}{700} = -11.959 \text{ kJ/K}$$ **Ans.**

For oil, change in entropy

$$(dS)_{\text{oil}} = mC_p \cdot \ln\left(\frac{T_f}{T_1}\right)$$
$$= 150 \times 2.5 \times \ln\left(\frac{315.38}{300}\right) = 18.748 \text{ kJ/K}$$ **Ans.**

∴ Net change of entropy for the system comprising steel ball and oil.

$$(dS)_{\text{universe}} = (dS)_{\text{steel ball}} + (dS)_{\text{oil}}$$
$$= -11.959 + 18.748 = 6.789 \text{ kJ/K}$$ **Ans.**

Entropy increases due to irreversibility.

Problem 13: *m_1, kg of water at 'T_1' is isobarically and adiabatically mixed with m_2 kg of water which is at T_2 ($T_1 > T_2$). Show that for equal masses of water, the change in entropy of the mixture is given by*

$$(dS)_{univ} = 2\,mC_p \cdot \log_e\left(\frac{T_1 + T_2}{2\sqrt{T_1 T_2}}\right)$$

and also prove that the value of change in entropy is positive.

Solution: Let T_f – final equilibrium temperature of the mixture.

By making energy balance, we can write

Heat lost by m_1 kg of water = Heat gained by m_2 kg of water

$$m_1 C_p \cdot (T_1 - T_f) = m_2 C_p\,(T_f - T_2)$$

$$m_1C_pT_1 - m_1\ C_p\ T_f = m_2C_pT_f - m_2C_pT_2$$
$$m_1C_pT_1 + m_2C_pT_2 = m_2C_pT_f + m_1C_pT_f$$

$$\therefore \qquad T_f = \frac{\cancel{C_p}\,[m_1T_1 + m_2T_2]}{\cancel{C_p}\,[m_1 + m_2]}$$

For m_1 kg of water whose temperature decreases from 'T_1' to 'T_f', then

$$(dS)_1 = m_1C_p \ln \frac{T_f}{T_1} = m_1C_p \cdot \ln \frac{\dfrac{m_1T_1 + m_2T_2}{m_1 + m_2}}{T_1}$$

Similarly for m_2 kg of water whose temperature increases from T_2 to T_f.

$$(dS)_2 = m_2\ C_p \ln \frac{T_f}{T_2} = m_2C_p \cdot \ln \frac{\dfrac{(m_1T_1 + m_2T_2)}{(m_1 + m_2)}}{T_2}$$

$\therefore$ For the universe, change in entropy can be written as

$$(dS)_{\text{universe}} = (dS)_1 + (dS)_2$$

$$= m_1C_p \ln \frac{m_1T_1 + m_2T_2}{T_1(m_1 + m_2)} + m_2C_p \ln \frac{m_1T_1 + m_2T_2}{T_2(m_1 + m_2)}$$

The given condition is $m_1 = m_2 = m$

$$= mC_p \cdot \ln \frac{\cancel{m}(T_1 + T_2)}{T_1 \times 2\cancel{m}} + mC_p \ln \frac{\cancel{m}(T_1 + T_2)}{T_2 \cdot 2\cancel{m}}$$

$$= mC_p \ln \frac{T_1 + T_2}{2T_1} + mC_p \ln \frac{T_1 + T_2}{2T_2}$$

The above expression can be written as

$$= mC_p \cdot \log_e \left[\frac{T_1 + T_2}{2T_1} \times \frac{T_1 + T_2}{2T_2}\right] \text{ or}$$

$$= mC_p \log_e \left[\frac{T_1 + T_2}{2\sqrt{T_1T_2}}\right]^2$$

$$= 2\ mC_p \cdot \ln \frac{T_1 + T_2}{2\sqrt{T_1T_2}} \quad \text{hence proved} \qquad \textit{Ans.}$$

- In the above equation, $\left(\dfrac{T_1 + T_2}{2}\right)$ – arithmatic mean always greater than $\sqrt{T_1T_2}$ – geometric mean.

Therefore the above equation yields a positive value and this satisfies the principle of increase of entropy.

Problem 14: *A piston-cylinder machine contains 0.05 m^3 of Nitrogen at 1 bar and 300 k. Determine change in entropy and work done during isothermal reversible compression of the gas during which pressure rises to 4 bar.*

Solution: Assume Nitrogen as a perfect gas

gas constant $$R = \frac{R_{mol}}{\text{molecular mass}} = \frac{8314}{28}$$

$$R = 296.93 \text{ J/kgK and}$$

$$m = \frac{P_1V_1}{RT_1} = \frac{1 \times 10^5 \times 0.05}{296.93 \times 300}$$

$$\boxed{m = 0.0561 \text{ kg}}$$

Change in entropy for the isothermal process 1-2;

$$S_2 - S_1 = mR \log_e \frac{P_1}{P_2}$$

$$= 0.0561 \times 296.93 \cdot \ln \frac{1}{4}.$$

$$= -23.092 \text{ J/k} = -0.023092 \text{ kJ/K}$$ *Ans.*

$\therefore$ $$Q_{1\text{-}2} = T \cdot dS = T \cdot (S_2 - S_1)$$

$$= 300\,(-0.023092) = -6.9276 \text{ kJ}$$

For Isothermal process

Work done = heat interaction

$\therefore$ $$WD = -6.9276 \text{ kJ}$$ *Ans.*

or $$\text{Work done} = P_1V_1 \ln \frac{V_2}{V_1} = P_1V_1 \ln \frac{P_1}{P_2}$$

$$= 1 \times 10^5 \times 0.05 \ln \frac{1}{4} = -6.9314$$ *Ans.*

Problem 15: *0.004 m^3 of Nitrogen is contained in a rigid cylinder at 1 bar and 300 K. It is heated reversibly until the temperature becomes 380 K. Determine heat supplied and increase in entropy. Take $\gamma = 1.4$.*

Solution: We know that $R = \frac{8314}{28} = 296.93$ J/kgK

and $$m = \frac{P_1V_1}{RT_1} = \frac{1 \times 10^5 \times 0.004}{296.93 \times 300}$$

$$\boxed{m = 4.4903 \times 10^{-3} \text{ kg}}$$

$$C_v = \frac{R}{\gamma - 1} = \frac{296.93}{1.4 - 1} = 742.32 \text{ J/kgK}$$

∴ For a constant volume process 1–2,

$$Q_{1-2} = mC_v\,(T_2 - T_1)$$
$$= 4.4903 \times 10^{-3} \times 742.32\,(380 - 300)$$
$$= 266.659 \text{ J}$$
$$\boxed{Q = 0.266 \text{ kJ}}$$

Ans.

Increase in entropy

$$S_2 - S_1 = mC_v \cdot \ln \frac{T_2}{T_1}$$
$$= 4.4903 \times 10^{-3} \times 742.32 \times \ln \frac{380}{300}$$
$$\boxed{S_2 - S_1 = 0.7879 \text{ J/k}}$$

Ans.

Problem 16: *A piston cylinder machine contains 0.01 m³ of steam at a pressure of 10 bar. The steam has dryness fraction of 0.95 and after fully resisted expansion it reaches a pressure of 2 bar by following the law PV = constant. Determine:*

(i) The initial and final temperature
(ii) The work done
(iii) The increase in entropy and heat transfer

Solution: Given

$$V_1 = 0.01 \text{ m}^3$$
$$P_1 = 10 \text{ bar}$$
$$x = 0.95$$
$$P_2 = 2 \text{ bar}$$
$$PV = C$$

(i) $P_1V_1 = P_2V_2 = C$

$$V_2 = \frac{P_1V_1}{P_2} = \frac{10 \times 0.01}{2}$$
$$\boxed{V_2 = 0.05 \text{ m}^3}$$

At 10 bar

$$V_f = 0.0011274 \text{ m}^3/\text{kg}$$
$$V_{fg} = V_g - V_f$$
$$= 0.19430 - 0.0011274$$
$$V_{fg} = 0.19317 \text{ m}^3/\text{kg}$$

At $P_1 = 10$ bar, $Ts = 179.88°\text{C}$ (from steam table)

WKT, $V_1 = V_f + x\,V_{fg}$

$$= 0.0011274 + 0.95\,(0.19317)$$
$$V_1 = 0.1846 \text{ m}^3/\text{kg}$$

$$\therefore \quad m = \frac{V_1}{v_1} = \frac{0.01}{0.1846}$$
$$m = 0.05417 \text{ kg}$$
$$m = \frac{V_2}{v_2}$$
$$v_2 = \frac{V_2}{m} = \frac{0.05}{0.05417}$$

$V_2 = 0.923 \text{ m}^3/\text{kg}$

At $P_2 = 2$ bar and $V_2 = 0.923 \text{ m}^3/\text{kg}$.

($V_2 > V_g$), the steam is superheated. Now to find temperature 'T_2',

at $V_2 = 0.923$

At $P = 2$ bar, $V_g = 0.88540 \text{ m}^3/\text{kg}$, $T_s = 120.23$ °C,

$P = 2$ bar, $V_{sup} = 0.95954 \text{ m}^3/\text{kg}$, $T_{sup} = 150$°C.

Now at $P = 2$ bar and $V = 0.923 \text{ m}^3/\text{kg}$

$$T_2 = 120.23 + \frac{0.923 - 0.88540}{0.95954 - 0.88540}(150 - 120.23)$$

$$\boxed{T_2 = 135.32°\text{C}}$$

∴ The initial and final temperatures are 179.88°C and 135.32°C, respectively. *Ans.*

(ii) To find work done during the process 1-2

$$W_{1\text{-}2} = \int_1^2 PdV \qquad PV = C$$

$$= \int_1^2 \frac{C}{V} dV \qquad P = \frac{C}{V}$$

$$= C \int_1^2 \frac{1}{V} dV$$

$$= C(\log_e V_2 - \log_e V_1) \qquad C = PV$$

$$= P_1V_1(\log_e V_2 - \log_e V_1)$$

$$= 10 \times 10^5 \times 0.01 (\log_e 0.05 - \log_e 0.01)$$

$$\boxed{W_{1\text{-}2} = 16.09 \text{ kJ}}$$ *Ans.*

(iii) The increase in entropy

$$S_1 = S_f + xS_{fg}$$

$$= 2.1382 + 0.95 \times 4.4447$$

$$S_1 = 6.3606 \text{ kJ/kgK}$$

At 10 bar

$S_f = 2.1382$ kJ/kgK

$S_{fg} = 4.4447$ kJ/kgK

At $P_2 = 2$ bar, $T_s = 120.23$°C, $S_g = 7.1268$ kJ/kgK

$P_2 = 2$ bar, $T_{sup} = 150$°C, $S_{sup} = 7.2794$ kJ/kgK

At $P_2 = 2$ bar and $T_2 = 135.32$ °C, $S_2 = ?$

$$S_2 = 7.1268 + \frac{135.32 - 120.23}{150 - 120.23}(7.2794 - 7.1268)$$

$$\boxed{S_2 = 7.2041 \text{ kJ/kgK}}$$

∴ The increase in entropy

$$S_2 - S_1 = 7.2041 - 6.3606$$

$$\boxed{S_2 - S_1 = 0.8435 \text{ kJ/kgK}}$$ *Ans.*

or $(S_2 - S_1)\, m = 0.8435 \times 0.05417$

$$\boxed{(S_2 - S_1) = 0.0456 \text{ kJ/K}}$$ *Ans.*

Heat transfer, $$Q = \left(\frac{T_1 + T_2}{2}\right) dS$$

$$= \left[\frac{(179.88 + 273) + (135.32 + 273)}{2}\right] [0.0456]$$

$$= [452.88 + 408.32]\ [0.0456]$$

$$\boxed{Q = 19.63 \text{ kJ}}$$ **Ans.**

Problem 17: *0.8 kg of air at a pressure of 1 bar and temperature of 310 k is contained in a Piston-Cylinder machine. The work done on the air to compress it is 12 kJ and pressure and temperature of air changes to 2.8 bar and 320 k. If the system exchanges heat only with a reservoir maintained at 310 k, determine change in entropy of air, change in entropy of the reservoir and heat transfer from the air.*

Take, $C_p = 1.005$ *kJ/kgK and* $R = 0.287$ *kJ/kgK for air.*

Solution: We know that, in terms of pressure and temperature, the change in entropy of air can be expressed as

$$S_2 - S_1 = mC_p \cdot \log_e \frac{T_2}{T_1} + mR \log_e \frac{P_1}{P_2}$$

$$= 0.8 \times 1.005 \log_e \frac{320}{310} + 0.8 \times 0.287 \log_e \frac{1}{2.8}$$

$$\boxed{S_2 - S_1 = -0.2108 \text{ kJ/K}}$$

$$S_2 - S_1 = -0.2108 \text{ kJ/K}$$

$$du = mC_v\, dT = m\,(C_p - R)\, dT$$

$$= 0.8\ (1.005 - 0.287)\ (320 - 310)$$

$$du = 5.744 \text{ kJ}$$

also we know that

$$\delta Q = du - \delta W = 5.744 - 10$$

$$\delta Q = -4.256 \text{ kJ}$$

The air rejects 4.256 kJ of heat and reservoir receives this heat energy.

∴ Entropy change for reservoir

$$(ds)_{reservoir} = \frac{\delta Q}{T} = \frac{4.256}{310}$$

$$(ds)_{rev} = 0.0137 \text{ kJ/K}$$

∴ The net entropy change

$$= (ds)_{air} + (ds)_{rev}$$

$$= -0.2108 + 0.0137 = -0.1971 \text{ kJ/K}$$

∴ For a composite system say universe, negative entropy change is impossible and hence the claim is false.

Problem 18: *During the polytropic expansion of 1 kg of air the pressure reduces from 8 bar to 1 bar. The temperature of air is 350 K. Determine: (i) The specific volume and temperature after expansion, (ii) change of internal energy, work done and heat interaction, (iii) increase in entropy.*

Take, $R = 287$ *J/kgK,* $\gamma = 1.4$ *and* $n = 1.2$

Solution:

(i) From Ideal gas equation we can write $PV = mRT$

$$\therefore \quad V_1 = \frac{RT_1}{P_1} = \frac{287 \times 350}{8 \times 10^5}$$

$$\boxed{V_1 = 0.1255 \text{ m}^3/\text{kg}}$$

Also we know that,

$$\left(\frac{P_1}{P_2}\right) = \left(\frac{V_2}{V_1}\right)^n$$

$$\left(\frac{V_2}{V_1}\right) = \left(\frac{P_1}{P_2}\right)^{1/n}$$

$$V_2 = 0.1255 \left(\frac{8}{1}\right)^{1/1.2}$$

$$\boxed{V_2 = 0.709 \text{ m}^3/\text{kg}}$$ ***Ans.***

Also, $$\frac{T_2}{T_1} = \left(\frac{P_2}{P_1}\right)^{\frac{n-1}{n}}$$

$$T_2 = 350 \left(\frac{1}{8}\right)^{\frac{1.2-1}{1.2}}$$

$$\boxed{T_2 = 247.487 \text{ k}}$$ ***Ans.***

or

$$PV = mRT$$

$$P_2V_2 = mRT_2$$

$$T_2 = \frac{1 \times 0.709 \times 10^5}{1 \times 287}$$

$$\boxed{T_2 = 247.038 \text{ k}}$$ ***Ans.***

(ii) Change in internal energy

$$du = C_v\, dT = \left(\frac{R}{\gamma - 1}\right) dT$$

$$= \left(\frac{287}{1.4 - 1}\right)(247.48 - 350)$$

$$du = -73558.1 \text{ J/kg}$$

$$\boxed{du = -73.55 \text{ kJ/kg}}$$ ***Ans.***

Work done during polytropic process 1-2 is given by

$$W_{1\text{-}2} = \frac{P_1V_1 - P_2V_2}{n-1} = \frac{R\,(T_1 - T_2)}{(n-1)}$$

$$= \frac{287\,(350 - 247.48)}{1.2 - 1}$$

$$W_{1\text{-}2} = 147116.2 \text{ J/kg}$$

$$\boxed{\delta W = 147.116 \text{ kJ/kg}}$$ *Ans.*

'δ_q' Heat interaction during the process 1-2

$$\delta q = \frac{\gamma - n}{\gamma - 1} \times WD = \frac{1.4 - 1.2}{1.4 - 1} \times 147.116$$

$$\boxed{\delta q = 73.55 \text{ kJ/kg}}$$ *Ans.*

or from first law of thermodynamics

$$\delta q = \delta W + du$$

$$= 147.116 - 73.55$$

$$\boxed{\delta q = 73.56 \text{ kJ/kg}}$$ *Ans.*

(iii) Increase in entropy

$$S_2 - S_1 = C_v \ln \frac{T_2}{T_1} + R \ln \frac{V_2}{V_1} \qquad \left| \; C_v = \frac{R}{\gamma - 1} \right.$$

$$= 717.5 \ln \frac{247.48}{350} + 287 \ln \left(\frac{0.709}{0.1255}\right) \qquad \left| \; = \frac{287}{1.4 - 1} \right.$$

$$= -248.68 + 496.95 \qquad \left| \; = 717.5 \text{ J/kgK} \right.$$

$$= 248.27 \text{ J/kgK}$$

$$\boxed{S_2 - S_1 = 0.248 \text{ kJ/kgK}}$$ *Ans.*

Problem 19: *Two identical bodies of constant heat capacity are at same initial temperature T_0. A refrigerator operates between these two, so that one body is cooled to temperature T_2. For no change of phase and constant pressure of two bodies, Show that* $W_{min} = C_p \left[\frac{T_0^2}{T_2} + T_2 - 2T_0\right]$.

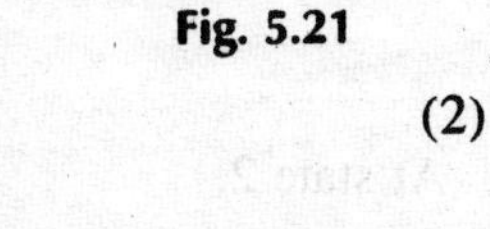

Fig. 5.21

Solution: Let T_0-initial temperature of finite bodies A and B. A refrigerator operates between A and B, cools the body A so that the temperature of it decreases from T_0 to T_2 and temperature of body 'B' changes to T_2^1.

$\therefore$ In body A, temperature decreases from T_0 to T_2

$\therefore \qquad Q = C_p\,(T_0 - T_2) \qquad (1)$

where C_p = heat capacity of A and B. In body B,
Heat received by body B

$$= Q + W = C_p\,(T_2^1 - T_0) \qquad (2)$$

$\therefore$ (2) – (1) gives

$$\not{Q} + W - \not{Q} = W = C_p\,(T_2^1 - T_0) - C_p\,(T_0 - T_2)$$

$$= C_p\,(T_2^1 - T_0 - T_0 + T_2) = C_p\,(T_2^1 + T_2 - 2T_0)$$

Now, for body A, entropy change

$$dS_A = \int_{T_0}^{T_2} C_p \frac{dT}{T} = C_p \log_e \frac{T_2}{T_0} \text{ (negative)}$$

for body B, entropy change

$$dS_B = \int_{T_0}^{T_2^1} C_p \frac{dT}{T} = C_p \log_e \frac{T_2^1}{T_0} \text{ (positive)}$$

In the refrigerator (cycle) entropy change is equal to zero for the combined system

$$(dS)_{\text{universe}} = (dS)_A + (dS)_B = C_p \ln \frac{T_2}{T_0} + C_p \ln \frac{T_2^1}{T_0}$$

According to entropy principles, we can write,

$$(dS)_{\text{univ}} \geq 0$$

$$\therefore \quad C_p \ln \frac{T_2}{T_0} + C_P \ln \frac{T_2^1}{T_0} \geq 0$$

$$C_p \ln \frac{T_2 T_2^1}{T_0^2} \geq 0$$

By keeping C_p, T_2 and T_0 constant, 'W' will be minimum when T_2^1 is minimum

$$\therefore \quad C_p \ln \frac{T_2 T_2^1}{T_0^2} = 0 = \ln 1.$$

$$T_2^1 = \frac{T_0^2}{T_2}$$

$$\therefore \quad W_{\min} = C_p \ln \left[\frac{T_0^2}{T_2} + T_2 - 2T_0 \right]$$

Problem 20: *One kg of superheated steam at 0.2 MPa and 200°C contained in a piston-cylinder machine is kept at ambient conditions of 300 K till the steam is condensed to saturated liquid at constant pressure. Calculate the change in entropy of the universe.*

Solution: At State1, $P_1 = 0.2$ MPa, and 200°C,

$P_1 = 2$ bar, $T_S = 120.23°C$

200°C > 120.23, the steam is superheated. From superheated steam tables at 2 bar and 200 °C.

$h_1 = 2870.5$ kJ/kg

$S_1 = 7.5072$ kJ/kgK

At state 2, $P_2 = 2$ bar, the liquid is saturated.

$hf_2 = 504.7$ kJ/kg and $Sf_2 = 1.5301$ kJ/kgK

$$\therefore \quad (\Delta S)_{\text{steam}} = S_2 - S_1 = 1.5301 - 7.5072$$

$$= -5.9771 \text{ kJ/kgK}$$

for a constant pressure process

$$q = \Delta h$$

Therefore, $q = h_2 - h_1 = 504.7 - 2870.5$

$$\boxed{q = -\,2365.8 \text{ kJ/kg}}$$

Entropy change of the surroundings

$$(\Delta S)_{sur} = \frac{dQ}{T} = \frac{+2365.8}{300}$$

$$= 7.886 \text{ kJ/kgK}$$

for the universe, net change in entropy

$$(\Delta S)_{univ} = (\Delta S)_{sys} + \Delta S_{sur}$$

$$= -\,5.9771 + 7.886$$

$$\boxed{(\Delta S)_{univ} = 1.9089 \text{ kJ/kgK}}$$ ***Ans.***

as $\Delta S_{univ} > 0$, the process is irreversible and possible.

Problem 21: *10 gms of water at 20°C is converted into ice at –10°C at constant atmospheric pressure. Assuming the specific heat of liquid water to remain constant at 4.2 kJ/kgK and that of ice to be half of this value and taking latent heat of fusion of ice at 0°C to 335 J/g, calculate the total entropy.* *(VTU, Aug. 2000)*

Solution:

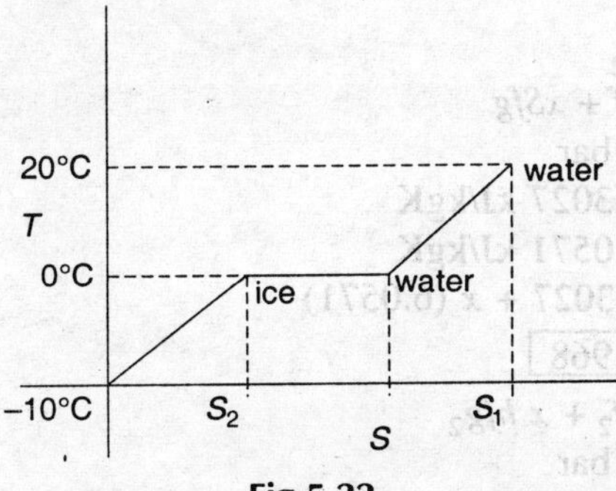

Fig 5.22

Total entropy change = (entropy change of water) + (entropy change of ice)

$$= mC_p \cdot \ln\left(\frac{T_f}{T_1}\right)_{water} + \left[\left(\frac{mhf_g}{T}\right) + mC_P \ln\left(\frac{T_f}{T_1}\right)\right]_{ice}$$

$$= 10 \times 4.2 \ln'\left(\frac{273}{293}\right) + \left[\frac{10 \times 335}{273} + 10 \times 2.1 \ln\left(\frac{263}{273}\right)\right]$$

$$-\,2.969 + [12.27 - 0.783]$$

$$\boxed{\text{Total entropy change} = 8.518 \text{ J/gK}}$$ ***Ans.***

Problem 22: *A cylinder contains 1 kg of steam at 4 bar and 200°C. During a reversible expansion process the steam pressure decreases to 1 bar and expansion is carried out in such a way that initial and final entropies remains same*

(i) Determine heat interaction during expansion when the work done is 80% of that for a reversible adiabatic process, between the same end states.

(ii) When the initial state and final pressure is same during a second process, which is adiabatic the steam does work equal to that in the first irreversible process. Determine increase in entropy during this process.

Solution: Given:

$$P_1 = 4 \text{ bar}, \; T_1 = 200° \text{ C}, \; t_s = 143.63°\text{C}$$

$$P_2 = 1 \text{ bar}, \; S_1 = S_2$$

(i) For reversible adiabatic process we can write

$$T\cancelto{0}{dS} - dW_x = \Delta\left(h + \cancelto{0}{\frac{V^2}{2g}} + \cancelto{0}{\frac{g}{gc}Z}\right)$$

$$\therefore \quad -dW_x = \Delta h$$

At state 1, the steam is in the superheated region.

At 4 bar and t_{sup} = 200°C, from superheated steam table,

$$h_1 = 2860.4 \text{ kJ/kg}$$

$$S_1 = 7.1708 \text{ kJ/kgK}$$

and given that

$$S_1 = S_2$$

$$\therefore \quad S_2 = Sf + xSfg$$

At $\quad P_2 = 1$ bar

$$Sf = 1.3027 \text{ kJ/kgK}$$

$$Sfg = 6.0571 \text{ kJ/kgK}$$

$$\therefore \quad 7.1708 = 1.3027 + x\,(6.0571)$$

$$\boxed{x = 0.968}$$

and $\quad h_2 = hf_2 + x\,hfg_2$

At $\quad P_2 = 1$ bar

$$hf_2 = 417.5 \text{ kJ/kg}$$

$$hfg_2 = 2257.9 \text{ kJ/kg}$$

$$\therefore \quad h_2 = 417.5 + 0.968\,(2257.9)$$

$$\boxed{h_2 = 2603.14 \text{ kJ/kg}}$$

$$\therefore \quad -dW_R = (h_2 - h_1)$$

$$dW_R = h_1 - h_2$$

$$= 2860.4 - 2603.14$$

$$dWR = 257.26 \text{ kJ/kg}$$

Given $\quad dW_x = 0.8\,dW_R$

$$= 0.8\,(257.26)$$

$$\boxed{dW_x = 205.80 \text{ kJ/kg}} \qquad \textit{Ans.}$$

According to first law, for irreversible process,

$$Q - W = \Delta h$$

$$Q = W + \Delta h$$
$$= -257.26 + 205.80$$
$$\boxed{Q = -51.46 \text{ kJ/kg}}$$ **Ans.**

(ii) $T_1 = 200 + 273 = 473$ k

At $P_2 = 1$ bar, $T_s = 99.63°C$

$= 372.63$ k

$$T \cdot dS = dQ$$

$$dS = \frac{dQ}{T}$$

$$S_2 - S_1 = \frac{Q}{\frac{T_1 + T_2}{2}}$$

$Q = +51.46$ kJ/kg

$$= \frac{51.46}{\frac{473 + 372.63}{2}}$$

$$\boxed{S_2 - S_1 = 0.121 \text{ kJ/kgK}}$$ **Ans.**

Problem 23: *In a vapour compression refrigerator, the working fluid enters the compressor at 1.2 bar and is delivered to the condenser as dry saturated vapour at 6.5 bar. At the exit of condenser the working fluid is a saturated liquid at the same pressure and is throttled to a pressure of 1.2 bar. Then it flows through the evaporator and then comes back to compressor. The compression process may be assumed to be reversible and adiabatic and throttling process to be adiabatic. By neglecting changes in kinetic energy and elevation, determine*

(i) Dryness fraction of the working fluid entering the compressor and hence shaft work done per unit mass of refrigerant.
(ii) Dryness fraction of the fluid after throttling.
(iii) Heat transfer per unit mass to the refrigerant in the evaporator
(iv) COP of the refrigerator.

For the working fluid in the refrigerator cycle, the property values are given below.

P bar	T_s °C	Specific enthalpy (kJ/kg)		Specific entropy (kJ/kgK)	
		Sat liquid	Sat Vapour	Sat liquid	Sat Vapour
1.2	–20	30.1	455.2	0.124	1.803
6.5	30	108.6	478.7	0.406	1.627

Solution: The *T-S* diagram for refrigeration cycle is as shown in figure.

$P_1 = 1.2$ bar

$P_2 = 6.5$ bar

(i) $S_1 = S_2$

$S_2 = 1.627$ kJ/kgK $= S_1$

$\therefore$ $S_1 = S_f + x\, S_{fg}$

$1.627 = 0.124 + x \cdot (1.679)$

$\boxed{x = 0.895}$ **Ans.**

$S_{fg} = S_g - S_f$

$= 1.803 - 0.124$

$= 1.679$ kJ/kgK

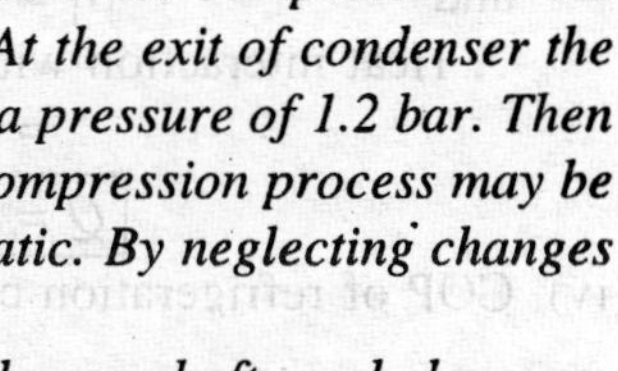

Fig. 5.23

$h_2 = 478.7$ kJ/kg

$h_1 = h_f + x\, h_{fg}$ $\quad$ | $h_{fg} = h_g - h_f = 425.1$ kJ/kg

$= 30.1 + 0.895\ (425.1)$

$\boxed{h_1 = 410.56 \text{ kJ/kg}}$

Compression work $= h_2 - h_1$

$= 478.7 - 410.56$

$\boxed{= 68.14 \text{ kJ/kg}}$ *Ans.*

(ii) Given $h_3 = 108.6$ kJ/kg (from table)

During throttling, enthalpy remains constant, i.e. $h_3 = h_4$.

Also $h_4 = hf + x\, hfg$

$108.6 = 30.1 + x\ (425.1)$

after throttling $\boxed{x = 0.184}$ *Ans.*

(iii) ∴ $T_1 = -20°C$

$= 253$ k

$S_4 = Sf + xSfg$

$= 0.124 + 0.184\ (1.678)$

$\boxed{S_4 = 0.4329 \text{ kJ/kgK}}$

and $S_1 = 1.627$ kJ/kgK

∴ Heat interaction with refrigerant $= Q = TdS = T\ (S_1 - S_2)$

$= 253\ (1.627 - 0.4329)$

$\boxed{Q = 302.10 \text{ kJ/kg}}$ Ans.

(iv) COP of refrigeration cycle

$$= \frac{Q}{W} = \frac{302.10}{68.14} = 4.43$$

$\boxed{\text{COP} = 4.43}$ *Ans.*

Problem 24: *In a closed system, air is at a pressure of 1 bar, temperature of 300 K and volume of 0.025 m³. The system executes the following processes during the completion of a thermodynamic cycle.*

1-2; Constant volume heat addition till pressure reaches 3.8 bar.

2-3; Constant pressure cooling of air.

3-1; Iso thermal heating to initial state.

Determine change in entropy in each process. Take $C_v = 0.718$ kJ/kgK, $R = 287$ J/kgK

Solution:

The *T-S* and *P-V* diagrams for the cycle 1-2-3-1 are as shown in figure. From the Ideal gas equation we can write, $PV = mRT$

$$m = \frac{PV}{RT} = \frac{P_1V_1}{RT_1} = \frac{1 \times 10^5 \times 0.025}{287 \times 300}$$

$\boxed{m = 0.029 \text{ kg}}$

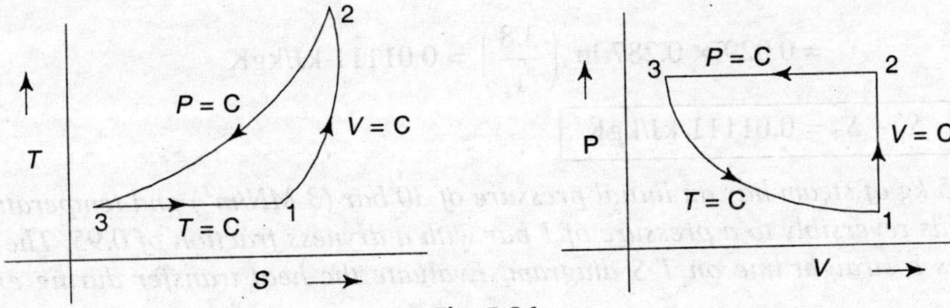

Fig. 5.24

Consider the process 1-2

$$\frac{P_2 \cancel{V_2}}{P_1 \cancel{V_1}} = \frac{\cancel{m}R\,T_2}{\cancel{m}R\,T_1}$$

$$\frac{P_2}{P_1} = \frac{T_2}{T_1}$$

$$\therefore \qquad T_2 = T_1\left(\frac{P_2}{P_1}\right) = 300\left(\frac{3.8}{1}\right)$$

$$\boxed{T_2 = 1140\text{ k}}$$ *Ans.*

$\therefore$ Change in entropy

$$S_2 - S_1 = mC_v \ln \frac{T_2}{T_1}$$

$$= 0.029 \times 0.718 \ln \left(\frac{1140}{300}\right) = 0.02779 \text{ kJ/kgK}$$

$$\boxed{S_2 - S_1 = 0.02779 \text{ kJ/kgK}}$$ *Ans.*

Consider the process 2-3,

$$T_3 = T_1 = 300 \text{ k}$$

$$\therefore \qquad S_3 - S_2 = mC_p \cdot \ln \frac{T_3}{T_2}$$

$$= m\,(C_v + R) \ln \frac{T_3}{T_2} = 0.029\,(0.718 + 0.287) \ln \frac{300}{1140}$$

$$= -\,0.0389 \text{ kJ/kgK}$$

$$\boxed{S_3 - S_2 = -\,0.0389 \text{ kJ/kgK}}$$ *Ans.*

Consider the process 3-1,

$$P_3 = P_2 = 3.8 \text{ bar}$$

$$\therefore \qquad S_1 - S_3 = mR \ln \frac{P_3}{P_1}$$

$$= 0.029 \times 0.287 \ln \left(\frac{3.8}{1}\right) = 0.01111 \text{ kJ/kgK}$$

$$\boxed{S_1 - S_3 = 0.01111 \text{ kJ/kgK}}$$ ***Ans.***

Problem 25: *4.5 kg of steam has an initial pressure of 30 bar (3 MN/m²) and temperature of 300°C. The steam expands reversibly to a pressure of 1 bar with a dryness fraction of 0.95. The expansion of steam appears as a straight line on T-S diagram. Evaluate the heat transfer during expansion and work done.*

Solution:

At state 1, $P_1 = 30$ bar, $T_1 = 300°C$.

$T_S = 233.84°C$ $(\because T_1 > T_S)$

From super heated steam table.

$$\left.\begin{aligned} V_1 &= 0.081159 \text{ m}^3/\text{kg} \\ h_1 &= 2995.1 \text{ kJ/kg} \\ S_1 &= 6.5422 \text{ kJ/kgK} \\ u_1 &= h_1 - P_1V_1 \end{aligned}\right\} \text{At } P_1 = 30 \text{ bar and } T_{sup1} = 300°C$$

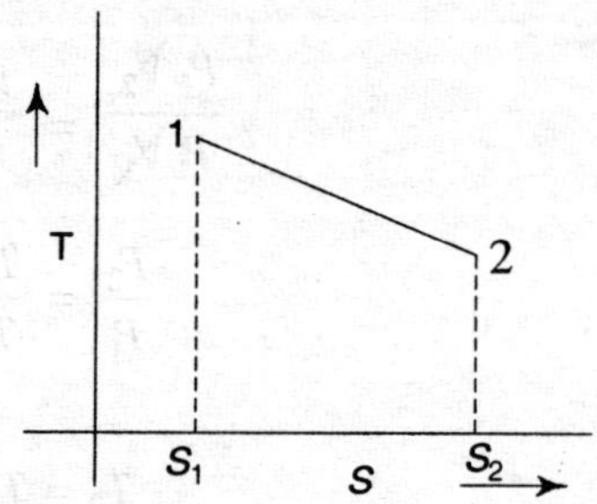

Fig. 5.25

$$= 2995.1 - \frac{30 \times 10^5}{1000} \times 0.081159$$

$$= 2751.62 \text{ kJ/kg}$$

At state 2,

$$S_2 = Sf_2 + x_2\, Sfg_2$$

At $P_2 = 1$ bar,

$$\left.\begin{aligned} Sf_2 &= 1.3027 \text{ kJ/kgK} \\ Sfg_2 &= 6.0571 \text{ kJ/kgK} \end{aligned}\right\} \text{From steam table}$$

$$S_2 = 1.3027 + 0.95 \times 6.0571$$

$$\boxed{S_2 = 7.0569 \text{ kJ/kgK}}$$

The corresponding saturation temperature is $T_s = T_2 = 99.63$ (from table).

also $h_2 = hf_2 + x_2\, hfg_2$

$$\left.\begin{aligned} hf_2 &= 417.5 \text{ kJ/kg} \\ hfg_2 &= 2257.9 \text{ kJ/kg} \end{aligned}\right\} \text{from steam table}$$

$\therefore$ $h_2 = 4175 + 0.95 \times 2257.9$

$h_2 = 2562.5$ kJ/kg

$\therefore$ $u_2 = h_2 - P_2V_2$

$$= 2562.5 - \frac{1 \times 10^5}{1000} \times 1.6090$$

$$\boxed{u_2 = 2401.6 \text{ kJ/kg}}$$

$V_2 = x_2 Vg_2$

$= 0.95 \times 1.6937$

$V_2 = 1.6090 \text{ m}^3/\text{kg}$

Heat transfer

$$Q = T \cdot dS$$

$$= \left(\frac{T_1 + T_2}{2}\right) dS$$

$$= \left[\frac{(300+273)+(99.63+273)}{2}\right][7.0569 - 6.5422]$$

$$\boxed{Q_{rev} = 243.35 \text{ kJ/kg}}$$ *Ans.*

For 4.5 kg of steam

$$Q = 243.35 \times 4.5$$

$$\boxed{Q = 1095.075 \text{ kJ}}$$ *Ans.*

This is the quantity of heat received by steam

$$Q = \Delta u + W$$

$$W = Q - \Delta u = Q - (u_2 - u_1)$$

$$= 243.35 - (2401.6 - 2751.62)$$

$$\boxed{W = 593.37 \text{ kJ/kg}}$$ *Ans.*

∴ Work done by 4.5 kg of steam is

$$WD = 593.37 \times 4.5$$

$$\boxed{WD = 2670.165 \text{ kJ}}$$ *Ans.*

Problem 26: *A heat engine is supplied with 300 kJ/s of heat at 290°C and heat rejection take place at 8.5°C. The results are as follows:*

(i) 215 kJ/s are rejected

(ii) 150 kJ/s are rejected

(iii) 75 kJ/s are rejected

Specify which of the above data represents a reversible, irreversible or impossible results.

(VTU, Aug. 2001)

Solution: According to Clausius inequality for the cyclic process, we can write

$$\oint \frac{dQ}{T} = \frac{Q_s}{Ts} - \frac{Q_r}{Tr}$$

(i) $\oint \frac{dQ}{T} = \frac{300}{563} - \frac{215}{281.5} = -0.2309 < 0.$

Hence the cycle is irreversible.

(ii) $\oint \frac{dQ}{T} = \frac{300}{563} - \frac{150}{281.5} = 0.$

The cycle is reversible.

(iii) $\oint \frac{dQ}{T} = \frac{300}{563} - \frac{75}{281.5} = 0.2664 > 0.$

The cycle is impossible.

Problem 27: *A reversible engine entracts 75 kW of energy from a reservoir at 750 k and produces 15 kW of work. The engine rejects heat to two reservoirs at 650 k and 580 k respectively. Determine quantity of heat rejected to each sink.*

Solution:

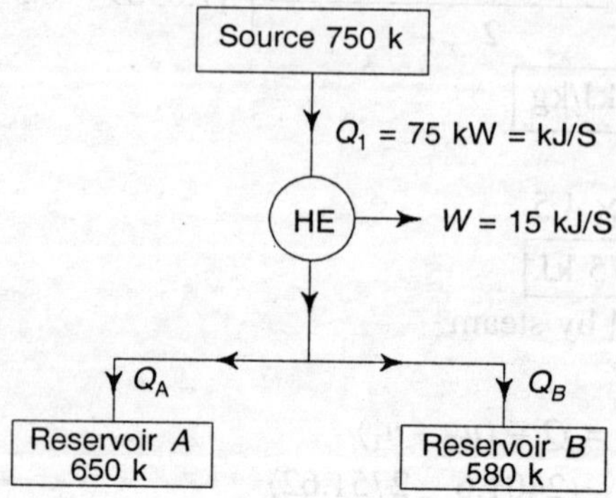

Fig. 5.26

Let Q_A – heat rejected to sink A

Q_B – heat rejected to sink B

$\therefore$
$$W = Q_1 - (Q_A + Q_B) = Q_1 - Q_A - Q_B$$
$$15 = 75 - Q_A - Q_B$$
$$Q_A + Q_B = 60$$
$$\boxed{Q_B = 60 - Q_A}$$

For a reversible engine, entropy change for the cycle is zero.

$$\oint \frac{dQ}{T} = 0.$$

$$\frac{75}{750} - \left[\frac{Q_A}{650} + \frac{60 - Q_A}{580}\right] = 0$$

$$\frac{75}{750} - \frac{Q_A}{650} - \frac{60 - Q_A}{580} = 0$$

$$\frac{75}{750} - \frac{Q_A}{650} - \frac{60}{580} + \frac{Q_A}{580} = 0$$

After solving the Eqn, we get

$$\boxed{Q_A = 18.27 \text{ kJ/s}}$$
$$Q_A = 1096.2 \text{ kJ/min}$$
$$Q_B = 60 - Q_A$$
$$= 60 - 18.27 = 41.73 \text{ kJ/s}$$
$$\boxed{Q_B = 2503.8 \text{ kJ/min}}$$

Ans.

Problem 28: *A reversible engine, shown below, during a cycle of operation draws 5 MJ from the 400 k reservoir and does 840 kJ of work. Find the amount and direction of heat interaction with other reservoirs.*

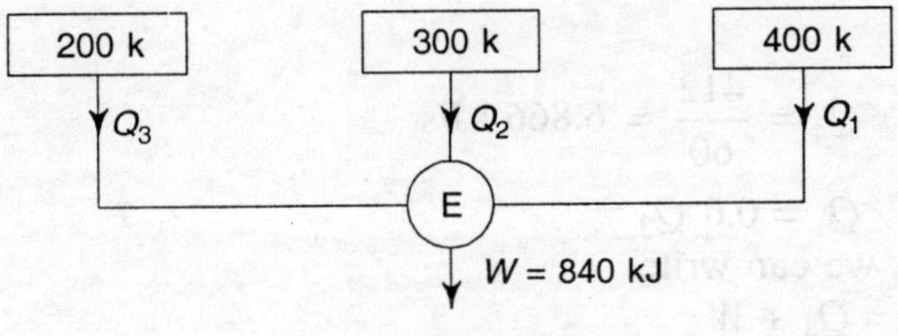

Fig. 5.27

Solution: Let Q_1, Q_2 and Q_3 are the heat interactions with reservoirs at 400 k, 300 k and 200 k respectively.

$$W = Q_1 - (\dot{Q}_2 + Q_3)$$

$$840 = 5 \times 10^3 - (Q_2 + Q_3)$$

$$\therefore \quad (Q_2 + Q_3) = 4160 \text{ kJ}$$

$$\therefore \quad Q_2 = (4160 - Q_3) \text{ kJ}$$

We know that

$$\oint \frac{dQ}{T} = 0, \text{ for a reversible engine.}$$

$$\therefore \quad -\frac{Q_3}{200} - \frac{Q_2}{300} + \frac{5 \times 10^3}{400} = 0$$

$$\therefore \quad -\frac{Q_3}{200} - \frac{(4160 - Q_3)}{300} + \frac{5 \times 10^3}{400} = 0$$

$$-5 \times 10^{-3}\, Q_3 - 13.866 + 3.33 \times 10^{-3}\, Q_3 + 12.5 = 0$$

$$\therefore \quad Q_3 = -1.366/1.67 \times 10^{-3}$$

$$Q_3 = -817.96 \text{ kJ}$$

$$\boxed{Q_3 = -0.817 \text{ MJ}}$$ *Ans.*

$$Q_2 = 4160 - Q_3$$

$$= 4160 + 817.96 = 4977.96 \text{ kJ}$$

$$\boxed{Q_2 = 4.97 \text{ MJ}}$$ *Ans.*

The engine receives $Q_2 = 4.97$ MJ of heat from a reservoir at 300 k and rejects $Q_3 = 0.817$ MJ to a reservoir at 200 k.

Problem 29: *A reversible engine operates between 3 heat reservoirs 1000 K, 800 K and 600 K and rejects heat to a reservoirs at 300 K. The engine develops 10 kW and rejects 412 kJ/min. If heat supplied by the reservoir at 1000 K is 60% of heat supplied by the reservoir at 600 K, find the quantity of heat supplied by each reservoir.* *(VTU, March 2001)*

Solution:

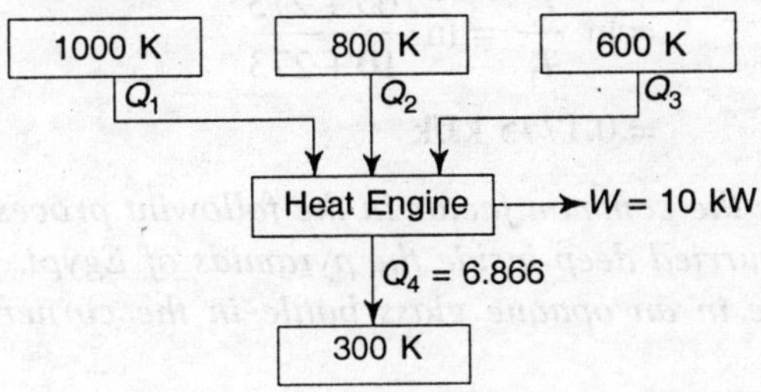

Heat rejected is given by

$$Q_4 = \frac{412}{60} = 6.866 \text{ kJ/s}$$

Also, $Q_1 = 0.6\ Q_3$

By making energy balance, we can write

$$Q_1 + Q_2 + Q_3 = Q_4 + W$$

$$0.6\ Q_3 + Q_2 + Q_3 = 6.866 + 10$$

$$Q_2 + 1.6\ Q_3 = 16.866 \quad (1)$$

By making entropy balance,

$$\frac{Q_1}{T_1} + \frac{Q_2}{T_2} + \frac{Q_3}{T_3} = \frac{Q_4}{T_4}$$

$$\frac{0.6 Q_3}{1000} + \frac{Q_2}{800} + \frac{Q_3}{600} = \frac{6.866}{300}$$

$$2.266 \times 10^{-3}\ Q_3 + 1.25 \times 10^{-3}\ Q_2 = 0.0228 \quad (2)$$

from (1) and (2), we can write

$$Q_2 + 1.6\ Q_3 = 16.866$$

$$1.25 \times 10^{-3}\ Q_2 + 2.266 \times 10^{-3}\ Q_3 = 0.0288$$

After solving, we get

$$\boxed{Q_3 = 6.75 \text{ kW}} \text{ and} \quad \textit{Ans.}$$

$$\boxed{Q_2 = 6.066 \text{ kW}} \quad \textit{Ans.}$$

$$\therefore \quad Q_1 = 0.6\ Q_3 = 0.6 \times 6.75$$

$$\boxed{Q_1 = 4.05 \text{ kW}} \quad \textit{Ans.}$$

Problem 30: *One kg of air is heated reversibly from an initial state of one atmosphere and 10°C to 1 atmosphere and 65°C. For reversible heating of air at constant pressure* $\frac{dQ}{dT}$ *= 1.0 kJ/k, determine change in entropy.* *(VTU, March 2000)*

Solution: Since $dQ = dT$

$$ds = \frac{dQ}{T} = \frac{dT}{T}$$

$\therefore$ change in entropy

$$DS = S_2 - S_1 = \int_{T_1}^{T_2} \frac{dT}{T}$$

$$= \ln \frac{T_2}{T_1} = \ln \frac{65 + 273}{10 + 273}$$

$$= 0.1775 \text{ kJ/k} \quad \textit{Ans.}$$

Problem 31: *Explain what is the common factor in the followint processes*

(i) A mummy (a corpse) burried deep inside the pyramids of Egypt.

(ii) An allopathic medicine in an opaque glass bottle in the corner of your shelf, well past it's expiry date.

(iii) In a perfectly insulated rigid container, an unstable Uranium 235 decays to become lead. *(VTU, Feb. 2004)*

Solution: Each of the above mentioned examples, constitutes an Isolated System with internal entropy generation.

Problem 32: *A rigid tank contains air at 35°C and is stirred by a paddle wheel which does 500 kJ of work on the air. During the stirring process, the temperature of air remains constant because of heat transfer to surroundings at 15°C. Estimate the change in entropy of air in the tank and change in entropy of the surroundings.* *(VTU Feb. 2004)*

Solution: For the system of air in the rigid tank, the initial and final states are same.

$$\therefore \quad S_1 = S_2$$

$$(\Delta S)\text{universe} = 0$$

By First Law of thermodynamics

$$Q - W = 0 \qquad (\text{Since } T_1 = T_2)$$

$$Q = W = -500 \text{ kJ}$$

for the surroundings

$$Q = +500 \text{ kJ}$$

$$\Delta S_{\text{surrounding}} = \frac{Q_{\text{surrounding}}}{T_{\text{surrounding}}} = \frac{500}{288.15}$$

$$\boxed{\Delta S_{\text{surroundings}} = 1.735 \text{ kJ/kgk}} \qquad \textit{Ans.}$$

Problem 33: *A turbine expands air adiabatically from 6 bar 400 K, 200 m/s to 1 bar, 290 K, 50 m/s. Determine the actual work output and the maximum possible work output for the above operating conditions. Also find the irreversibility of the process. Take surrounding atmosphere to be at 1 bar and 280 K.* *(VTU, Feb. 2004)*

Solution: The flow process is adiabatic

$$W_{\text{act}} = (h_1 - h_2) + \left(\frac{V_1^2 - V_2^2}{2}\right)$$

$$= C_p (T_1 - T_2) + \left(\frac{V_1^2 - V_2^2}{2}\right)$$

$$= 1004.5\,(400 - 290) + \left(\frac{200^2 - 50^2}{2}\right)$$

$$= \boxed{129.24 \text{ kJ/s}} \qquad \textit{Ans.}$$

Maximum possible work–Reversible work

$$= (h_1 - h_2) + T_0(S_1 - S_2) + \frac{V_1^2 - V_2^2}{2}$$

$$= 129240 + 280\left[C_p \ln\frac{T_2}{T_1} - R\ln\frac{P_2}{P_1}\right]$$

$$= 129240 + 280\left[1004.5 \ln \frac{290}{400} - 287 \ln\left(\frac{1}{6}\right)\right]$$

$$= \boxed{182.80 \text{ kJ/s}}$$ *Ans.*

$$\text{Irreversibility} = 182.80 - 129.24$$

$$= \boxed{53.56 \text{ kJ/s}}$$ *Ans.*

5.19 REVIEW QUESTIONS

1. State and prove Clausius theorem.
2. Derive Clausius inequality and hence prove that entropy is a property.
3. Derive an expression for entropy.
4. Define entropy and show that entropy is a property of the system.
5. Explain the principle of increase of entropy.
6. State Clausius inequality and show that the cycle is reversible, irreversible or impossible.
7. How the entropy change of a reversible process estimated? Will it be different for an irreversible process between the same end states.
8. Show that transfer of heat through a finite temperature difference is irreversible.
9. Show that the adiabatic mixing of two fluids is irreversible.
10. What is the maximum work obtainable from two finite bodies at temperatures T_1 and T_2.
11. Determine the maximum work obtainable by using one finite body at temperature T and a thermal energy reservoir at temperature T_0, $T > T_0$.

CHAPTER 6

Pure Substances

6.1 INTRODUCTION

The thermodynamic systems comprise fluids which do not change chemically or exhibit significant electrical, magnetic or capillary effects. This chapter is devoted to explain nature of such simple systems. The system may consists of several working fluids and steam is one such working fluid used extensively in steam engines, steam turbines in atomic power plants for power generation, process industries and in heat exchanger as a heat carrying agent. The desired effect can be obtained during state change of steam which changes steam properties. The chemical composition of steam does not change in all the phases i.e. water (liquid phase) and ice (solid phase) and also it is freely available in nature. When a substance changes it's chemical composition, it cannot be considered as pure substance. But water is considered as a pure substance which is homogeneous in chemical aggregation and invariable in chemical aggregation i.e. it is a substance with chemical homogeneity and constant chemical composition.

One of the important features of the pure substance is that its state can be represented as a point on a plane diagram and state changes can be recognised by the curves.

6.2 DEFINITION OF PURE SUBSTANCE

A pure substance is one that has a homogeneous and invariable chemical composition, i.e. the substance retains its chemical composition even during a change of phase.

A pure substance may exist in three different states (phases), i.e. solid (ice), liquid (water) and gas (steam or water vapour). A system consists of liquid and vapour phases of water is also a pure substance.

When heat is added to ice, it melts and phase transformation occurs from solid to liquid. The transformation in the opposite direction is known as "freezing". Further heat addition to water results in phase transformation from liquid to gas phase and is known as "vapourisation". The phase transformation from gas to liquid is known as "condensation".

Some examples of pure substances are system comprising steam, steam and water, Oxygen + Nitrogen gas (air); Air + fuel vapour, Gaseous combustion products etc. . .

On the other hand, air + liquid air, air and fuel vapour with liquid fuel, air and fuel vapour with combustion products in a system are not considered as pure substances.

6.3 THE TWO PROPERTY RULE

The two property rule states that "The state of a pure substance of given mass can be fixed by specifying two properties, provided that the system is in equilibrium and gravity, motion, electricity, magnetism and capillarity are without significant effects."

The two property rule applies only to liquid and gaseous systems. For solids, this rule cannot be applied as it has positive or negative pressures according to the direction of measurement. When the system comprising a solid, it requires more than two properties to specify its state.

Some systems which are not actually pure substances obeys two property rule.

Suppose (x, y, z) be the three properties of a system, we can write

$$z = f(x, y)$$

If P, V and t are pressure, specific volume and temperature of a particular substance, then we can write

$$PV = (a + bt)$$

$$P = \frac{a + bt}{V}$$ where a and b are called constants.

6.4 THE ENERGY OF A PURE SUBSTANCE : INTERNAL ENERGY

The internal energy of a pure substance can be expressed by U by using two property rule. In case, the system is of unit mass the letter 'u' will be used.

For a pure substance E and U are related as E = 'U' + terms accounting for gravity, motion, capillarity, electricity and magnetism.

In the absence of above terms, the change in internal energy of a pure substance can be written by using first law of thermodynamics

$$Q - W = \Delta U$$

If the system is of mass m,

$$\frac{Q - W}{m} = \Delta u$$

6.5 ENTHALPY OF A PURE SUBSTANCE

It is the sum of internal energy and PV work and is expressed as 'H'. It is the amount of heat energy contained in a given mass of steam

$$H = U + PV$$

For unit mass of system

The specific enthalpy; $h = u + PV$

$$\therefore \quad \Delta h = h_2 - h_1 = (u_2 - u_1) + (P_2V_2 - P_1V_1)$$

$$\Delta h = \Delta u + W$$

6.6 SPECIFIC HEAT AT CONSTANT VOLUME AND CONSTANT PRESSURE

The specific heat at constant volume 'C_v' is the rate of change with temperature of the specific internal energy of the system when the specific volume is held constant.

$$\therefore \quad C_v = \left(\frac{\partial u}{\partial t}\right)_v$$

$\frac{\partial u}{\partial t}$ → rate of change of 'u' wr to 't'.

Characteristics of 'C_v'

(i) It is a property i.e. a function of the state.

(ii) 'C_v' can be taken as constant over a range of temperature and pressures.

For example, 'C_v' for water may be taken as constant and equal to 4186 J/kgK over a range of temperature and pressure.

(iii) The specific heat is defined in terms of the two other properties, *u* and *t*. From first law of thermodynamics we can write

$$q = W + u$$

$\therefore$ In differential form, $\delta q = du + \delta W$

$$\delta q = du + pdv, \text{ by assuming only displacement work}$$

$$q = (\Delta u)_{v-1}, \text{ for a constant volume process and after integration}$$

$$q = \int_1^2 C_v \, dT$$

The specific heat at constant pressure 'C_p' is the rate of specific enthalpy with respect to temperature, when the pressure of the system is held constant.

$$\therefore \quad C_p = \left(\frac{\partial h}{\partial t}\right)_p$$

From Ist law,

$$\delta q = du + \delta W = du + PdV$$

$$h = u + PV$$

$$dh = du + PdV + VdP$$

$$dh = \delta q + VdP$$

$$\therefore \quad \delta q = dh - VdP$$

for constant pressure process $VdP = 0$

$$(\delta q)_P = (dh)_P$$

After integration

$$q = \int_1^2 C_p \, dT$$

Characteristics of 'C_p'

(i) It is a property and changes with state of the substance.

(ii) It is the heat transfer required to raise temperature of unit mass of substance through one degree at constant pressure

(iii) By using two property rule, '*h*' can be expressed as function of *t* and *P*.

$$\therefore \quad dq = dh - VdP$$

$$= \frac{\partial h}{\partial t} dt + \frac{\partial h}{\partial p} dP - VdP$$

For constant pressure process

$$\frac{\partial h}{\partial P} dP = VdP = 0$$

$$\therefore \quad dq = \left(\frac{\partial h}{\partial t} dt\right)_p$$

$\boxed{dq = C_p}$ for unit temperature rise.

6.7 P-V DIAGRAM FOR A PURE SUBSTANCE

Consider a piston-cylinder arrangement contain unit mass of ice at –10°C and 1 atm. If the ice is heated slowly at constant pressure, there is state change of water as temperature is increased and can be plotted on a *P-V*-coordinate. The distinct regimes of heating are:

1-2: The ice increases its temperature from –10°C to 0°C, at which ice would start melting. The volume of ice also increases from V_1 to V_2.

2-3: At constant temperature of 0°C, ice melts into water. At state 2, melting of ice begins and at state 3, melting process ends. The volume decreases from V_2 to V_3.

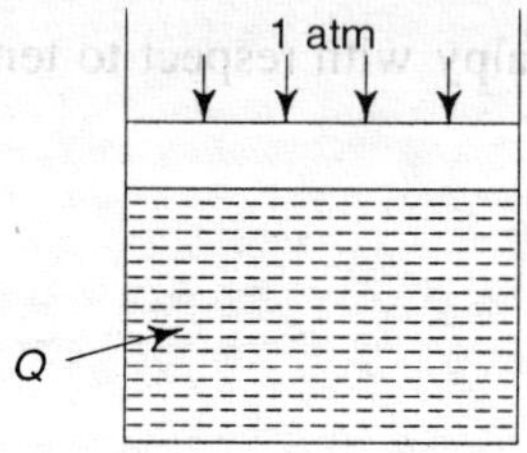

Fig. 6.1 Heating of water at constant pressure

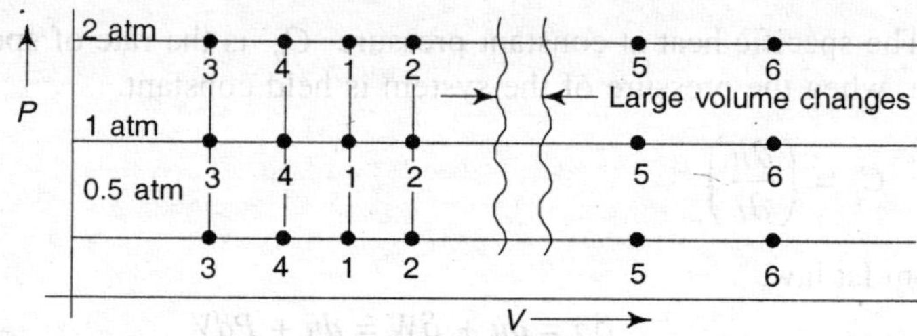

Fig. 6.2 Changes in volume of water during heating at constant pressure

3-4: Further addition of heat at the same pressure, increases temperature of water from 0°C to 100°C. The volume of water increases because of thermal expansion.

4-5: The phase change from liquid to vapour (vapourisation or boiling) occurs at a constant temperature of 100°C. (the pressure is being constant at 1 atm). The boiling of water starts at 4 and ends at 5 during which there is large increase in volume.

5-6: The vapour at state '5' is heated to superheated condition i.e. to state '6'. The volume of vapour increases from V_5 to V_6.

In regime 1-2, water existed in solid phase, 3-4 is the liquid phase and beyond state 5, water exists in the vapour phase. Between 2-3, the ice (solid) absorbs latent heat of fusion and changes into liquid phase. Between 4-5, liquid changes into vapour by absorbing latent heat of vapourization. Both these processes occur at constant temperature and pressure.

The states 2, 3, 4 and 5 are known as saturation states. It is the state from which a phase change may occur without a change of pressure or temperature.

Saturated Solid State: At state 2, a solid changes into liquid at constant pressure and temperature. State '2' represents end of solid phase or it is the state at which ice begins to melt.

Saturated Liquid States: States 3 and 4 are both saturated liquid states. At state 3, the liquid is saturated with respect to solidification and at state '4', the liquid is saturated with respect to vapourisation.

Saturated Vapour State: From state 5, the vapour may condense into liquid without a change of pressure or temperature.

Similar regimes of heating would have been obtained with similar saturation states 2, 3, 4 and 5 by considering heating of ice at –10°C and at a constant pressure of 2 atm. and the state changes were plotted on *P-V* diagram.

By considering heating of ice at different pressures, all the state changes may similarly be plotted on the *P-V* coordinates. Then join all the solid states 2 at various pressures by a line. Similarly join all the saturated states 3, 4 and 5 together as shown in Fig. 6.3.

For a pure substance other than water, the state changes are as shown in Fig. 6.4. In this case during melting of substance, volume increases.

In the Figs. 6.3 and 6.4, the saturated solid line is the line which passes through all the saturated solid states 2. Similarly, the saturated liquid lines passes through the states 3 and 4 i.e. saturated liquid states with respect to solidification and vapourization respectively. The saturated vapour line is one which passes through all saturated vapour states 5. The saturated liquid line with respect to vapourization and saturated vapour line meet at the critical state and the shape of the curve is known as saturation or vapour dome.

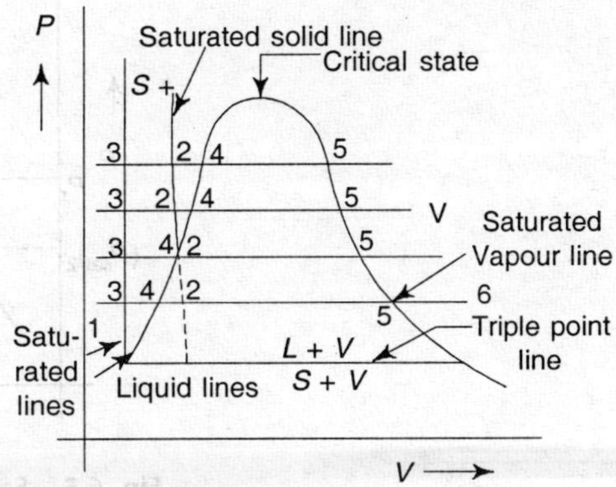

Fig. 6.3 P-V diagram of water, whole volume decreases on melting

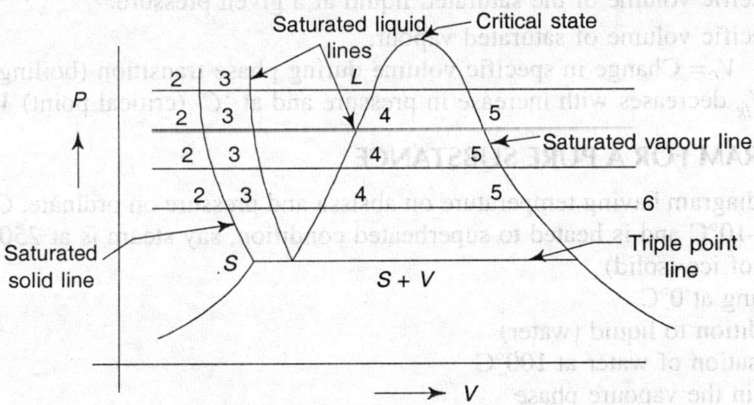

Fig. 6.4 P-V diagram of a pure substance other than water whose volume increases on melting

The solid region (S) exists to the left of saturated solid line. The solid-liquid mixture ($S + L$) region exists between saturated solid line and saturated liquid line with respect to solidification. The compressed liquid region is present bet-ween two saturated liquid lines. The vapour dome consists of liquid-vapour mix-ture ($L + V$) region between saturated liquid and saturated vapour lines and vapour region exists to the right of saturated vapour line. The triple point is a line on *P-V* diagram where all three phases (solid, liquid, gas) exists in equilibrium.

The substance cannot exist in the liquid phase at a pressure below the triple point line. After heating, the substance changes from solid to vapour by extracting the latent heat of sublimation from the surroundings and is known as "sublimation". Hence this region represents solid-vapour ($S + V$) mixture region.

For a substance, phase change occurs at constant temperature and pressure. The saturation temperature is the temperature at which phase transformation occurs from liquid to vapour at a given pressure. The saturation temperature is the function of pressure. Similarly at constant temperature, the liquid will boil or condense only at a particular pressure, called the saturation pressure which is a function of temperature.

In the Fig. 6.5 shown, $(t_{sat})_{(1)}$ is the saturation temperature at P_1 and $(P_{sat})_2$ is the saturation pressure at given temperature t_2. The saturation temperature increases with increase in pressure up to critical point. Between points '*A*' & '*B*', boiling of liquid take place. At '*A*' the substance is in the liquid form only (saturated) and at '*B*' the substance is in the vapour form only (saturated). The vapour content gradually increases from *A* to *B* as the liquid changes its state.

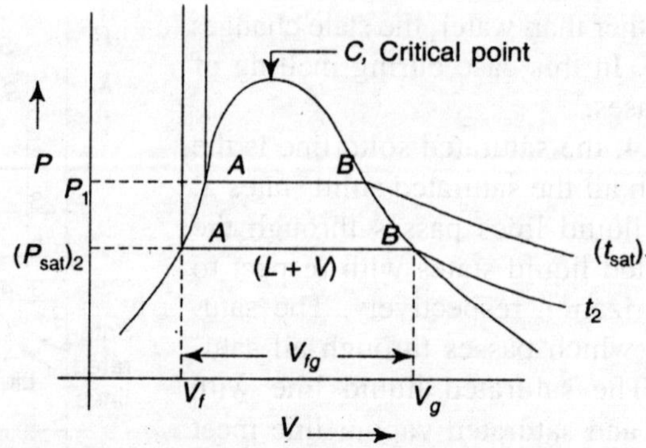

Fig. 6.5 Saturation pressure and temperature

V_f – Specific volume of the saturated liquid at a given pressure.
V_g – Specific volume of saturated vapour.
$V_{fg} = V_g - V_f$ = Change in specific volume during phase transition (boiling or condensation).
The value of V_{fg} decreases with increase in pressure and at 'C' (critical point) V_{fg} becomes zero.

6.8 P-T DIAGRAM FOR A PURE SUBSTANCE

It is the property diagram having temperature on absissa and pressure on ordinate. Consider heating of ice at 1 atm and –10°C and is heated to superheated condition, say steam is at 250°C.

1-2: Heating of ice (solid)
2-3: Ice melting at 0°C
3-4: Heat addition to liquid (water)
4-5: Vapourisation of water at 100°C
5-6: Heating in the vapoure phase

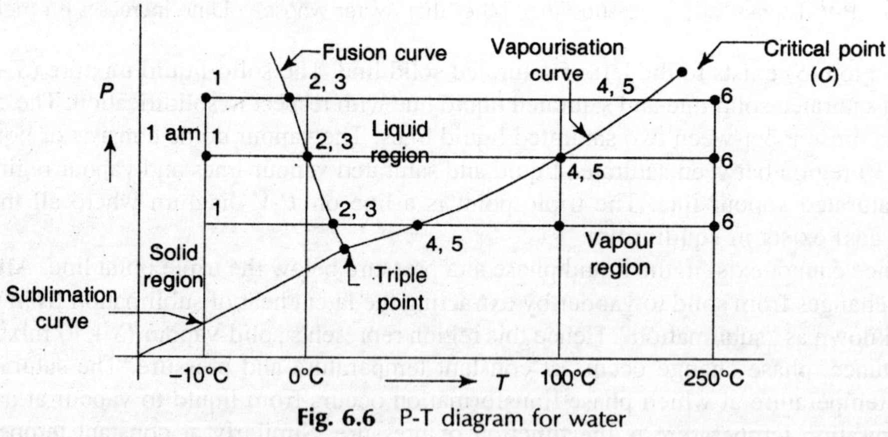

Fig. 6.6 P-T diagram for water

By joining points 2, 3 at different pressures, a fusion curve will be obtained. Similarly, by joining points 4, 5 at different pressures, vapourisation curve will be obtained. This curve represents vapourisation or condensation (on cooling) at different temperatures and pressures. The vapour pressure of a solid at different temperatures will give the sublimation curve.

The fusion curve, vapourisation curve and sublimation curve meet at a point called triple point. For water, the triple point is at 4.58 mm of Hg and 273.16 K. For carbon dioxide, it is at 3885 mm Hg (about 5 atm.) and 216.55 K. When solid CO_2 ('dry ice') is exposed to 1 atm. pressure, it gets transformed into vapour directly.

6.9 P-V-T SURFACE

Consider a system in which heat is added to the system causes its temperature to increase and changes in the volume. The pressure of the system is maintained constant, when the weight on the piston remain constant as shown in Fig 6.9.1. The heat addition process is repeated for several different loads and the properties *P-V* and *T* are measured. These datas are used for construction of *P-V-T* surface.

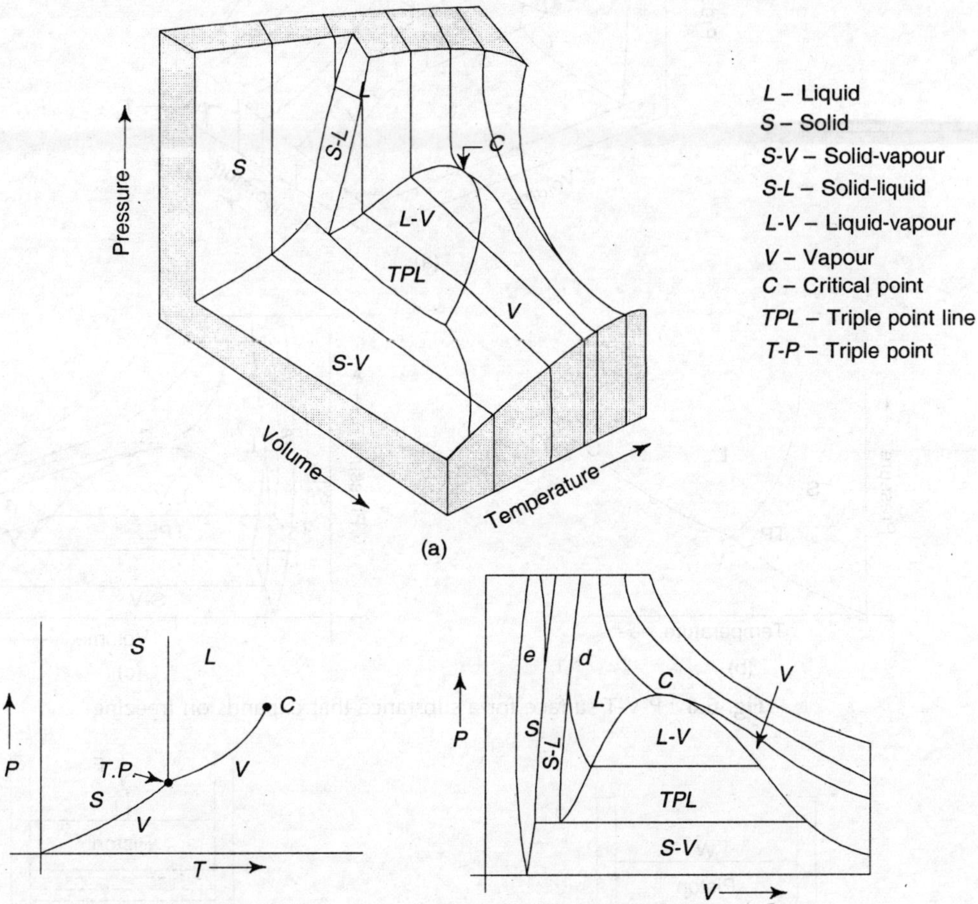

Fig. 6.7 P-V-T surface for a substance that contracts on freezing

In the Fig. 6.9.2, it is shown that the system temperature is maintained constant by using a heat reservoir. By varying force applied to the piston, volume can be changed corresponding to the change in pressure of the system. The properties *P-V* and *T* are measured at different temperatures and datas are used for the construction of *P-V-T* surface.

P-V-T surface is the three dimensional variation describing three thermodynamic properties *P-V* and *T*. It is the graphical representation of the *P-V-T* equation of state also known as thermal equation of state. At least two properties are required to define state. Therefore, on *P-V-T* surface, by knowing any two properties, the third property can be seen and continuous variation of these properties is available. Here, pressure, volume and temperature are taken on mutually perpendicular axis and each possible equilibrium state is represented by a point on the surface i.e. any point on the *P-V-T* surface represents an equilibrium state of the substance.

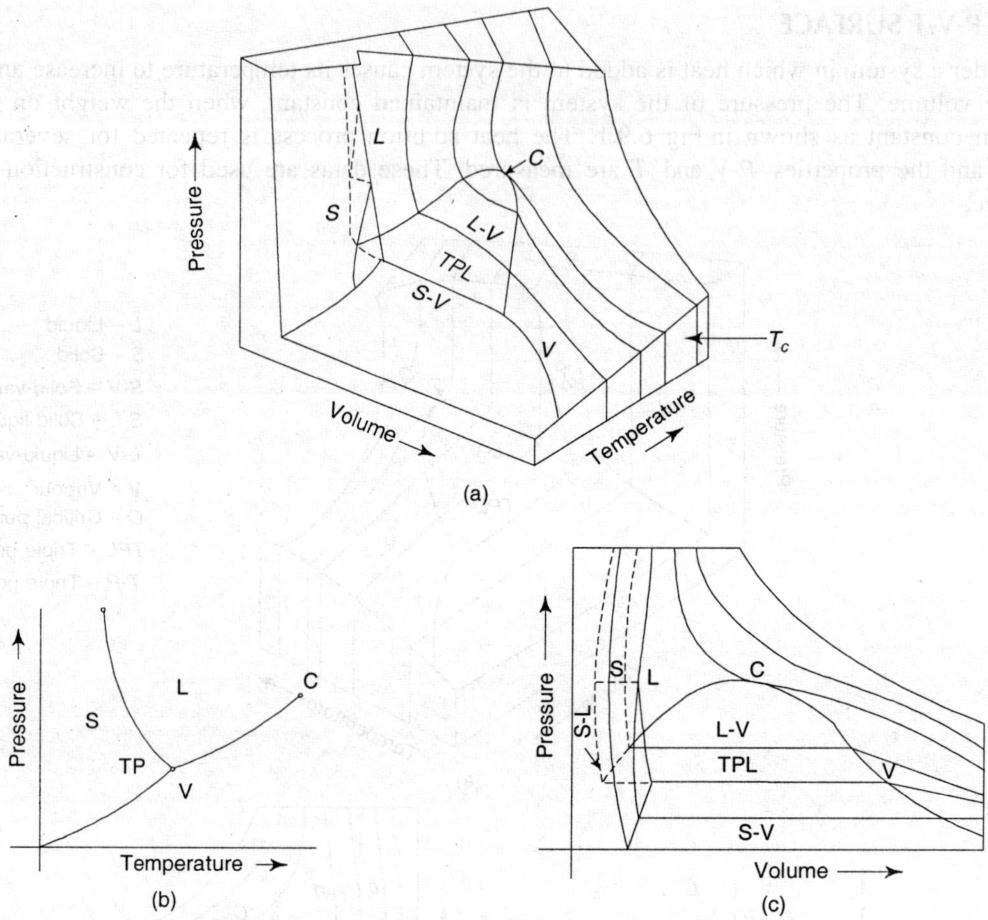

Fig. 6.8 P-V-T surface for a substance that expands on freezing

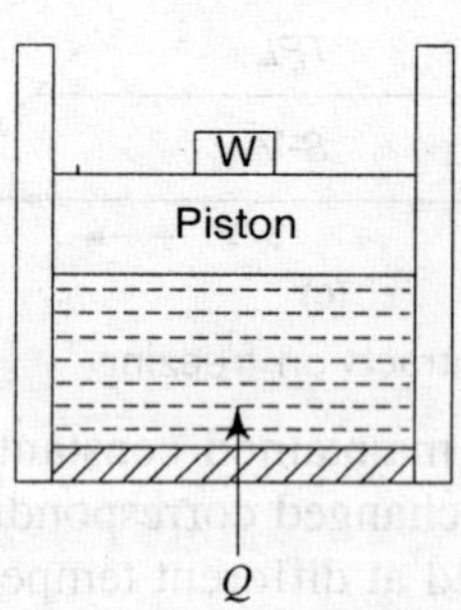

Fig. 6.9.1 Variation of T with V at cons-tant P

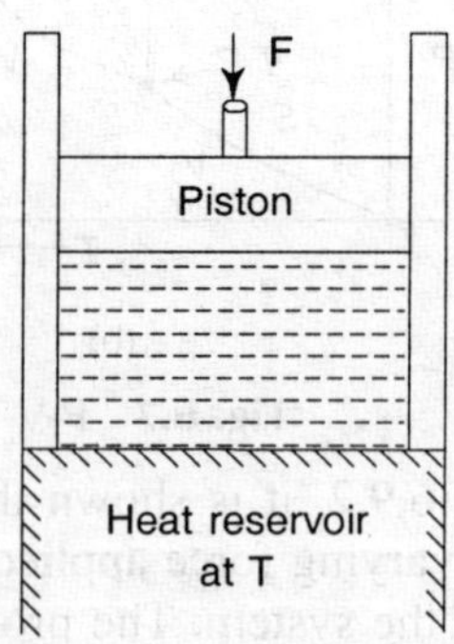

Fig. 6.9.2 Variation of P with V at cons-tant T

The regions of the surface that represent a single phase i.e. solid, liquid and vapour phases are shown on *PVT* surface and these surfaces are being curved. The two phase regions (*S-L*, *S-V* and *L-V*) are ruled surfaces i.e. they are strai-ght lines parallel to *V* (specific volume) axis. It means that, in two phase region, constant pressure and constant temperature lines are same, though the specific volume may change. The triple point appears as the triple line on *P-V-T* surface.

6.10 FORMATION OF STEAM AT CONSTANT PRESSURE

The transformation of one kg of ice into superheated steam at constant pressure is shown on a temperature-enthalpy diagram as in Fig. 6.10. Assume atmospheric pressure and temperature of ice is below 0°C before supplying the heat.

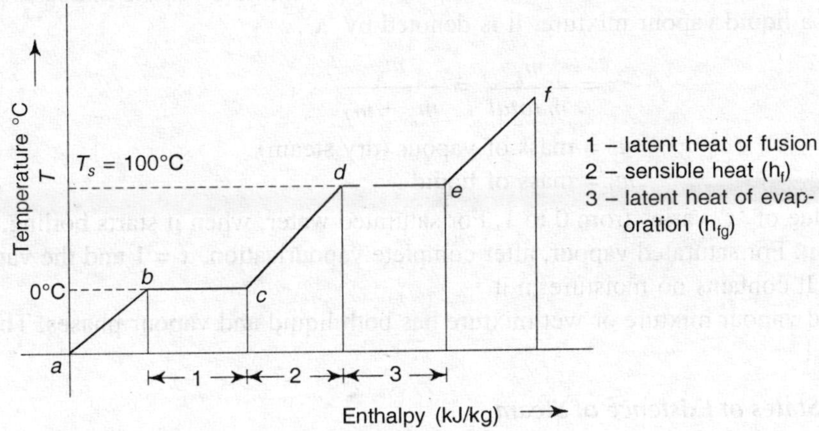

Fig. 6.10 Phase transformation

ab: Heat addition at constant pressure, during which temperature of ice increases till it reaches 0°C.

bc: Heat addition causes melting of ice at 0°C, and is continued till all the ice gets melted. The heat added during this process is known as 'latent heat of fusion' or enthalpy of fusion. (The heat required to convert one kg of ice in to 1 kg water).

cd: Further heat addition increases water temperature to reach 100°C. The amount of heat absorbed by water to heat it from freezing point to boiling point (say from 0°C to T_s = 100°C) is known as sensible heat or total heat of water.

de: The heat addition after point '*d*', is used to vapourise the water without increasing its temperature. The temperature at which water starts boiling is known as saturation temperature (T_s) of water. The heat required to vapourise unit mass of water at saturation temperature 'T_s' to unit mass of dry saturated steam is known as latent heat of evaporation or enthalpy of evaporation (internal latent heat). The point '*e*' represents the saturated condition of steam.

***ef*:** If the heating of dry steam is continued, temperature starts increasing and after point '*e*', the steam condition is known as "superheated steam". The '*ef*' represents heating of dry steam above its saturation temperature 'T_s' and is known as superheating. 'T_{sup}' is the superheat temperature of the steam, above the saturation temperature.

The quantity of heat required to increase the temperature from T_s to T_{sup} at constant pressure is known as superheat and (T_{sup} – T_s) is "degree of superheat".

Effect of Pressure on Saturation Temperature

The boiling temperature of water increases with increasing pressure. The boiling temperature of water at a particular pressure is known as saturation temperature. It is the temperature at which water starts boiling. The corresponding pressure is known as "saturation pressure".

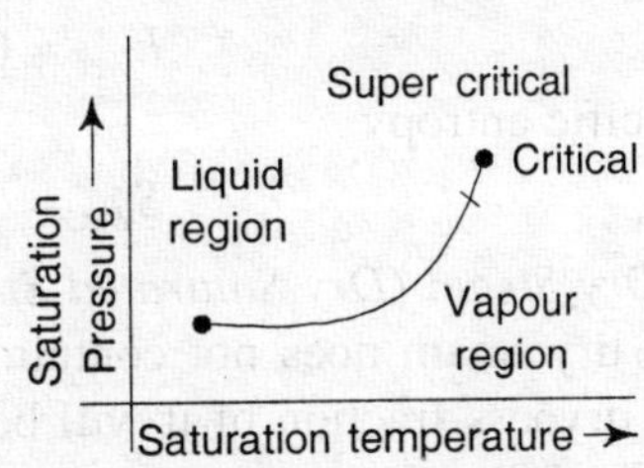

Fig. 6.11 Effect of pressure on saturation temperature

6.11 DRYNESS FRACTION (QUALITY OF STEAM)

Dryness fraction is the basic parameter used to know the state of substance in liquid-vapour mixture region (wet region). It is defined as the ratio of mass of pure vapour and total mass of the substance. Suppose 'x' kg is the mass of vapour in 1 kg of liquid-vapour mixture, then $(1 - x)$ kg be the mass of liquid. 'x' is known as the quality of liquid vapour mixture which indicates the mass fraction of vapour in a liquid vapour mixture. It is denoted by 'x'.

$$x = \frac{m_g}{m\ total} = \frac{m_g}{m_g + m_f}$$

m_g = mass of vapour (dry steam)

m_f = mass of liquid

The value of 'x' varies from 0 to 1. For saturated water, when it starts boiling, $x = 0$. It contains no vapour in it. For saturated vapour, after complete vapourisation, $x = 1$ and the vapour is said to be dry saturated. It contains no moisture in it.

A liquid vapour mixture or wet mixture has both liquid and vapour phases. The value of 'x' ranges from 0 to 1.

Different States of Existence of Steam

The steam can exist in any of the three different states as wet steam, dry steam or superheated steam.

1. *Wet Steam or Two Phase Mixture*

The wet steam contains finely divided water particles suspended in the state of equilibrium. It means both liquid and vapour phases co-exist. The suspended water particles will be at boiling temperature and has not yet absorbed the latent heat and evaporated. A wet steam may have different proportions of water particles and dry steam. The quality of wet steam is expressed in terms of dryness fraction 'x'.

For a given saturation temperature and pressure, we can write the following relations for the wet steam.

Specific volume;

$$V_{\text{wet}} = (1 - x)\, V_f + x\, V_g$$

$$= V_f - x\, V_f + x\, V_g = V_f + x\,(V_g - V_f)$$

$$V_{\text{wet}} = V_f + x\, V_{fg} \qquad (\because V_g - V_f = V_{fg})$$

$$V_{\text{wet}} \approx x\, V_g \text{ for } V_f << V_g$$

Specific internal energy

$$u_{\text{wet}} = (1 - x)\, u_f + x\, u_g$$

$$u_{\text{wet}} = u_f + x\, u_{fg}$$

Specific enthalpy

$$h_{\text{wet}} = (1 - x)\, h_f + x\, h_g = h_f + x\, h_{fg}$$

Specific entropy

$$S_{\text{wet}} = (1 - x)\, S_f + x\, S_g = S_f + x\, S_{fg}$$

2. *Dry Steam (Dry Saturated Steam)*

The dry steam does not contain any water particles suspended in it at the saturation temperature T_s and dryness fraction of it will be unity.

The following expressions are used for dry steam when the saturation pressure and temperature are fixed

Specific volume; $V_g = V_f + V_{fg}$
Specific internal energy; $u_g = u_f + u_{fg}$
Specific enthalpy; $h_g = h_f + h_{fg}$
Specific entropy; $S_g = S_f + S_{fg}$

h_f = Sensible heat
h_{fg} = Latent heat of evaporation

V_g – Specific volume of dry steam – m³/kg
V_f – Specific volume of saturated water – m³/kg
V_{fg} – Change in specific volume during evaporation – m³/kg
u – Specific internal energy – kJ/kg
h – Specific enthalpy – kJ/kg
h_g, h_f, h_{fg} – Specific enthalpy of saturated steam, saturated water, and latent heat of evaporation of saturated water in kJ/kg, respectively.
S_g, S_f, S_{fg} – Specific entropy of saturated steam, saturated water and entropy change during evaporation in kJ/kgK respectively.

$$u_{fg} = h_{fg} + Pv_{fg}$$

$$S_{fg} = \frac{h_{fg}}{T}$$

3. *Vapour Phase or Superheated Steam*

When dry saturated steam is heated at constant pressure beyond its saturation condition, the temperature of steam increases and becomes superheated steam. The corresponding temperature of steam is known as superheated temperature and is denoted by T_{sup}. The superheated steam behaves like a perfect gas.

Specific volume;

From Charle's law, between dry saturated and superheated conditions, we can write,

$$\frac{V_g}{T_{sat}} = \frac{V_{sup}}{T_{sup}}$$

$$\therefore \quad V_{sup} = V_g \left(\frac{T_{sup}}{T_{sat}}\right) \text{ m}^3/\text{kg}$$

Total enthalpy or Total heat of superheated steam

H_{sup} = heat of dry steam + amount of superheat at same constant pressure kJ/kg

$$H_{sup} = H_g + C_{ps}(T_{sup} - T_{sat})$$
$$= H_f + H_{fg} + C_{ps}(T_{sup} - T_{sat})$$

H_g – enthalpy of dry steam
C_{ps} – specific heat of superheated steam
T_{sup} – superheated temperature
T_{sat} – saturation temperature

6.12 METHODS OF DETERMINATION OF DRYNESS FRACTION OF STEAM

The quality of steam may be determined by using different types of calorimeters. They are:

1. Barrel or tank calorimeter
2. Separating calorimeter
3. Throttling calorimeter

4. Separating and throttling calorimeter
5. Electrical calorimeter

1. Barrel or Tank Calorimeter

In this calorimeter, a known mass of steam is made to pass through the water contained in the bucket calorimeter where the steam is completely condensed. The heat lost by steam is equated to the heat gained by water. The mass of calorimeter, the mass of calorimeter with water before and after mixing the steam are obtained separately. The temperature of water before and after mixing the steam are recorded by using mercury thermometer. The pressure gauge gives pressure of the steam. In this case losses due to condensation and radiation are neglected, hence the value of dryness fraction measured is not accurate.

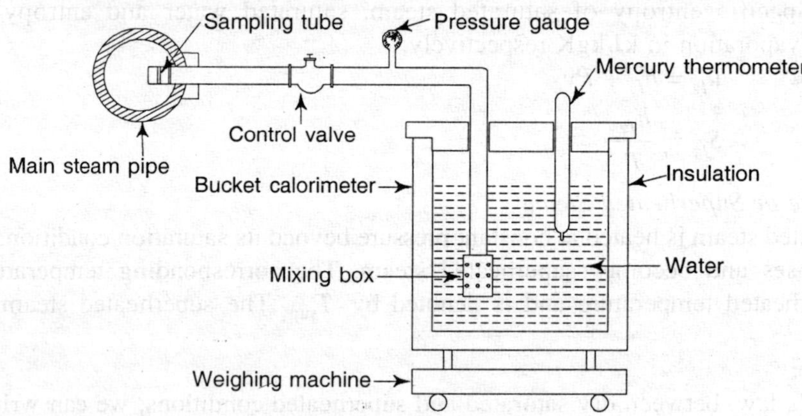

Fig. 6.12 Tank calorimeter

Let T_s – saturation temperature
P_s – gauge pressure
P_{abs} – absolute pressure of steam $-P_{atm} + P_s$
m_c – mass of calorimeter alone
m_1 – mass of calorimeter with water
m_2 – mass of calorimeter with water and condensed steam
m_s – mass of steam condensed
h_{fg} – latent heat of evaporation of steam at P_{abs}
T_1, T_2 – temperature of water before and after mixing the steam
x – dryness fraction of steam

By making energy balance

Heat lost by steam = Heat gained by water and calorimeter (By neglecting losses)

$$m_s\,[x\,h_{fg} + C_{ps}\,(T_s - T_2)] = (m_1 - m_2)\,(T_2 - T_1)\,C_{pW} + m_c\,C_{pc}\,(T_2 - T_1)$$

$$\therefore\ m_s\,[x\,h_{fg} + C_{ps}\,(T_s - T_2)] = (m_w\,C_{pW} + m_c\,C_{pc})\,(T_2 - T_1)$$

where $(m_1 - m_2) = m_w$ = mass of water before mixing

C_{pc}, C_{pW}, C_{ps} – specific heat of calorimeter, water and steam respectively.

$$\therefore \qquad x = \frac{(m_w\,C_{pW} + m_c\,C_{pc})\,(T_2 - T_1) - m_s\,C_{ps}\,(T_s - T_2)}{m_s\,h_{fg}}$$

2. Separating Calorimeter

This calorimeter is used when the steam is very wet. In this type, a known mass of wet steam is taken through sampling tube. The moisture is separated mechanically from the steam. The steam is made to pass through perforated trays or tubes where the water separates due to inertia of droplets. Then the outgoing steam is condensed in the bucket calorimeter and gets condensed as explained earlier.

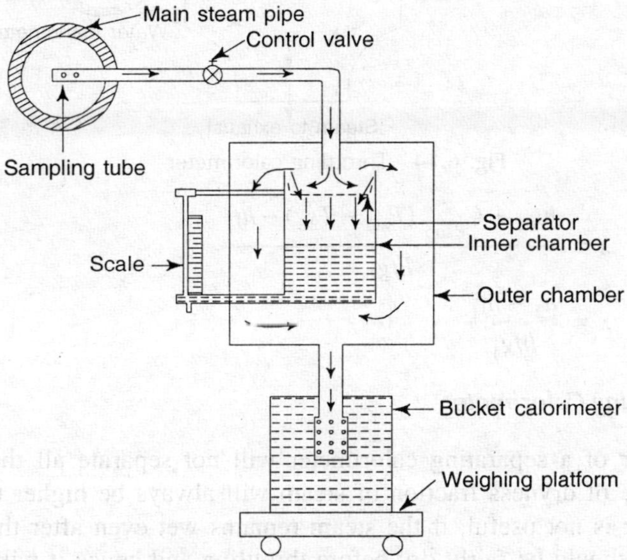

Fig. 6.13 Separating Calorimeter

Let m_W = mass of water separated

m_s = mass of steam condensed

$$\therefore \quad x = \frac{m_s}{m_w + m_s}$$

This formula is used with an assumption that all the moisture contents are removed in separator and only dry steam enters bucket calorimeter. Hence the value of 'x' obtained with this equipment is not accurate.

3. Throttling Calorimeter

This calorimeter is used to measure the quality of steam, when it is nearly dry. In this type, steam from the sampling tube is passed through a throttle valve which is a narrow or restricted passage. During throttling, steam will expand from high pressure to low pressure without doing any work. The enthalpy of steam remain constant before and after throttling and steam reaches superheated state after throttling. The pressure and temperature of steam after throttling are measured by using a manometer and mercury in glass thermometer.

Consider wet mixture being at state 1, after throttling it gets superheated and attains a new state 2.

Enthalpy 'h_1' before throttling = Enthalpy 'h_2' after throttling

$$\therefore \quad h_1 = h_2$$

$$hf_1 + x\, hfg_1 = hg_2 + C_{psup}\,(T_{sup} - Ts_2)$$

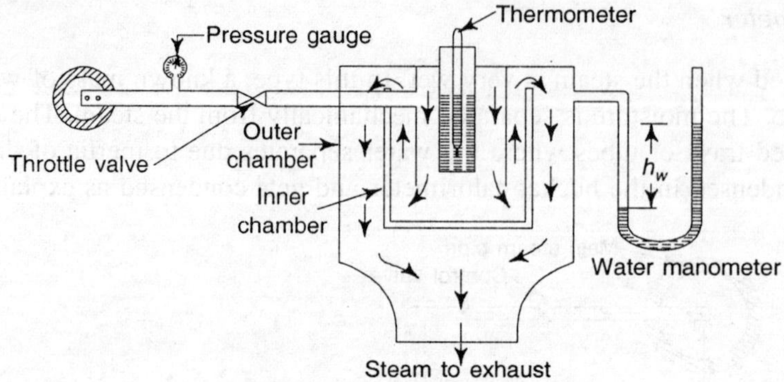

Fig. 6.14 Throttling calorimeter

$$\therefore \qquad x = \frac{hg_2 + C_{p_{sup}}\,(T_{sup} - Ts_2) - hf_1}{hfg_1}$$

$$x = \frac{h_2 - hf_1}{hfg_1}$$

4. Separating and Throttling Calorimeter

The mechanical separator of a separating calorimeter will not separate all the water particles and hence the measured value of dryness fraction of steam will always be higher than the actual value. The throttling calorimeter is not useful, if the steam remains wet even after throttling. For accurate determination, the steam should be fairly dry before throttling and hence it will be superheated after throttling. A combined separating and throttling calorimeter overcomes these difficulties and measures dryness fraction of too wet steam (x = 0.6 to 0.9) accurately.

The arrangement of the combined type is as shown in Fig 6.15. The steam from the sampling tube is first passes through the separating calorimeter where it is dried up to a certain extent. The steam which is less wet, then passes through the throttling calorimeter, where it expands in the throttle valve and comes out in the superheated form. Then the steam condenses in a condenser and condensate is collected in a collector. The mass of condensate, mass of water separated in the separating calorimeter and temperature and pressure of the final superheated steam are recorded.

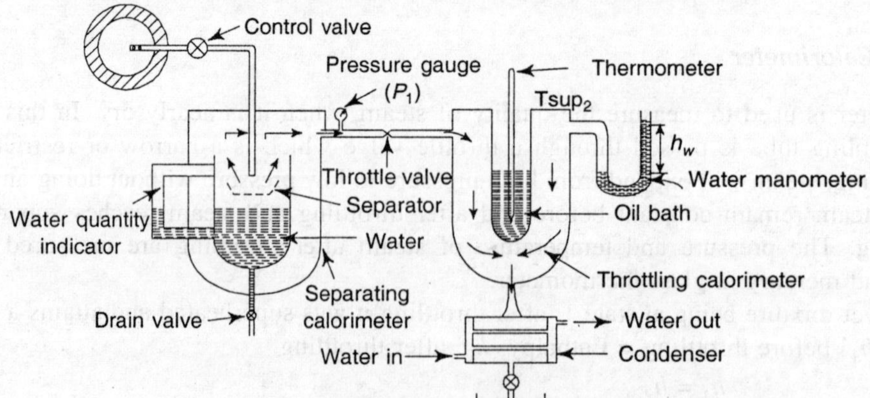

Fig. 6.15 Separating and Throttling calorimeter

Let m_c – mass of condensate from condenser

m_w – mass of water separated in the separating calorimeter

m – mass of steam sample = $m_c + m_w$

m_d – mass of dry steam which comes out from separating calorimeter

$= x_2 \times m_c$

x – actual dryness fraction of steam sample

x_1 – $\dfrac{m_c}{m_c + m_w}$ apparent dryness fraction of steam assuming the steam from separating calorimeter is completely dry

x_2 – actual dryness fraction of steam comes out from separating calorimeter

Amount of water initially in the steam

$$= (1 - x)\,(m_c + m_w) \qquad (1)$$

Amount of water separated in separating calorimeter

$$= (1 - x_1)\,(m_c + m_w) \qquad (2)$$

Amount of water carried by steam into the throttling calorimeter

$$= (1 - x_2)\,m_c \qquad (3)$$

∴ Amount of water initially in the steam = Amount of water separated in separating calorimeter + Amount of water carried by steam

$$\therefore \quad (1) = (2) + (3)$$

$$(1 - x)\,(m_c + m_w) = (1 - x_1)\,(m_c + m_w) + (1 - x_2)\,m_c$$

$$\therefore \quad (1 - x) = (1 - x_1) + (1 - x_2)\,\frac{m_c}{m_c + m_w}$$

But $$x_1 = \frac{m_c}{m_c + m_w}$$

$$\therefore \quad (1 - x) = (1 - x_1) + (1 - x_2)\,x_1$$

$$\cancel{1} - x = \cancel{1} - \cancel{x_1} + \cancel{x_1} - x_1 x_2$$

$$\therefore \quad \boxed{x = x_1 x_2}$$

5. Electrical Calorimeter

The principle of this calorimeter is similar to that of a throttling calorimeter. In this type, the wet steam is superheated by electrical heating and not by throttling. The wet steam is first made to pass through an electric heater and gets superheated. The pressure and temperature of superheated steam are recorded. The heat added in a heater is determined by using the relation

$$Q = VI$$

Let m_1 – mass of steam sample taken in 't' secs.

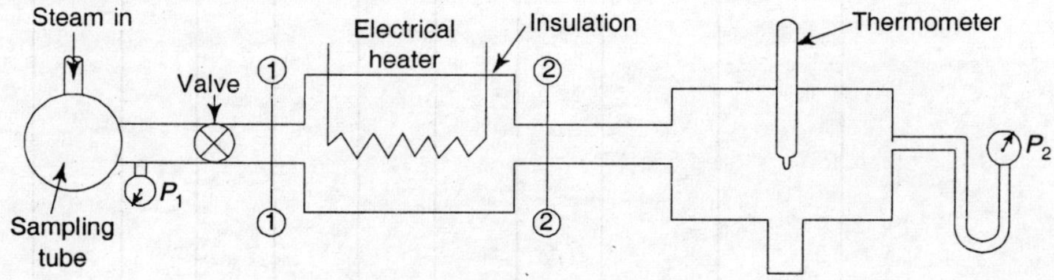

Fig. 6.16 Electrical calorimeter

h_1 & h_2 – enthalpies of steam before and after throttling

From first law at sections 1-1 and 2-2, we can write

$$Q = m_1 (h_2 - h_1) = m_1 [h_{sup} - (hf_1 + x\, hfg_1)]$$

$$\therefore \quad x = \frac{h_{sup} - (h_2 - h_1) - hf_1}{hfg_1}$$

$$x = \frac{h_{sup} - \left(\frac{Q}{m_1}\right) - hf_1}{hfg_1}$$

6.13 STEAM TABLES

Steam being pure substance has its unique and constant properties at different pressures and temperatures. The steam generation at different pressures has been studied experimentally and steam properties at different conditions have been estimated and tabulated for future uses. These are tabulated either on the temperature basis or pressure basis. Steam table is a tabular presentation of properties like specific enthalpy, entropy, internal energy and specific volume at different saturation pressures and temperatures.

Table 6.1 Pressure based steam table (for saturated water and dry saturated steam)

Pressure bar	Temperature T_s°C	Specific Volume m³/kg		Enthalpy kJ/kg			Entropy kJ/kgK			Internal energy kJ/kg		
P	T_s	V_f	V_g	h_f	h_{fg}	h_g	s_f	s_{fg}	s_g	u_f	u_{fg}	u_g

Table 6.2 Temperature based steam table (for saturated water and dry saturated steam)

Temperature °C	Pressure bar	Specific volume m³/kg		Enthalpy kJ/kg			Entropy kJ/kgK			Internal energy kJ/kg		
T_s	P	V_f	V_g	h_f	h_{fg}	h_g	s_f	s_{fg}	s_g	u_f	u_{fg}	u_g

6.14 EXTERNAL WORK OF EVAPORATION

At a given pressure, saturated water absorbs latent heat and evaporates into dry steam. This evaporation results in large increase in the volume of water from V_f to V_g. Thus the latent heat absorbed, not only causes phase change, but also do some external work due to large increase in volume, and piston inside the cylinder moves up.

External work of evaporation is the amount of external work due to large increase in volume when water changes into steam.

For wet steam;

External work = $x \cdot PV_g$ kJ/kg

for dry saturated steam;

External work = PV_g kJ/kg

for superheated steam;

External work = PV_{sup} kJ/kg

6.15 INTERNAL ENERGY OF STEAM

It is the actual heat energy stored in the steam. We know that, the latent heat of steam is equal to internal latent heat plus external work of evaporation, the total heat or enthalpy of dry saturated steam consists of sensible heat, internal latent heat and external work of evaporation. The amount of heat corresponds to external work of evaporation is not stored in the steam in the form of energy and is utilised in doing external work.

Thus the internal energy of steam consists of sensible heat and internal latent heat. Therefore, internal energy of steam is defined as the difference between the enthalpy of steam and external work of evaporation. It is denoted by 'u' and is expressed in kJ/kg.

For the wet steam

Internal energy = $u_{wet} = h_f + x\,h_{fg} - x \cdot PV_g$ kJ/kg

For dry steam; $u_g = h_g + PV_g$ kJ/kg

For superheated steam;

$u_{sup} = h_{sup} - PV_{sup}$ kJ/kg

6.16 MOLLIER DIAGRAM

It is the h-s diagram, i.e. enthalpy-entropy diagram for steam. Fig. 6.17 shows phase equilibrium diagram of a pure substance for all phases on the h-s co-ordinates, showing the saturated solid line, saturated liquid lines and saturated vapour line, the various phases and the transition zones.

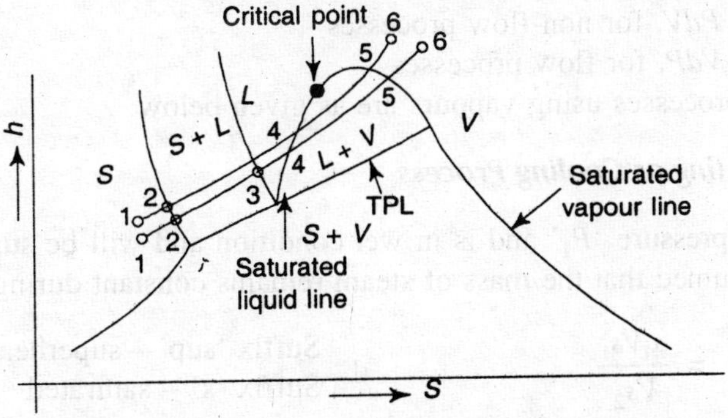

Fig. 6.17 Phase equilibrium diagram (h-s diagram)

The Fig. 6.18 is the Mollier diagram showing only the liquid and vapour phases. The saturation temperature increases with increase in pressure and hence increases slope of the isobar. The constant pressure lines diverge from one another, and the critical isobar is a tangent at the critical point. Different significant lines such as saturated liquid line, saturated vapour line, isobaric line, isothermal lines, constant specific volume lines, constant dryness fraction lines are shown on the Mollier diagram.

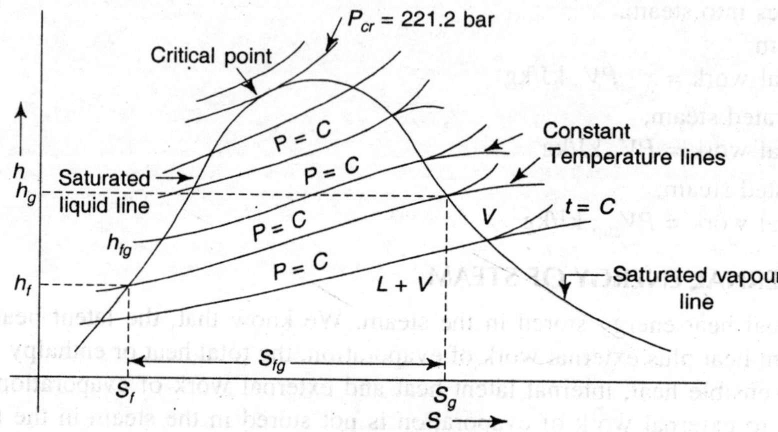

Fig. 6.18 *h-s diagram for water*

6.17 PROCESSES EXECUTED BY PURE SUBSTANCES

When a system involving pure substance execute a change of state, has to satisfy, conditions of two property rule to represent the process on diagrams. The pure substance must be in equilibrium and effects due to gravity, motion, capillarity, electricity and magnetism must be absent. When these conditions are satisfied, a given point on the diagram represents a state of the system and all the properties of system can be determined at that point (state).

To represent a process on the diagram, the path of change of state is the succession of states passed through by the system. If all these states are in equilibrium, they may be represented by a point on the diagram. All these states merge continuously, one into the next, the corresponding points form a continuous line joining initial and final state to represent a process.

The ideal gas equation $PV = mRT$ is not used for vapours, but basic equations derived from Ist and IInd law of thermodynamic can be used for gases as well as vapours.

For vapours ρ, we use the basic equations,

$$\delta Q = du + PdV, \text{ for non-flow processes}$$

$$\delta Q = dh - VdP, \text{ for flow processes}$$

The different non-flow processes using vapours are as given below.

1. Constant Volume Heating or Cooling Process

The steam is initially at pressure 'P_1' and is in wet condition and will be superheated and reaches a pressure of 'P_2'. It is assumed that the mass of steam remains constant during superheating of steam.

$$\frac{V_{\text{sup}\,2}}{Vs_2} = \frac{T_{\text{sup}\,2}}{Ts_2} = \frac{x_1 Vs_1}{Vs_2}$$ Suffix 'sup' – superheatd

Suffix 's' – saturated

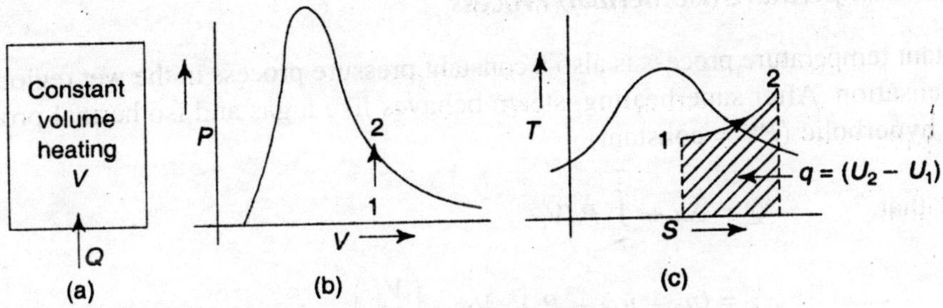

Fig. 6.19 Constant volume heating

After heating, if the steam remains wet, 'x_2' is the final dryness fraction and is given by

$$x_2 Vs_2 = x\, Vs_1$$

$$x_2 = x_1 \cdot \frac{Vs_1}{Vs_2}$$

From I[st] law of Thermodynamic for non-flow process

$$Q = \Delta u + \int_1^2 Pdv\ ; \quad \int_1^2 Pdv = 0 \text{ for constant volume process}$$

$$= \Delta u$$

$$= u_2 - u_1$$

$$Q = (h_2 - P_2\, V_{sup2}) - (h_1 - P_1 x_1\, Vs_1)$$

If the steam is wet after heating, then

$$Q = (h_2 - P_2\, x_2\, Vs_2) - (h_1 - P_1\, x_1\, Vs_1)$$

The analysis can also be applied for cooling process.

2. *Constant Pressure Heating or Cooling*

The wet steam is heated to superheated condition similar to generation of steam in boilers. From I[st] law for non-flow process

$$Q = \Delta u + \int_1^2 PdV = (u_2 - u_1) + (V_2 - V_1)\, P$$

$$Q = (u_2 + P_2 V_2) - (u_1 + P_1 V_1)$$

$$Q = (h_2 - h_1)$$

Thus for a constant pressure heating or cooling process, heat added or removed is the change in enthalpy between the given conditions.

The work done $\quad W = P(V_2 - V_1)$

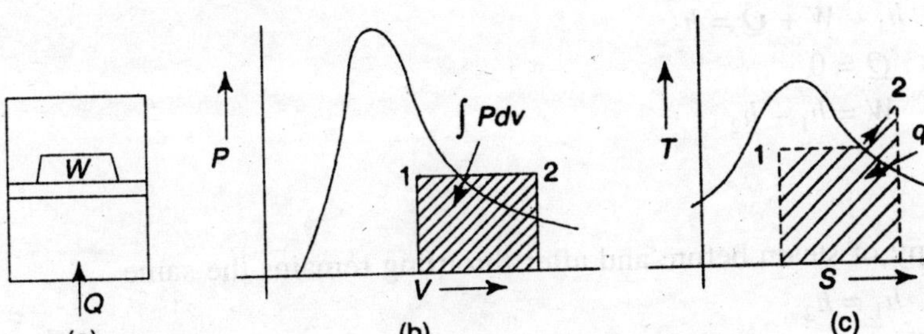

Fig. 6.20 Constant pressure process

3. Constant Temperature (Isothermal) Process

The constant temperature process is also a constant pressure process in the wet region during evaporation and condensation. After superheating, steam behaves like a gas and isothermal process in superheated region is hyperbolic (PV = constant)

We know that, $$Q = \Delta u + \int_1^2 PdV$$

$$= (u_2 - u_1) + P_1V_1 \log_e\left(\frac{V_2}{V_1}\right)$$

$$= (h_2 - P_2V_2) - (h_1 - P_1V_1) + P_1V_1 \log_e\left(\frac{V_2}{V_1}\right)$$

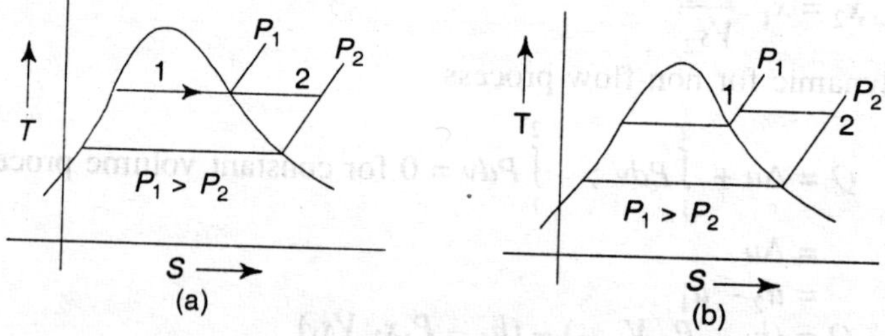

Fig. 6.21 Hyperbolic expansion

4. Isentropic or Reversible Adiabatic Expansion

During isentropic expansion 1-2, entropy remains constant.

$S_1 = S_2$, during process 1-2 (Ideal)

This condition is used to find 'x_2', i.e. quality of steam after expansion

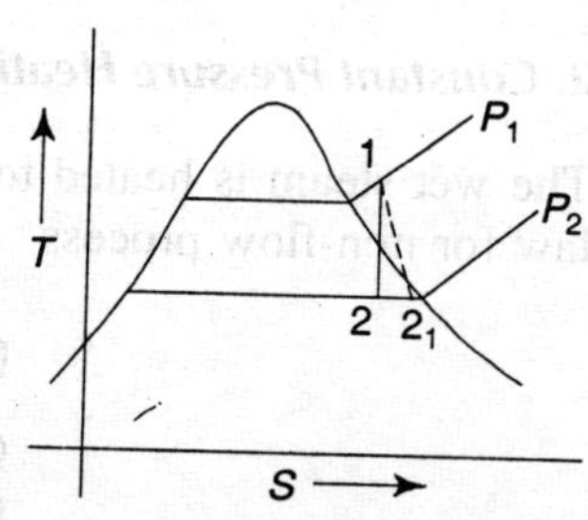

Fig. 6.22

We know that, $$Q = \Delta u + \int_1^2 PdV$$

$$Q = 0, \text{ (process is adiabatic)}$$

$$\therefore \quad 0 = \Delta U + W$$

$$W = (u_1 - u_2)$$

By applying SFEE for reversible adiabatic process

$$u_1 + P_1V_1 - W + Q = u_2 + P_2V_2$$

$$h_1 - W + Q = h_2$$

$$Q = 0$$

$$W = h_1 - h_2$$

5. Throttling Process

We know that, enthalpy of steam before and after throttling remains the same

$$h_1 = h_2$$

If state '2' is in superheated region after throttling

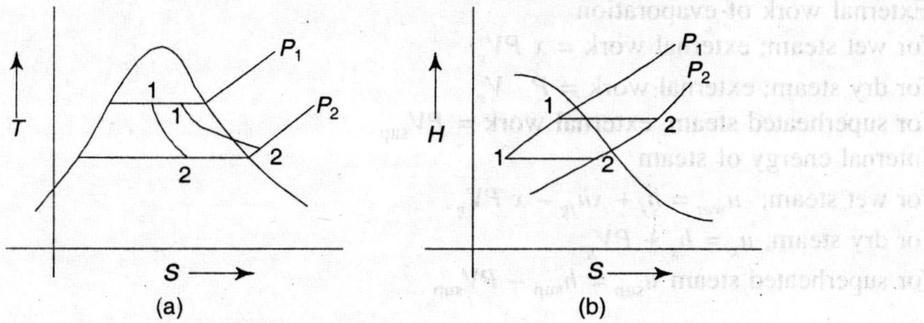

(a) (b)

Fig. 6.23

$$hf_1 + x_1\, hfg_1 = hf_2 + hfg_2 + C_{ps}\,(T_{sup} - T_s)$$

If the steam is wet after throttling

$$hf_1 + x_1\, hfg_1 = hf_2 + x_2\, hfg_2$$

6.18 FORMULAE USED IN SOLVING PROBLEMS

(1) Two property rule $= P = \dfrac{a+bt}{V}$

(2) Internal energy $\Delta u = \dfrac{Q-W}{m}$

(3) Enthalpy $\Delta h = h_2 - h_1 = (u_2 - u_1) + (P_2V_2 - P_1V_1)$

$$\Delta h = \Delta u + W$$

(4) $C_v = \left(\dfrac{\partial u}{\partial t}\right)_v$ and $C_p = \left(\dfrac{\partial h}{\partial t}\right)_p$

(5) $x = \dfrac{m_g}{m_{total}} = \dfrac{m_g}{m_g + m_f}$

(6) For wet steam

$$V_{wet} = V_f + x\, V_{fg} \approx x\, V_g$$
$$u_{wet} = u_f + x\, u_{fg}$$
$$h_{wet} = h_f + x\, h_{fg}$$
$$s_{wet} = s_f + x\, s_{fg}$$

(7) For dry steam

$$V_g = V_f + V_{fg}$$
$$u_g = u_f + u_{fg}$$
$$h_g = h_f + h_{fg}$$
$$s_g = s_f + s_{fg}$$

(8) For superheated steam

$$V_{sup} = V_g \left(\frac{T_{sup}}{T_{sat}}\right)$$

$$H_{sup} = H_g + C_{ps}\,(T_{sup} - T_{sat})$$

(9) External work of evaporation
for wet steam; external work = $x\,PV_g$
for dry steam; external work = $P \cdot V_g$
for superheated steam, external work = PV_{sup}

(10) Internal energy of steam
for wet steam; $u_{wet} = h_f + xh_{fg} - x\,PV_g$
for dry steam, $u_g = h_g + PV_g$
for superheated steam $u_{sup} = h_{sup} - PV_{sup}$

6.19 NUMERICAL PROBLEMS

Problem 1: *A fluid system consists of 2.5 kg of a pure substance with an initial pressure of 700×10^3 N/m^2 and temperature 200°C, the corresponding value of specific volume is 0.2 m^3/kg. During a process, the system reaches final pressure and specific volume of 700×10^3 N/m^2 and 0.2 m^3/kg, where pressure and specific volume are independent properties.*

(i) Determine final temperature of substance and the increase in its specific internal energy.

(ii) During the process, the work done by the system is 12,000 Nm. Determine magnitude and direction of the heat transfer.

Solution : Given data:

$$m = 2.5 \text{ kg}, P_1 = 700 \times 10^3 \text{ N/m}^2, t_1 = 200°\text{C}$$
$$V_1 = 0.2 \text{ m}^3/\text{kg}, P_2 = 700 \times 10^3 \text{ N/m}^2, V_2 = 0.2 \text{ m}^3/\text{kg}$$

(i) Since

$$P_1 = P_2 = 700 \times 10^3 \text{ N/m}^2$$
$$V_1 = V_2 = 0.2 \text{ m}^3/\text{kg} \qquad (PV = mRT)$$
$$\boxed{t_1 = t_2 = 200°\text{C}} \qquad \textit{Ans.}$$
$$\left[\frac{P_1 V_1}{mRT_1} = \frac{P_2 V_2}{mRT_2}\right] \qquad [\therefore\; t_1 = t_2]$$

(ii) W.D. = 12,000 Nm

$$\frac{Q - W}{m} = du = 0$$
$$Q - W = 0$$
$$\therefore \quad \boxed{Q = W = 12{,}000 \text{ Nm} = 12 \text{ kJ}} \qquad \textit{Ans.}$$

Problem 2: *A mass of 0.3 kg of a pure substance at a pressure of 90×10^3 N/m^2 and a temperature 40°C occupies a volume of 120 cm^3. If the internal energy of the substance is 30.4×10^3 J, determine the specific enthalpy of the substance.*

Given: *$m = 0.3$ kg, $P = 90 \times 10^3$ N/m^2, $T = 40°$C,*
$V = 120 \times 10^{-3}$ m$^3_1$ $U = 30.4 \times 10^3$ J, h = ?

Solution :

$$u = \frac{U}{m} = \frac{30.4 \times 10^3}{0.3} = 101.33 \times 10^3 \text{ J}$$

$$V = \frac{120 \times 10^{-3}}{0.3} = 400 \times 10^{-3} \text{ m}^3/\text{kg}$$

$$h = u + PV = 101.33 \times 10^3 + (90 \times 10^3 \times 400 \times 10^{-3})$$

$$h = 137330 \text{ J/kg} = 137.33 \text{ kJ/kg}$$

$\therefore$ $\boxed{h = 137.33 \text{ kJ/kg}}$ *Ans.*

Problem 3: *A piston-cylinder arrangement contains pure substance. It undergoes a fully resisted, constant pressure process in which the only work done is the displacement work. Show that the heat transfer during the process is equal to the increase in enthalpy of the substance.*

Solution :

We know that, $Q = u + W$

$$dQ = du + dW$$

But $dW = Pdv$ for constant pressure process

$$h = u + PV$$

$$dh = du + Pdv + vdp$$

But $du = dQ - dW$

$\therefore$ $dh = dQ - dW + Pdv + vdp$

$$dh = dQ - \cancel{Pdv} + \cancel{Pdv} + 0$$

$\boxed{dh = dQ}$ hence proved.

Problem 4: *In a cylinder-piston arrangement containing pure substance, the paddle rotation causes the substance to undergo a fully-resisted constant pressure process. Assuming no heat transfer, show that the stirring work done on the substance is equal to the increase in enthalpy of the substance.*

Solution : Pressure is constant and heat transfer is zero.

$$dQ = du + dw$$

$$0 = du + dw$$

$$du = -dw = -(dwd + dws)$$

dwd = displacement work

dws = stirring work which is done on the system

We know that, $h = u + PV$

$$dh = du + Pdv + vdp = du + Pdv + 0$$

$$dh = -dwd + dws + Pdv = -\cancel{Pdv} + dws + \cancel{Pdv}$$

$\boxed{dh = dws}$ *Ans.*

Problem 5: *The properties of oxygen gas are expressed by the relation*

$$PV = 260\,t + 71 \times 10^3$$

$$t = 1.52\,u - 273$$

where P is in N/m², V – m³/kg, t –°C and u – kJ/kg

(i) Determine specific heat at constant volume and constant pressure in J/kgK

(ii) Show that, for any process $\Delta u = C_v\,\Delta t$ and $\Delta h = C_p\,\Delta t$ per unit mass of oxygen.

Solution :

(i) We know that, $C_v = \left(\dfrac{\partial u}{\partial t}\right)_v$

Given: $t = 1.52u - 273$

$$u = \frac{t + 273}{1.52}$$

$$\frac{\partial u}{\partial t} = \frac{1}{1.52}[1 + 0] \quad = C_v = 0.657 \text{ kJ/kgK}$$

We know that, $C_p = \left(\frac{\partial h}{\partial t}\right)_p$

also, $$h = u + PV = \frac{t + 273}{1.52} + \frac{260t + 71 \times 10^3}{1000}$$

$$\therefore \quad C_p = \frac{\partial h}{\partial t} = \frac{1}{1.52}(1 + 0) + \frac{260}{1000} + 0 = \frac{1}{1.52} + \frac{260}{1000}$$

$$C_p = 0.917 \text{ kJ/kg°C}$$

$$\therefore \quad \boxed{C_v = 0.657 \text{ kJ/kg°C}} \ \& \ \boxed{C_p = 0.917 \text{ kJ/kg°C}}$$ *Ans.*

(ii) $\therefore \ \Delta u = C_v \Delta t$

But $u = \dfrac{t + 273}{1.52}$ 'u' is a function of 't' only

$$C_v = \frac{\Delta u}{\Delta t}$$

$$\therefore \quad \Delta u = C_v \, \Delta t$$

also we know that,

$$C_p = \frac{\Delta h}{\Delta t}$$

'h' is a function of t

$$\therefore \quad \Delta h = C_p \cdot \Delta t$$

Problem 6: *In the above system (mentioned in problem 5), unit mass of oxygen at a pressure of 600 × 10³ N/m² and temperature of 280°C, contained in a rigid vessel. Heat transfer to the oxygen increases its enthalpy by 28 kJ/kg. Determine final temperature, heat and final pressure.*

Given: $P_1 = 600 \times 10^3 \text{ N/m}^2$, $t_1 = 280°C$, $dh = 28$ kJ/kg

Solution : We know that, $h_1 = u_1 + P_1V_1$

$$= \frac{t_1 + 273}{1.52} + \frac{260t_1 + 71 \times 10^3}{1000} \text{ kJ/kg}$$

$$h_1 = \frac{280 + 273}{1.52} + \frac{260 \times 280 + 71 \times 10^3}{1000} = 507.62 \text{ kJ/kg}$$

$$h_2 = u_2 + P_2V_2 = \frac{t_2 + 273}{1.52} + \frac{t_2(260) + 71 \times 10^3}{1000} \text{ kJ/kg}$$

Given: $$h_2 - h_1 = 28 \text{ kJ/kg}$$

$\therefore \qquad h_2 = 507.62 + 28 = 535.62$ kJ/kg

$\therefore \qquad 535.62 = 0.657\ t_2 + 179.6 + 0.26\ t_2 + 71$

$285.02 = 0.917\ t_2$

$\boxed{t_2 = 310.81\ °C}$ **Ans.**

also, $$\frac{Q - W}{m} = u_2 - u_1 \qquad \text{(from first law)}$$

$$w = 0, = \frac{310.81 + 273}{1.52} - \frac{280 + 273}{1.52}$$

$\boxed{Q = 20.27 \text{ kJ/kg}}$ **Ans.**

$V_1 = V_2$

Given: $P_1V_1 = 260\ t_1 + 71 \times 10^3$

$P_2V_2 = 260\ t_2 + 71 \times 10^3$

$$\therefore \qquad \frac{P_2 \not{V_2}}{P_1 \not{V_1}} = \frac{260 \times 310.81 + 71 \times 10^3}{260 \times 280 + 71 \times 10^3}$$

$P_2 = 1.055 \times P_1 = 1.055 \times 600 \times 10^3$

$\boxed{P_2 = 633.42 \times 10^3 \text{ N/m}^2}$ **Ans.**

Problem 7: *The properties of a certain gas are related by $PV = 310\ (t + 273)$ and*

$u = u_0 + 0.84\ t$

where P – N/m^2, V – m^3/kg, t – °C, u – kJ/kg

A cylinder-piston machine contains 0.02 m^3 of this gas at a pressure of 330 $\times$ 10^3 N/m^2 and a temperature of 85°C. The work done by the gas is 2900 Nm, during expansion of gas and heat transfer from the gas is 1.9 $\times$ 10^3 J.

(i) Determine the temperature of gas after expansion.

(ii) For the adiabatic process between the same end states, determine work done by the gas.

Given: $PV = 310\ (t + 273)$ and

$u = u_0 + 0.84\ t$

(i) $Q = W + du$

$$\frac{Q - W}{m} = du = u - u_0$$

$$\therefore \qquad u - u_0 = \frac{-1.9 \times 10^3 - 2900}{m} = \frac{-4800}{m} \text{ J/kg}$$

But given, $P_0V_0 = 310\ (t + 273)$

$$V_0 = \frac{310\,(85 + 273)}{330 \times 10^3} = 0.336 \text{ m}^3\text{/kg}$$

$$m = \frac{V}{v} = \frac{0.02}{0.336} = 0.0595 \text{ kg}$$

$$\therefore \qquad u - u_0 = \frac{-4800}{0.0595} = -80.67 \text{ kJ/kg}$$

$$u = u_0 + 0.84\ t$$

$\therefore$ $$u - u_0 = 0.84\ t$$

$\therefore$ $$0.84\ t = -80.67$$

$$\boxed{t = -96.03\ °C}$$

$\therefore$ The rise in temperature

$$= 85 - 96.03 = \boxed{-11.03\ °C}$$ **Ans.**

(ii) For an adiabatic process, $Q = 0$

$$u - u_0 = 0.84\ t = 0.84\ (-96.03) = -80.66\ \text{kJ/kg}$$

$$\frac{Q - W}{m} = u - u_0 = -80.66$$

$\therefore$ $$-W = m\ (-80.66)$$

$$= 0.0595\ (-80.66)$$

$$\boxed{W = 4.79\ \text{KNm}}$$ **Ans.**

Problem 8: *Determine specific enthalpy of saturated steam at a pressure of 128 kN/m².*

Solution : The specific enthalpy (h_g) values of saturated steam are available at 1.2 bar and 1.4 bar in the steam table.

$\therefore$ For 120 KN/m² $= \dfrac{120 \times 10^3}{10^5}$ bar = 1.2 bar

At 1.2 bar; $h_g = 2683.4$ kJ/kg (1 N/m² = 10^{-5} bar)

Similarly 140 KN/m² $= \dfrac{140 \times 10^3}{10^5}$ bar = 1.4 bar

At 1.4 bar; $h_g = 2690.3$ kJ/kg

(refer steam table corresponding to properties of saturated water and dry saturated steam)

At 128 KN/m², the value of 'h_g' can be obtained by linear interpolation.

$\therefore$ $$h_g = \frac{128 - 120}{140 - 120} \times (2690.3 - 2683.4) + 2683.4$$

$$\boxed{h_g = 2686.16\ \text{kJ/kg}}$$ **Ans.**

Problem 9: *Determine the values of h, v and u for P = 1.5 × 10⁶ N/m², t = 429 °C.*

Solution : At $P = 1.5 \times 10^6$ N/m², 't' values are available in 400°C and 500°C etc. in the steam table.

$\therefore$ At $P = 1.5 \times 10^6\ \text{N/m}^2 - \dfrac{1.5 \times 10^6}{10^5} = 15$ bar

At 15 bar and $t_{sup} = 400°C$, $h = 3256.6$ kJ/kg

At 15 bar and $t_{sup} = 500°C$. $h = 3472.8$ kJ/kg

(from steam table corresponding to properties of superheated steam)

$\therefore$ At $P = 15$ bar and $t = 429°C$, $h = ?$

$$h = \frac{429 - 400}{500 - 400} \times (3472.8 - 3256.6) + 3256.6$$

$$\boxed{h = 3319.298\ \text{kJ/kg}}$$ **Ans.**

Similarly for '*V*',

At $P = 15$ bar, $t = 400$ °C, $V = 0.20292$ m^3/kg

$P = 15$ bar, $t = 500$ °C, $V = 0.23503$ m^3/kg

At $P = 15$ bar, $t = 429$°C, $V = ?$

$$V = \left(\frac{429 - 400}{500 - 400}\right) \times (0.23503 - 0.20292) + 0.20292$$

$$\boxed{V = 0.21223 \text{ m}^3/\text{kg}}$$ ***Ans.***

$\therefore$ $u = h - PV$

$= 3319.3 \times 10^3 - 1.5 \times 10^6 \times 0.21223$

$$\boxed{u = 3000.95 \text{ kJ/kg}}$$ ***Ans.***

Problem 10: *Determine the value of specific enthalpy at $P = 1.68 \times 10^6$ N/m^2, $t = 429$°C.*

Solution : Determine the values of '*h*' at $t = 429$°C for $P = 1.5 \times 10^6$ N/m^2 and 2.0×10^6 N/m^2

At $t = 429°C,\ P = \dfrac{1.5 \times 10^6}{10^5} = 15$ bar, $h = 3319.29$ kJ/kg (refer previous problem)

At 20 bar, $t = 400$ °C, $h = 3248.7$ kJ/kg

20 bar, $t = 500$ °C, $h = 3467.3$ kJ/kg

At 20 bar & $t = 429$ °C, $h = ?$

$$h = \left(\frac{429 - 400}{500 - 400}\right)(3467.3 - 3248.7) + 3248.7$$

$$\boxed{h = 3312.09 \text{ kJ/kg}}$$ ***Ans.***

At $t = 429$°C, $P = 15$ bar, $h = 3319.29$ kJ/kg

$t = 429$ °C, $P = 20$ bar, $h = 3312.09$ kJ/kg

$t = 429$ °C, $P = \dfrac{1.68 \times 10^6}{10^5} = 16.8$ bar, $h = ?$

$$h = \frac{16.8 - 15}{20 - 15}(3312.09 - 3319.29) + 3319.29$$

$$\boxed{h = 3316.69 \text{ kJ/kg}}$$ ***Ans.***

Problem 11: ***Determine condition of steam, whether it is wet, dry or superheated, when***

(i) steam pressure is 10 bar and has specific volume of 0.175 m³/kg,

(ii) steam pressure is 20 bar and temperature is 230°C,

(iii) steam has a pressure of 20 bar and if 2750 kJ/kg of heat is required to generate the steam from water at 0°C.

Solution :

(i) At $P = 10$ bar, $V_g = 0.19430$ m^3/kg

(from steam table for properties of saturated water and dry saturated steam)

The given specific volume for steam is

$$V = 0.175 \text{ m}^3/\text{kg which is less than } 'V_g'$$

∴ The steam is wet

$$\therefore \text{ Dryness fraction, } x = \frac{\text{The given sp. volume of steam at 10 bar}}{\text{Sp. volume of dry sat. steam at 10 bar}}$$

$$= \frac{0.175}{0.19430} = 0.90$$

$$\boxed{x = 0.9}$$ *Ans.*

(ii) At 20 bar, $t_s = 212.37°C$

Given: $t = 230°C$, which is greater than 't_s'.

∴ The given steam is superheated

The degree of superheat = $t_{sup} - t_{sat}$

$= 230 - 212.37 = 17.63°C$ *Ans.*

(iii) At 20 bar, $h_g = 2797.2$ kJ/kg

The given value is less than 'h_g'

∴ The given steam is wet, and

$$h = h_f + x\, h_{fg}$$

$$2750 = 908.6 + x\,(1888.7)$$

$$\boxed{x = 0.97}$$ *Ans.*

(At 20 bar, $h_f = 908.6$ kJ/kg, $h_{fg} = 1888.7$ kJ/kg from steam table).

Problem 12: *Throttling calorimeter has steam entering at 10 MPa and leaving out at 0.05 MPa and 100°C. Determine dryness fraction of the steam.*

Solution: During throttling

$$h_1 = h_2$$

h_2 = enthalpy of steam at state '2'.

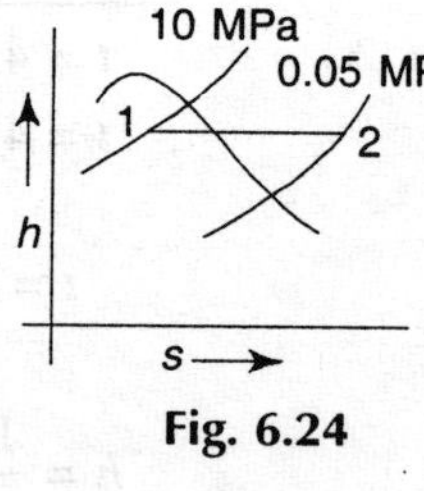

Fig. 6.24

[1 N/m^2 = 1 Pa = 1 × 10^{-5} bar]

0.05×10^6 Pa = 0.5 bar

At $P = 0.5$ bar, $t_s = 81.35°C$

∴ 100°C > 81.35 , the steam at state 2 will be superheated

at $P = 0.5$ and $t_{sup} = 100°C$, $h_2 = 2682.6$ kJ/k

(from steam table for super heated steam)

At state 1, before throttling

$P = 10$ MPa = 100 bar; $h_f = 1408.1$ kJ/kg

$h_{fg} = 1319.7$ kJ/kg

$$\therefore \quad h_1 = hf_1 + x_1\, hfg_1 = h_2$$

$$2682.6 = 1408.1 + x_1\,(1319.7)$$

$$\boxed{x_1 = 0.965}$$

The dryness fraction of wet stream at state 1 is 0.965 *Ans.*

Problem 13: *Steam is available at 10 bar and 0.9 dry. Determine final dryness fraction of steam, when*

(i) there is heat loss of 120 kJ from the steam at constant pressure,

(ii) the steam temperature decreases to 170°C,

(iii) In a turbine, steam expands to 4 bar and equivalent work done is 210 kJ/kg.

Solution :

(i) At $P = 10$ bar, $h_{fg} = 2013.6$ kJ/kg

h_{fg} = latent heat of evaporation from steam table

$\therefore$ Increase in wetness due to removal of 120 kJ of heat

$$= \frac{120}{2013.6} = 0.0595$$

$\therefore$ final dryness fraction = 0.9 – 0.0595

$$= \boxed{0.8405}$$ *Ans.*

(ii) At $P = 10$ bar, $t_s = 179.88$ °C

When temperature decreases to 170°C the pressure also decreases and becomes equal to saturation pressure corresponding to reduced temperature of 170°C.

final pressure = 7.9202 bar (corresponding to 170°C)

Sp. volume at 10 bar = 0.194 m^3/kg

Sp. volume at 7.92 bar = 0.24 m^3/kg

Since volume remains constant

$$0.9 \times 0.194 = x \times 0.24$$

$$\therefore \quad \boxed{x = 0.73}$$ *Ans.*

(iii) At $P = 10$ bar, $hf_1 = 762.6$ kJ/kg

$hfg_1 = 2013.6$ kJ/kg

$\therefore$ At state 1,

$$h_1 = hf_1 + x_1\, hfg_1$$

$$= 762.6 + 0.9\,(2013.6)$$

$$h_1 = 2574.84 \text{ kJ/kg}$$

At 4 bar, $hf_2 = 604.7$ kJ/kg

$hfg_2 = 2132.9$ kJ/kg

$\therefore$ At state 2,

$$h_2 = hf_2 + x_2\, hfg_2$$

But $h_2 = h_1 -$ work done

$\therefore$ $h_2 = 2574.84 - 210$

$h_2 = 2364.84$ kJ/kg

$\therefore$ $2364.84 = 604.7 + x_2\,(2132.9)$

$$\boxed{x_2 = 0.825}$$ *Ans.*

Problem 14: *Determine specific volume, enthalpy, internal energy and entropy of wet steam at 20 bar and dryness fraction 0.85.*

Solution: Specific volume of wet steam

$$V_{\text{wet}} = x\, V_g$$

At 20 bar, $V_g = 0.0995\ m^3/kg$

$$h_f = 908.6\ kJ/kg$$
$$h_{fg} = 1888.7\ kJ/kg$$
$$S_f = 2.4468\ kJ/kgK$$
$$S_{fg} = 3.8909\ kJ/kgK$$

$\therefore$ $V_{wet} = 0.85 \times 0.0995$

$$\boxed{V_{wet} = 0.0845\ m^3/kg}$$ *Ans.*

$$h = h_f + x\,h_{fg}$$
$$= 908.6 + 0.85\,(1888.7)$$
$$\boxed{h = 2513.99\ kJ/kg}$$ *Ans.*

External work $PV = 20 \times 10^5 \times 0.0845$ or *x.vg.p.*

$$= \boxed{169\ kJ}$$ *Ans.*

Internal energy $u = h - PV$

$$= 2513.99 - 169$$
$$= \boxed{2344.99\ kJ/kg}$$ *Ans.*

Entropy, $s = sf + x\ sfg$

$$= 2.4468 + 0.85\,(3.8909)$$
$$\boxed{s = 5.754\ kJ/kgK}$$ *Ans.*

Problem 15: *A vessel of volume 0.04 m^3 contains a mixture of saturated water and saturated steam at a tempt. of 240°C, the mass of liquid present is 8 kg. For the mixture evaluate the pressure, the mass, the specific volume, the enthalpy.*

Given: $V = 0.04\ m^3,\ t = 240°C$

$$m_f = 8\ kg$$

Solution: At $t = 240°C$ (from steam table)

$$P = 33.478\ bar$$
$$V_f = 0.0012291\ m^3/kg$$
$$V_g = 0.059645\ m^3/kg$$
$$h_f = 1037.6\ kJ/kg$$
$$h_{fg} = 1764.6\ kJ/kg$$

The volume occupied by liquid $= V_f \times m_f$

$$= 0.0012291 \times 8 = 9.8328 \times 10^{-3}\ m^3$$

The volume occupied by vapour $= V - V_f = 0.04 - 9.8328 \times 10^{-3}$

$$= 0.030167\ m^3$$

$\therefore$ The mass of vapour, $m_g = \dfrac{0.030167}{0.059645} = 0.50577\ kg$ $\quad (V = m_g V_g)$

$\therefore$ The dryness fraction, $x = \dfrac{\text{mass of vapour}}{\text{total mass}}$

$$x = \frac{0.50577}{8 + 0.50577} = 0.05946$$

$\boxed{\text{Pressure} = 33.478 \text{ bar}}$ *Ans.*

The mass = 8 + 0.50577 = $\boxed{8.50577 \text{ kg}}$ *Ans.*

The specific volume of the mixture $V = \dfrac{V}{m_f + m_g} = \dfrac{0.04}{8.50577}$

$\boxed{= 4.7026 \times 10^{-3} \text{ m}^3/\text{kg}}$ *Ans.*

Enthalpy $h = h_f + xh_{fg}$

$= 1037.6 + 0.05946\ (1764.6)$

$\boxed{h = 1142.53 \text{ kJ/kg}}$ *Ans.*

Total enthalpy $= h \times m_{total}$

$= 1142.53 \times 8.50577$

$= 9718.09$ kJ

Problem 16: *A rigid vessel contains 1 kg of mixture of saturated water and saturated steam at a pressure of 140 × 10³ N/m². When the mixture is heated, the state passes through the critical point. Determine*

(i) The volume of the vessel,

(ii) The mass of liquid and of vapour in the vessel initially,

(iii) The temperature of the contents of the vessel when the pressure increases to 30 × 10⁶ N/m².

Solution :

(i) $m = 1$ kg, $P = 140 \times 10^3$ N/m²

'*V*' of the mixture at critical point

$V = 0.00317$ m³/kg

∴ Volume of vessel $= 0.00317 \times m = 0.00317$ m³ *Ans.*

(ii) $P = 140 \times 10^3 \text{ N/m}^2 = 1.4$ bar

$V_f = 0.0010513$ m³/kg (from steam table)

$V_g = 1.2363$ m³/kg ($V_{fg} = V_g + V_f$)

$V_{fg} = 1.2352$ m³/kg

$V = V_f + xV_{fg}$

$0.00317 = 0.0010513 + x\ (1.2352)$

$\boxed{x = 1.715 \times 10^{-3}}$

∴ $m_g = \dfrac{\text{Volume of vapour}}{\text{Sp. volume of vapour}} = \dfrac{xV_g}{V_g} = 0.001715$ kg

$m_f = m - m_g = \boxed{0.9982 \text{ kg}}$ *Ans.*

(iii) $P_2 = 30 \times 10^6$ N/m², $V = 0.00317$ m³/kg, t = ?

for $V_1 = 0.0028306$ m³/kg at 400°C

$V_2 = 0.0086808$ m³/kg at 500°C

[from steam table corresponding to properties of superheated steam]

∴ $t = \dfrac{0.00317 - 0.0028306}{0.0086808 - 0.0028306}(500 - 400) + 400$

$\boxed{t = 405.79\ °\text{C}}$ *Ans.*

Problem 17: *0.4 kg of steam with initial dryness of 0.8 is heated at a constant pressure of 8 bar till the volume is doubled. Determine the final temperature, heat and work transfer.*

Solution: The specific volume at state 1

$V_1 = x_1\ Vg_1 = 0.8 \times 0.24026$ (at P = 8 bar, $V_g = 0.24026$ m³/kg)

$V_1 = 0.192208\ \text{m}^3/\text{kg}$

$V_2 = 2 \times V_1 = 2 \times 0.192208 = 0.38441\ \text{m}^3/\text{kg}$

At 8 bar, $V_2 = 0.38441\ \text{m}^3/\text{kg}$

$t_1 = 400°\text{C},\ V_1 = 0.38416\ \text{m}^3/\text{kg}$

$t_2 = 500°\text{C},\ V_2 = 0.44317\ \text{m}^3/\text{kg}$

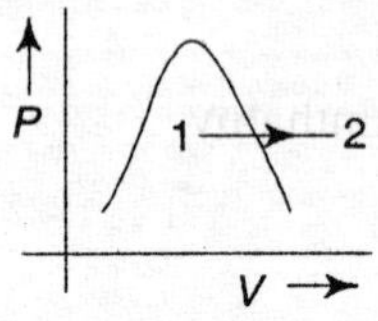

Fig. 6.25

By interpolation,

$$= \left(\frac{0.38441 - 0.38416}{0.44317 - 0.38416}\right)(500 - 400) + 400$$

$$\boxed{t = 400.5°\text{C}}$$

$$\boxed{\text{final temperature } t = 400.5°\text{C}}$$ *Ans.*

Work done

$$= m \cdot \int Pdv = mp\ (V_2 - V_1)$$

$$= \frac{0.4 \times 8 \times 10^5}{1000}\ (0.38441 - 0.192208)\ \text{kJ} \qquad \left[\cancel{kg} \times \frac{N}{\cancel{m^2}} \times \frac{\cancel{m^3}}{\cancel{kg}}\right] \quad \text{Nm: J}$$

$$\boxed{\text{W.D.} = 61.504\ \text{kJ}}$$ *Ans.*

$$Q = W + (u_2 - u_1)\ m$$

$$= mp\ (v_2 - v_1) + (u_2 - u_1)m$$

$$= m(h_2 - h_1)$$

at 8 bar from steam table

$hf_1 = 720.9$ kJ/kg

$hfg_1 = 2046.5$ kJ/kg

$$h_1 = hf_1 + x\ fg_1$$

$$h_1 = 720.9 + 0.8\ (2046.5)$$

$$\boxed{h_1 = 2358.1\ \text{kJ/kg}}$$

$h_2 = 3267.5$ kJ/kg (from superheated steam tables corresponding to P = 8 bar and t_{sup} = 400.5 °C

$$Q = 0.4\ (3267.5 - 2358.1)$$

$$\boxed{Q = 363.76\ \text{kJ}}$$ *Ans.*

Problem 18: *Steam at a pressure of 15 bar and 250°C expands according to the law $PV^{1.25} = C$ to a pressure of 1.5 bar. Evaluate the final conditions, work done, heat transfer and change in entropy. The mass of system is 0.8 kg.*

Solution: At $P_1 = 15$ bar, $t_s = 198.28°\text{C}$

Given: $t = 250°\text{C} > t_s$

The steam is in the super heated condition.

$$V_1 = V_{supl} = 0.15199\ \text{m}^3/\text{kg}$$

(from super heated tables at P = 15 bar and t_{sup} = 250°C)

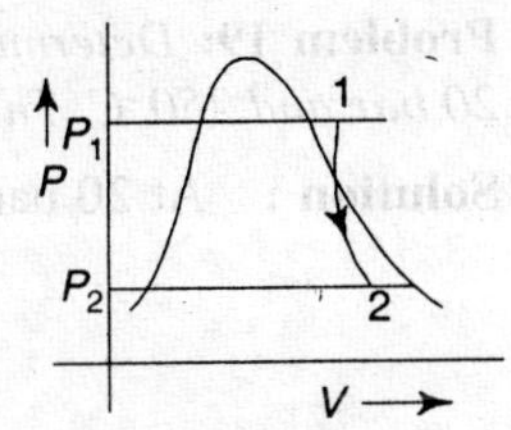

Fig. 6.26

$$\therefore \quad \frac{V_2}{V_1} = \left(\frac{P_1}{P_2}\right)^{1/n}$$

$$V_2 = 0.15199 \left(\frac{15}{1.5}\right)^{1/1.25}$$

$$V_2 = 0.9589 \text{ m}^3/\text{kg}$$

$$V_2 = x_2 \, Vg_2$$

$$0.9589 = x_2 \, (1.159)$$

$$\boxed{x_2 = 0.827}$$

$Vg_2 = 1.159$ (from steam table at $P_2 = 1.5$ bar)

The steam is wet at state 2 **Ans.**

At 1.5 bar, $t_s = 111.32$ (from steam table)

$$W = \left(\frac{P_1 V_1 - P_2 V_2}{n-1}\right) m$$

$$= 0.8 \left(\frac{15 \times 10^5 \times 0.15199 - 1.5 \times 0.9589 \times 10^5}{(1.25-1)}\right)$$

$$\boxed{W = 269.28 \text{ kJ}}$$ **Ans.**

$$u_1 = h_1 - P_1 V_1$$

$$u_1 = 2923.5 \times 10^3 - 15 \times 10^5 \, (0.15199)$$

$$u_1 = 2695.51 \text{ kJ/kg}$$

$$u_1 = 2695.51 \times 0.8 = 2156.408 \text{ kJ}$$

$$u_2 = h_2 - P_2 V_2$$

$$h_2 = hf_2 + x_2 \, hfg_2$$

$$h_2 = 466.9 + 0.827 \, (2226.4)$$

$$h_2 = 2308.13 \text{ kJ/kg}$$

$h_1 = 2923.5$ at 15 bar and 250°C from superheated steam tables

at $P = 1.5$ bar

$h_f = 466.9$ kJ/kg

$hfg = 2226.4$ kJ/kg

(from steam table)

$$u_2 = 2308.13 \times 10^3 - 1.5 \times 10^5 \, (0.9589)$$

$$= 2164.29 \text{ kJ/kg}$$

$$u_2 = 2164.29 \times 0.8 = 1731.43 \text{ kJ}$$

$$Q = W + (u_2 - u_1)$$

$$= 269.28 + (1731.43 - 2156.408)$$

$$\boxed{Q = -155.69 \text{ kJ}}$$ **Ans.**

$S_1 = 6.7099$ kJ/kgK from superheated steam tables at $P = 15$ bar

and $t_{sup} = 250°C$

$$S_2 = Sf_2 + x_2 \, Sfg_2$$

$$S_2 = 1.43295 + 0.827 \, (5.79115)$$

$$S_2 = 6.2222 \text{ kJ/kgK}$$

At 1.5 bar

$Sf_2 = 1.43295$ kJ/kgK

$Sfg_2 = 5.79115$ kJ/kgK

$$\therefore \quad S_2 - S_1 = 6.2222 - 6.7099$$

$$= -0.4877 \text{ kJ/kgK}$$

$$= -0.4877 \times 0.8 = \boxed{-0.39016 \text{ kJ/K}}$$ **Ans.**

Problem 19: *Determine the enthalpy, volume, internal energy and entropy of superheated steam at 20 bar and 250°C. The volume of water may be neglected and assume specific heat = 2.2 × 10³ J/kgK*

Solution : At 20 bar, from steam tables

$$V_g = 0.099549 \text{ m}^3/\text{kg}$$
$$h_f = 908.6 \text{ kJ/kg}$$
$$h_{fg} = 1888.7 \text{ kJ/kg}$$
$$S_f = 2.4468 \text{ kJ/kgK}$$
$$S_{fg} = 3.8909 \text{ kJ/kgK}$$
$$t_{sat} = 212.37°\text{C}$$

Total heat of 1kg of super heated steam

$$h_{sup} = h_f + h_{fg} + C_{ps}(t_{sup} - t_{sat})$$
$$= 908.6 + 1888.7 + 2.2\,(250 - 212.37)$$
$$\boxed{h_{sup} = 2880.08 \text{ kJ/kg}}$$ *Ans.*

The superheated steam behaves like a perfect gas, so from Charle's law

$$V_{sup} = V_g \frac{t_{sup}}{t_{sat}}$$
$$= 0.099549 \times \frac{250 + 273}{212.37 + 273}$$
$$\boxed{V_{sup} = 0.1072 \text{ m}^3/\text{kg}}$$ *Ans.*

External work= $PV = 20 \times 10^5 \times 0.1072$

$$= 214.4 \text{ kJ}$$

Internal energy $u = h_{sup} -$ External work

$$u = 2880.08 - 214.4$$
$$\boxed{u = 2665.68 \text{ kJ/kg}}$$ *Ans.*

The entropy of superheated steam

$$S_{sup} = S_f + S_{fg} + C_p s \log_e \frac{t_{sup}}{t_{sat}}$$
$$= 2.4468 + 3.8909 + 2.2 \ln \frac{250 + 273}{212.37 + 273}$$
$$\boxed{S_{sup} = 6.5019 \text{ kJ/kgK}}$$ *Ans.*

Problem 20: *A boiler produces steam at 9 bar and 0.90 dry which enters into a superheater to absorb 80 kJ/kg of heat at constant pressure. Subsequently the steam is throttled to a pressure of 0.1 bar. Evaluate the final condition of steam.*

Solution : Let 1 be the condition of steam before entering the superheater

$$\therefore \quad h_1 = hf_1 + x_1\, hfg_1$$
$$= 742.6 + 0.9\,(2029.5)$$
$$h_1 = 2569.15 \text{ kJ/kg}$$

$$\left\{\begin{array}{ll}\text{At 9 bar} & hf_1 = 742.6 \text{ kJ/kg} \\ & hfg_1 = 2029.5 \text{ kJ/kg}\end{array}\right\}$$

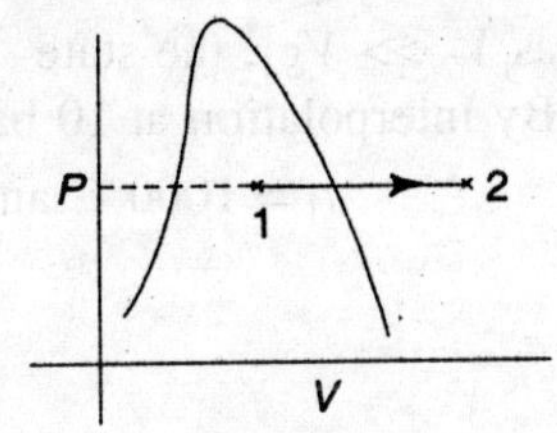

Fig. 6.27

At 2 the steam condition is superheated

$$h_2 = h_1 + Q$$

$$\therefore \quad h_2 = 2569.15 + 80$$

$$h_2 = 2649.15 \text{ kJ/kg}$$

$$(Q_2 = (h_2 - h_1))$$

During throttling enthalpy remains constant and after steam has been throttled to $P_2 = 0.1$ bar, $h_2 = h_3$
$= 2649.15$ kJ/kg

At 0.1 bar, $\quad h_g = 2584.8$ kJ/kg **(from steam table)**

$\therefore$ The heat of superheat $= h_3 - h_g$

$$= 2649.15 - 2584.8 = 64.35 \text{ kJ/kg}$$

Assume specific heat of superheated steam = 2.1 kJ/kgK

$$\therefore \quad \text{Degree of superheat} = \frac{64.35}{2.1} = 30.64°\text{C}$$ *Ans.*

Problem 21: *In a piston cylinder arrangement the steam at 1.0 MPa and 80% dryness fraction and 0.05 m³ volume is heated to increase the volume to 0.2 m³. Determine the quantity of heat added.*

Solution : At state 1, $P_1 = 1.0 \text{ MPa} = 1 \times 10^6 \text{ Pa} = \frac{1 \times 10^6}{105}$ bar

$$= 10 \text{ bar}$$

$$V_1 = 0.05 \text{ m}^3, \; x_1 = 0.85$$

At state 2, $\quad V_2 = 0.2 \text{ m}^3, \; P_2 = P_1 = 10$ bar

At $\quad P_1 = 10$ bar,

$$V_f = 0.0011274 \text{ m}^3/\text{kg}$$

$$V_g = 0.19430 \text{ m}^3/\text{kg}$$ {from steam tables}

$$h_f = 762.6 \text{ kJ/kg}$$

$$h_{fg} = 2013.6 \text{ kJ/kg}$$

Now at state 1, steam is wet

$$\therefore \quad V_1 = V_f + x \cdot V_{fg}$$

$$V_1 = 0.0011274 + 0.80\,(0.19317)$$

$$\boxed{V_1 = 0.1556 \text{ m}^3/\text{kg}}$$

$$\left|\begin{array}{l} V_g = V_f + V_{fg} \\ V_{fg} = V_g - V_f \end{array}\right.$$

$$\therefore \quad \text{Mass of steam} = \frac{V_{\text{all}}}{V_1} = \frac{0.05}{0.1556} = 0.3213 \text{ kg}$$

$$1l^{ly} \quad \text{Mass of steam} = \frac{V_2}{v_2} = \frac{0.2}{v_2}$$

$$\therefore \quad V_2 = \text{Specific volume at } 2 = \frac{0.2}{0.3213}$$

$$\boxed{V_2 = 0.6224 \text{ m}^3/\text{kg}}$$

as $V_2 >> Vg_1$, the state '2' is in the superheated region
By interpolation at 10 bar and $V_2 = 0.6641$ m³/kg

$$t_1 = 1000°C \text{ and } t_2 = 1100°C,\ V_1 = 0.6335 \text{ and } V_2 = 0.5871$$

$$t = \frac{0.6224 - 0.5871}{(0.6335 - 0.5871)}(1100 - 1000) + 1000$$

$$\therefore \quad \boxed{t = 1076.07°C}$$

$$h_1 = hf_1 + x\, hfg_1$$
$$= 762.6 + 0.8\,(2013.6) = 2373.48 \text{ kJ/kg}$$
$$u_1 = u_1 - P_1V_1 = 2373.48 \times 10^3 - 10 \times 10^5\,(0.1556)$$
$$\boxed{u_1 = 2217.88 \text{ kJ/kg}}$$

u_2 at 1 MPa (10 bar) and $t_{sup} = 1076.07°C$

$$u_2 = 4208.6 \text{ kJ/kg}$$
$$\therefore \quad Q = W + \Delta U = W + (U_2 - U_1)$$
$$W = P\,(V_2 - V_1)$$
$$= 10 \times 10^5(0.2 - 0.05) = 150 \text{ kJ}$$
$$\therefore \quad Q = 150 + (4208.6 - 2217.88)$$
$$Q = 2140.72 \text{ kJ/kg} = 2140.72 \times m = 2140.72 \times 0.3213$$
$$\boxed{Q = 687.81 \text{ kJ}}$$

Ans.

Problem 22: *Steam with an initial pressure of 15 bar and 250°C in a closed container of volume 0.12 m³ is cooled in such a way that the pressure falls to 5 bar. Determine the final temperature, heat exchange and change in entropy.*

Solution: At state '1', steam is in superheated region
At 15 bar and 250°C (from superheated steam table)

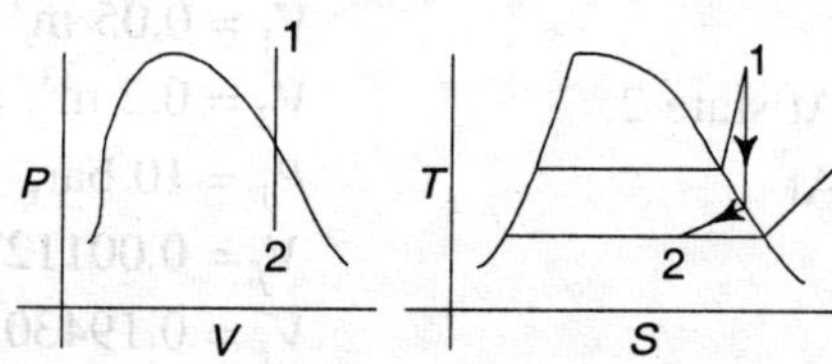

Fig. 6.28

$$V_1 = 0.15199 \text{ m}^3/\text{kg}$$

$$m = \frac{V}{v_1} = \frac{0.12}{0.15199} = 0.789 \text{ kg}$$

At $P_2 = 5$ bar, $Vg_2 = 0.37466$ m³/kg

$$t_{sat} = 151.85°C$$

Specific volume of wet steam

$$V_2 = x_2\, Vg_2$$

$$0.12 = x_2 \times 0.37466 \times (0.789) \qquad \left(m^3 = \frac{m^3}{kg} \times kg\right)$$

$$\therefore \quad \boxed{x_2 = 0.405}$$

∴ The final temperature is the saturation temperature and is equal to151.85°C. *Ans.*

$$Q = W + (u_2 - u_1)$$
$$W = 0 \qquad (\because V_1 = V_2)$$
$$\therefore \quad Q = u_2 - u_1$$

$u_1 = h_1 - P_1 V_1$
$= 2923.5 \times 10^3 - 15 \times 10^5 \; (0.15199)$
$_1 = 2695.51$ kJ/kg
$u_2 = h_2 - P_2 V_2$
$h_2 = hf_2 + x_2 \; hfg_2$
$h_2 = 640.1 + 0.405 \; (2107.4)$
$= 1493.59$ kJ/kg
$\therefore \quad u_2 = 1493.59 \times 10^3 - 5 \times 10^5 \; (0.1517)$
$u_2 = 1417.74$ kJ/kg
$\therefore \quad Q = u_2 - u_1 = 1417.74 - 2695.51$
$= -1277.77$ kJ/kg
$= -1277.77 \times 0.789$
$\boxed{Q = -1008.16 \text{ kJ}}$ **Ans.**

At 15 bar and 250°C,
$h_1 = 2923.5$ kJ/kg
(from super heated steam table)
At 5 bar
$hf = 640.1$ kJ/kg
$hfg = 2107.4$ kJ/kg
$V_2 = x_2 \times Va_2$
$= 0.405 \times 0.37466$
$= 0.1517 \text{ m}^3\text{/kg}$
$\boxed{\text{or } V_2 = V_{f2} + x_2 V_{fg2}}$

At state 1 $\quad S_1 = 6.7099$ kJ/kgK
At state 2, $\quad S_2 = Sf_2 + x_2 \; Sfg_2$
$S_2 = 1.8604 + 0.405 \; (4.9588)$
$\boxed{S_2 = 3.8687 \text{ kJ/kgK}}$

At 15 bar and 250°C
$S_{sup} = 6.7099$ kJ/kgK
At 5 bar
$Sf_2 = 1.8604$ kJ/kgK
$Sfg_2 = 4.9588$ kJ/kgK

$\therefore \quad S_2 - S_1 = 3.8687 - 6.7099$
$S_2 - S_1 = -2.8412$ kJ/kgK
$S_2 - S_1 = -2.8412 \times m = -2.8412 \times 0.789$
$\boxed{S_2 - S_1 = -2.241 \text{ kJ/K}}$ **Ans.**

Problem 23: *Steam expands isentropically from a pressure of 20 bar and 300°C to 2 bar. By taking mass of the system as 0.9 kg, determine end conditions and the work done.*

Solution : During isentropic expansion
$S_1 = S_2 = Sf_2 + x_2 \; Sfg_2$
$6.7696 = 1.5301 + x_2 \; (5.5967)$
$\boxed{x_2 = 0.936}$

At 20 bar and 300°C
$S_1 = 6.7696$ kJ/kgK
$h_1 = 3025.0$

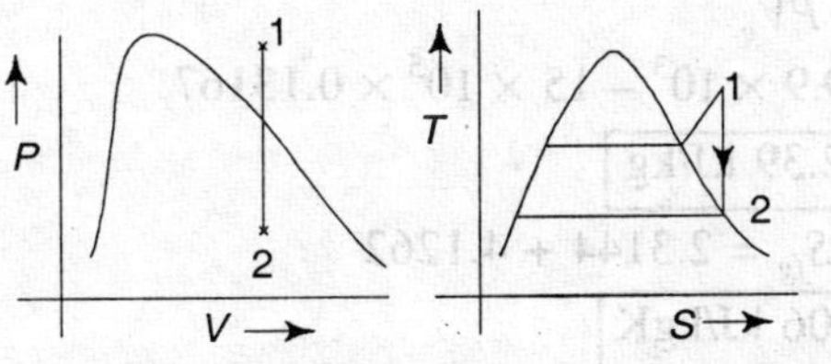

Fig. 6.29

We know that, $Q = 0$,
for isentropic process
$Q = W + (u_2 - u_1)$
$W = u_1 - u_2$

At 2 bar
$Sf_2 = 1.5301$ kJ/kgK
$Sfg_2 = 5.5967$ kJ/kgK
$t_{sat} = 120.23$ °C

$u_1 = m(h_1 - P_1V_1)$

$u_1 = 0.9\,(3025 \times 10^3 - 20 \times 10^5\,(0.1255) \times 10^3$

$\boxed{u_1 = 2496.6 \text{ kJ}}$

$u_2 = m(h_2 - P_2V_2)$

$h_2 = hf_2 + x_2\,hfg_2$

$= 504.7 + 0.936\,(2201.6)$

$\boxed{h_2 = 2565.39 \text{ kJ/kg}}$

At 20 bar and 300°C

$h_1 = 3025.0$ kJ/kg

$V_1 = 0.12550$ m³/kg

At 2 bar

$hf_2 = 504.7$ kJ/kg

$hfg_2 = 2201.6$ kJ/kg

$V_2 = Vf_2 + x_2\,Vfg_2$

or $V_2 = x_2Vg_2$

$= 0.936 \times 0.8854$

$V_2 = 0.8287$ m³/kg

$u_2 = 0.9\,(2565.39 \times 10^3 - 2 \times 10^5 \times 0.8287)$

$u_2 = 2159.68$ kJ

$\therefore \quad W = U_1 - U_2 = 2496.6 - 2159.68$

$\boxed{W = 336.92 \text{ kJ}}$ *Ans.*

Problem 24: *Determine the enthalpy, internal energy and entropy of unit mass of steam at 15 bar when (i) the steam is dry and saturated, (ii) steam is 20% wet, and (iii) when it is superheated to 300°C.*

Solution :

At 15 bar, from steam table

$t_s = 198.28$ °C

$V_g = 0.13167$ m³/kg

$V_f = 0.0011538$ m³/kg

$h_f = 844.6$ kJ/kg

$hfg = 1945.3$ kJ/kg

$S_f = 2.3144$ kJ/kgK

$S_{fg} = 4.1262$ kJ/kgK

(i)For dry saturated steam

$h_g = h_f + h_{fg} = 844.6 + 1945.3$

$= 2789.9$ kJ/kg (or we can take directly from steam table)

$\boxed{h_g = 2789.9 \text{ kJ/kg}}$ *Ans.*

Internal energy

$u_g = h_g - PV_g$

$u_g = 2789.9 \times 10^3 - 15 \times 10^5 \times 0.13167$

$\boxed{u_g = 2592.39 \text{ kJ/kg}}$ *Ans.*

Entropy $\quad S_g = S_f + S_{fg} = 2.3144 + 4.1262$

$\boxed{S_g = 6.4406 \text{ kJ/kgK}}$ *Ans.*

(ii)When the steam is 80% dry,

Enthalpy of wet steam

$h_{wet} = h_f + x\,h_{fg} = 844.6 + 0.8\,(1945.3)$

$\boxed{h_{wet} = 2400.84 \text{ kJ/kg}}$ *Ans.*

Internal energy

$$u_{wet} = h_{wet} - x \cdot P \cdot Vg$$
$$= 2400.84 \times 10^3 - (0.8 \times 15 \times 10^5 \times 0.13167)$$

$\boxed{u_{wet} = 2242.83 \text{ kJ/kg}}$ *Ans.*

Entropy $S_{wet} = S_f + x\, S_{fg} = 2.314 + 0.8\ (4.1262)$

$\boxed{S_{wet} = 5.61496 \text{ kJ/kgK}}$ *Ans.*

(iii) When the steam is superheated to 300°C

Enthalpy $h_{sup} = 3038.9$ kJ/kg

$V_{sup} = 0.16970 \text{ m}^3/\text{kg}$

$S_{sup} = 6.9207$ kJ/kgK

[(At 15 bar and 300°C from super heated steam tables)]

∴ The internal energy

$$u = h - PV = 3038.9 \times 10^3 - 15 \times 10^5 \times 0.16970$$

$\boxed{u = 2784.35 \text{ kJ/kg}}$ *Ans.*

Problem 25: *A boiler supplies steam at a pressure of 15 bar and 95% dry to a steam engine. The pressure of the steam when it flows through the pipe remains constant, but it loses 30 kJ/kg of heat due to poor insulation on the pipe. Determine the temperature and dryness fraction of the steam at the engine end of the pipe line.*

Solution : Let Suffix 1 – state of steam at boiler end
Suffix 2 – state of steam at the engine side

At 15 bar, $t_s = 198.28$°C

Pressure in the pipe line is constant

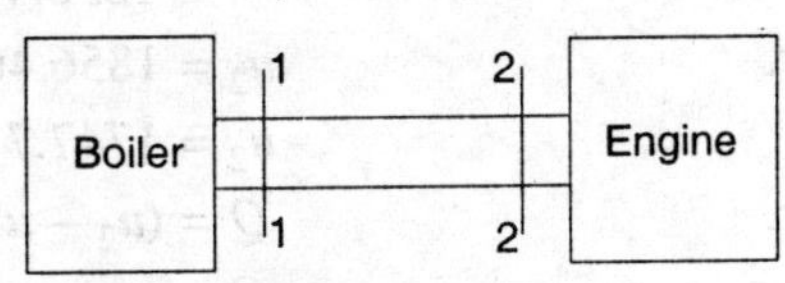

Fig. 6.30

∴ $hf_1 = hf_2 = 844.6$ kJ/kg

$hfg = hfg_2 = 1945.3$ kJ/kg

Now, $h_1 = hf_1 + x_1\, hfg_1$

$h_1 = 844.6 + 0.95\ (1945.3)$

$h_1 = 2692.64$ kJ/kg

$h_2 = hf_2 + x_2 .\ hfg_2$

But $h_2 = h_1 - 30 = 2692.64 - 30 = 2662.64$ kJ/kg

∴ $2662.64 = 844.6 + x_2\ (1945.3)$

∴ $\boxed{x_2 = 0.934}$ *Ans.*

At the engine side in the pipeline, steam is 93.4% dry and its temperature is 198.28°C.

Problem 26: *A pressure cooker has 1.5 kg of saturated steam at 5 bar. Determine quantity of heat which is to be rejected to reduce quality to 60% dry. Also determine pressure and temperature at new state* *(VTU, March 2001)*

Solution :

At 5 bar, $t_s = 151.8$°C, $Vg_1 = 0.375 \text{ m}^3/\text{kg}$

$hf_1 = 640.1$ kJ/kg, $hfg_1 = 2107.4$ kJ/kg

Volume of the cooker

$$V_1 = m \cdot Vg_1$$

$$1.5 \times 0.375 = 0.5625 \text{ m}^3$$

At state 1, specific internal energy

$$u_1 = h_1 - P_1V_1 = h_1 - P_1Vg_1$$

The cooker contains saturated steam, hence $x = 1$

$$\therefore \quad h_1 = hf_1 + hfg = 640.1 + 2107.4$$

$$= 2747.5 \text{ kJ/kg} \quad \text{(or refer '}h_g\text{' value from steam table)}$$

$$\therefore \quad u_1 = 2747.5 \times 10^3 - 5 \times 10^5 \times 0.375$$

$$u_1 = 2560 \text{ kJ/kg}$$

In the cooker volume remains same during cooling process

$$V_1 = V_2 = m\, x_2\, Vg_2$$

$$0.5625 = 1.5 \times 0.6 \times Vg_2$$

$$Vg_2 = 0.625 \text{ m}^3\text{/kg}$$

At $\quad Vg_2 = 0.625 \text{ m}^3\text{/kg}$,

the following steam properties are read from the steam table,

$$P_2 = 2.9 \text{ bar}, \; hfg_2 = 2166.6 \text{ kJ/kg}$$

$$t_s = 132.4 \text{ °C}$$

$$hf_2 = 556.5 \text{ kJ/kg}$$

∴ At state 2, internal energy

$$u_2 = h_2 - P_2V_2$$

$$h_2 = hf_2 + x_2\, hfg_2$$

$$= 556.5 + (0.6)(2166.6)$$

$$= 1856.46 \text{ kJ/kg}$$

$$V_2 = x_2\, Vg_2 = 0.6 \times 0.625$$

$$V_2 = 0.375 \text{ m}^3\text{/kg}$$

$$\therefore \quad u_2 = 1856.46 \times 10^3 - (2.9 \times 10^5 \times 0.375)$$

$$u_2 = 1747.71 \text{ kJ/kg}$$

$$\therefore \quad Q = (u_2 - u_1) + W$$

$$Q = u_2 - u_1 = 1747.71 - 2560$$

$$Q = -812.29 \text{ kJ/kg}$$

$W = 0$, during constant volume process

∴ The total heat transferred

$$= m \cdot Q = 1.5\,(-812.29)$$

$$\boxed{Q_{total} = -1218.44 \text{ kJ}}$$ *Ans.*

The negative sign implies that the heat is rejected from the system.

Problem 27: *Determine change in enthalpy and entropy of steam when it expands in an engine from initial condition of 10 bar and 300°C to 0.25 bar. The exhaust steam has a dryness fraction of 0.85.*

Solution: At state 1, before steam entering the engine, $P = 10$ bar, $t_s = 179.88$°C, $t_{sup} = 300$°C

∴ From superheated steam tables at 10 bar and 300°C

$$V_1 = 0.25798 \text{ m}^3\text{/kg}$$

$$h_1 = 3052.1 \text{ kJ/kg}$$

$$S_1 = 7.1251 \text{ kJ/kgK}$$

At 0.25 bar

$$h_f = 272 \text{ kJ/kg}$$

$$h_{fg} = 2346.4 \text{ kJ/kg}$$

$$S_f = 0.8933 \text{ kJ/kgK}$$

$$S_{fg} = 6.9390 \text{ kJ/kgK}$$
$$h_2 = hf_2 + x_2\, hfg_2$$
$$= 272 + 0.85\ (2346.4) = 2266.44 \text{ kJ/kg}$$
$$S_2 = Sf_2 + x_2\, Sfg_2 = 0.8933 + 0.85\ (6.9390)$$
$$S_2 = 6.7914 \text{ kJ/kgK}$$

$\therefore$ Change in enthalpy $= h_1 - h_2$
$$= 3052.1 - 2266.44 = \boxed{785.66 \text{ kJ/kg}}$$ *Ans.*

Change in entropy $= S_1 - S_2 = 7.1251 - 6.7914$
$$\boxed{S_1 - S_2 = 0.337 \text{ kJ/kgK}}$$ *Ans.*

Problem 28: *Two boilers 'A' and 'B' both operate at a pressure of 15 bar supplies steam in equal proportions to a common main. The boiler 'A' supplies steam in the superheated condition at 350°C and pressure in the main remains same (say at 15 bar). If the temperature of the resulting mixture in the steam main is 230°C, evaluate the quality of steam supplied by boiler 'B' which is without superheater. Also determine net change in entropy of the system. Take* C_{ps} *= 2.25 kJ/kgK for superheated steam.*

Solution: At 15 bar, $t_{sat} = 198.28°C$

In the main $t = 230°C > t_{sat}$

$\therefore$ It is superheated

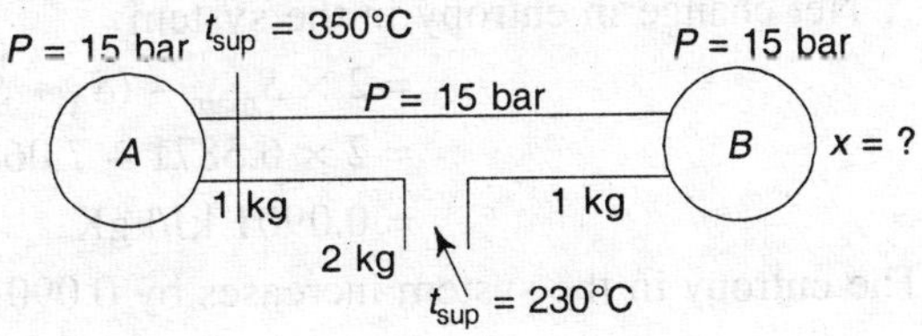

Fig. 6.31

$\therefore$ In the main, $h_{sup} = hf + hfg + C_{ps}\ (t_{sup} - t_{sat})$

At 15 bar $hf = 844.6$ kJ/kg

$$hfg = 1945.3 \text{ kJ/kg}$$
$$h_{sup} = 844.6 + 1945.3 + 2.25\ (230 - 198.28)$$
$$= 2861.27 \text{ kJ/kg}$$

Heat supplied by boiler A

$$= hf + hfg + C_{ps}\ (t_{sup} - t_{sat})$$
$$= 844.6 + 1945.3 + 2.25\ (350 - 198.28)$$
$$= 3131.27 \text{ kJ/kg}$$ (from table 3148.7 kJ/kg)

Let x = dryness fraction of steam in B

$\therefore$ Heat supplied by $B = hf + x\ hfg$
$$= 844.6 + x\ (1945.3)$$

By neglecting loss, the energy balance equation can be written as

Heat supplied by 'A' + Heat supplied by 'B'

= Heat in the main

$3131.27 + 844.6 + x\ (1945.3) = m \times 2861.27$ (where m = 2 kg in the main)

$\therefore\ 3975.87 + x\ (1945.3) = 2 \times 2861.27$

$$\boxed{x = 0.897}$$ *Ans.*

In the main, the dryness fraction of steam is 89.7%

Entropy of steam in the main

$$S_{main} = S_{sup} = S_f + S_{fg} + C_{ps} \log_e \frac{t_{sup}}{t_{sat}}$$ from steam table.

$$S_{sup} = 2.3144 + 4.1262 + 2.25 \ln\left(\frac{230+273}{198.28+273}\right)$$

$$= 6.5871 \text{ kJ/kgK}$$

From steam table At 15 bar,

$S_f = 2.3144$ kJ/kgK

$S_{fg} = 4.1262$ kJ/kgK

At 15 bar & 350°C

$S = 7.1044$ kJ/kgK

In the boiler *A*

$$S_A = S_{supA} = S_f + S_{fg} + C_{ps} \ln\left(\frac{t_{sup}}{t_{sat}}\right)$$

$$= 2.3144 + 4.1262 + 2.25 \ln\left(\frac{350+273}{198.28+273}\right)$$

$$= 7.0685 \text{ kJ/kgK}$$

Entropy in boiler '*B*',

$$S_B = S_f + x\, S_{fg}$$
$$= 2.3144 + (0.897)(4.1262)$$
$$S_B = 6.0156 \text{ kJ/kgK}$$

∴ Net change in entropy in the system

$$= 2 \times S_{main} - (S_A + S_B)$$
$$= 2 \times 6.5871 - 7.0685 - 6.0156$$
$$= 0.0901 \text{ kJ/kgK}$$ *Ans.*

The entropy in the system increases by 0.0901 kJ/kgK. *Ans.*

Problem 29: *A rigid vessel of 2 m^3 volume is filled with superheated steam at 20 bar and 350°C. The vessel is cooled until the steam is just dry saturated. Calculate the mass of steam in the vessel, the final pressure of steam and amount of energy transferred as heat to the surroundings.*

Solution: At state 1,

$$P_1 = 20 \text{ bar}, \; t_{sup} = 350°\text{C}$$
$$V_1 = 2 \text{ m}^3$$

From superheated steam table

$$V_1 = 0.13866 \text{ m}^3/\text{kg}$$

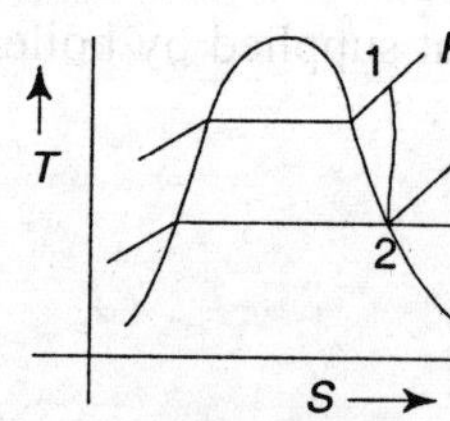

Fig. 6.32

∴ Mass of steam $\quad m = \dfrac{V_1}{V_1} = \dfrac{2}{0.13866}$

$$\boxed{m = 14.423 \text{ kg}}$$ *Ans.*

(i) For a rigid vessel, volume remains constant

∴ $\quad V_1 = V_2 = 0.13866$ m^3/kg

and at state 2, the condition of steam is dry and saturated

(ii) ∴ $\quad Vg_2 = V_2 = 0.13866$ m^3/kg

from steam table

At 15 bar $\quad Vg = 0.13167$

16 bar $\quad Vg = 0.14073$

By inter polation

$$P ; \frac{0.13866 - 0.13167}{0.14073 - 0.13167}(16 - 15) + 15$$

$$P_2 = 15.77 \text{ bar at } 0.13866 \text{ m}^3/\text{kg}$$

(iii) W.D. = 0, for a constant volume process

$\therefore \quad q = du + W$

$q = u_2 - u_1$

$u_2 = h_2 - P_2V_2$

$h_2 = hg = 2790.4$ kJ/kg (at 15.77 bar, by interpolation)

$u_2 = [2790.4 \times 10^3 - 15.77 \times 10^5 \times 0.13866]$

$u_2 = 2571.73$ kJ/kg

$u_1 = h_1 - P_1V_1$

$= (3138.6 \times 10^3 - 20 \times 10^5 \times 0.13866)$

$u_1 = 2861.28$ kJ/kg

$h_1 = 3138.6$ (at 20 bar & 350°C from superheated steam tables)

$\therefore \quad q_{1\text{-}2} = u_2 - u_1 = 2571.73 - 2861.28$

$q_{1\text{-}2} = -289.55$ kJ/kg

∴ Total energy transferred as heat to the surroundings

$Q_{1\text{-}2} = -289.55 \times 14.423$

$\boxed{Q_{1\text{-}2} = -4176.17 \text{ kJ}}$ *Ans.*

Problem 30: *During a polytropic expansion, steam changes its condition from 15 bar & 300°C to 1 bar. If the index of expansion is 1.2. Find (i) specific volume and temperature after expansion (ii) heat interaction & (iii) change in entropy.*

Solution: At state 1, i.e.

At 15 bar and 300°C, from superheated steam tables

$V_1 = 0.16970$ m³/kg

$h_1 = 3038.9$ kJ/kg

$S_1 = 6.9207$ kJ/kgK

$u_1 = 2784.35$ kJ/kg

$u_1 = h_1 - P_1V_1$

$= 3038.9 \times 10^3 - 15 \times 10^5 \times 0.16970$

$u_1 = 2784.35$ kJ/kg

For polytropic expansion

we can write, $\quad P_1V_1^n = P_2V_2^n = c$

$$V_2^n = \frac{P_1}{P_2} \cdot V_1^n = \left(\frac{15}{1}\right) (0.16970)^{1.2}$$

$V_2 = (1.7852)^{1/1.2}$

$V_2 = 1.6208$ m³/kg

At state 2, 1 bar,

$Vg_2 = 1.6937$ m³/kg

$hf_2 = 417.5$ kJ/kg

$hfg_2 = 2257.9$ kJ/kg

$Sf_2 = 1.3027$ kJ/kgK

$Sfg_2 = 6.0571$ kJ/kgK

$t_{sat} = 99.63$°C

Since $V_2 < Vg_2$, the steam is wet at state 2

$$\therefore \quad x_2 = \frac{V_2}{Vg_2} = \frac{1.6208}{1.6937} = 0.956$$

$$h_2 = hf_2 + x_2\, hfg_2$$
$$= 417.5 + 0.956\ (2257.9)$$
$$h_2 = 2576.05 \text{ kJ/kg}$$
$$u_2 = h_2 - P_2V_2$$
$$= 2576.05 \times 10^3 - 1 \times 10^5 \times 1.6208$$
$$\boxed{u_2 = 2413.97 \text{ kJ/kg}}$$

$$V_2 = x_2\, Vg_2$$
$$= 0.956 \times 1.6937$$
$$V_2 = 1.6208 \text{ m}^3\text{/kg}$$

$$S_2 = Sf_2 + x_2\, Sfg_2$$
$$= 1.3027 + 0.956\ (6.0571)$$
$$S_2 = 7.0932 \text{ kJ/kgK}$$

$$V_2 = x \cdot Vg_2$$

$$\text{Workdone} = \text{WD}_{1\text{-}2} = \frac{P_1V_1 - P_2V_2}{n-1}$$

$$= \frac{15 \times 10^5 \times 0.1697 - 1 \times 10^5 \times 1.6208}{1.2 - 1}$$

$$\text{WD}_{1\text{-}2} = 462.35 \text{ kJ/kg}$$

From first law of thermodynamics

$$q_{1\text{-}2} = (u_2 - u_1) + W_{1\text{-}2}$$
$$= (2413.97 - 2784.35) + 462.35$$
$$q_{1\text{-}2} = 91.97 \text{ kJ/kg}$$ *Ans.*

Entropy change $= S_2 - S_1 = 7.0932 - 6.9207$

$$\boxed{= 0.1725 \text{ kJ/kgK}}$$ *Ans.*

Problem 31: *Steam at 20 bar and 300°C is throttled to 10 bar, and then in a steam turbine expands isentropically to 1 bar. Determine isentropic heat drop and quality of steam at the exit of turbine. Use Mollier diagram (h-s chart).*

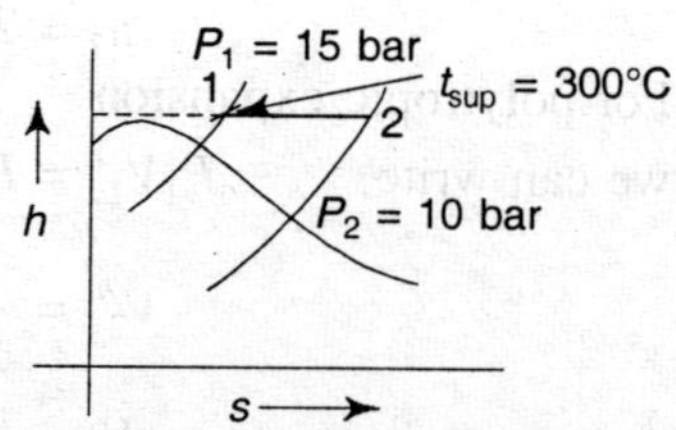

Fig. 6.33

Solution: *Procedure to locate state points on h-s diagram*

(i) At P_1 = 20 bar, t_{sup} = 300°C. Locate point 1.

(ii) During throttling, enthalpy remains constant, i.e. $h_1 = h_2$. Draw a horizontal line from 1 to 2 to meet P_2 line (P_2 = 10 bar) from the chart

$$h_1 = h_2 = 3020 \text{ kJ/kg}$$

(iii) During isentropic expansion in the turbine entropy remains constant ($S_2 = S_3$). ∴ from 2, drop a vertical to meet pressure line at 1 bar to represent state 3.

At state 3, h_3 = 2460 kJ/kg

$$x_3 = 0.88$$

∴ The isentropic heat drop in the turbine is

$$h_2 - h_3 = 3020 - 2460 = 560 \text{ kJ/kg}$$

and the condition of steam is 88% dry.

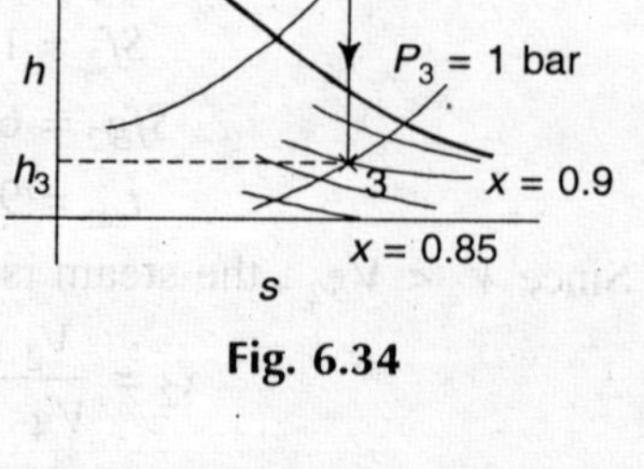

Fig. 6.34

Problem 32: *In a steam turbine, steam expands isentropically from 10 bar and 300°C to a pressure of 0.5 bar. Determine dryness fraction and work developed per kg of steam. If the actual drop is 90% of isentropic drop and flow rate of steam is 450 kg/min. Calculate power developed by the turbine by using h-s chart.*

Solution: Procedure

(i) Locate state 1, corresponding to 10 bar and 300°C on *h-s* chart.

(ii) Drop a vertical from state 1, to meet constant pressure line corresponding to 0.5 bar at state 2. Then 1-2 represents isentropic heat drop or isentropic expansion in the turbine.

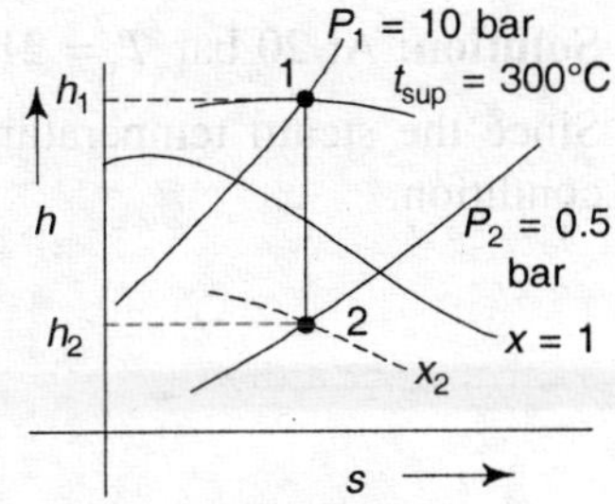

Fig. 6.35

From the chart $h_1 = 3040$ kJ/kg

$h_2 = 2440$ kJ/kg

$x_2 = 0.91$

The quality of steam after isentropic expansion in the turbine is 91% dry

Workdone per kg of steam

$$= h_1 - h_2 = (3040 - 2440)$$
$$= 600 \text{ kJ/kg}$$

Since actual heat drop is 90% of isentropic heat drop

$$(h_1 - h_2)_{actual} = 0.9\ (600) = 540 \text{ kJ/kg}$$

mass flow rate = 450 kg/min (given)

$$\therefore \quad (h_1 - h_2)_{actual} = 540 \times 450 \text{ kJ/min}$$
$$= 243000 \text{ kJ/min}$$
$$\div \text{ by } 60$$
$$\text{Power} = 4050 \text{ kJ/S} = \text{kW}$$

∴ The power developed by the turbine = 4050 kW *Ans.*

Problem 33: *If a certain amount of steam is produced at a pressure of 8 bar and dryness fraction of 0.8 calculate*

(i) External workdone during evaporation

(ii) External latent heat of steam. *(VTU, Aug. 2001)*

Given: $P = 8$ bar, $x = 0.8$

Solution:

(i) External work done during evaporation

$$= Px\ Vg = 8 \times 10^2 \times 0.8 \times 0.24$$
$$= \boxed{153.6 \text{ kJ}}$$ *Ans.*

(where $Vg = 0.24$ m^3/kg at $P = 8$ bar from steam table)

(ii) External latent heat of steam.

$$= xh + g - \text{External work done}$$
$$= 0.8 \times 2046.5 - 153.6$$
$$= \boxed{1483.6 \text{ kJ}}$$ *Ans.*

Problem 34: *Two boilers one with super heater and other without super heater are delivering equal quantities of steam into a common main. The pressure in the boiler is 20 bar. The temperature of steam from a boiler with a super heater is 350°C and temperature of the steam in the main is 250°C. Determine quality of steam supplied by the other boiler. Take C_{ps} = 2.25 kJ/kg. (VTU, Aug. 2001)*

Solution: At 20 bar $T_s = 212.42°C$.

Since the steam temperature in the main is 250°C > T_s, steam in the main is in the super heated condition.

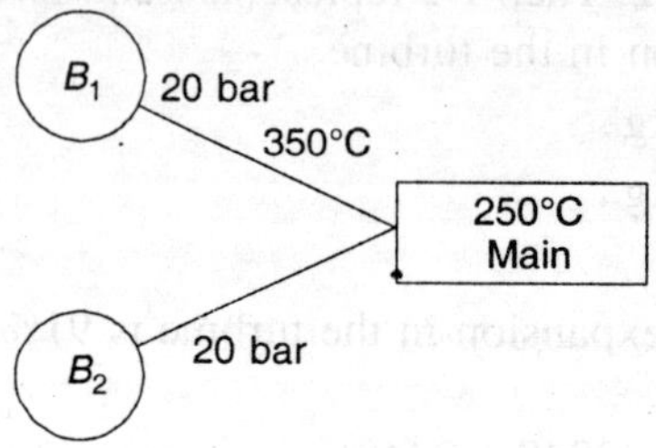

Fig. 6.36

Let B_1 — Boiler with super heater

B_2 — Boiler without super heater

In Boiler B_1, $T_1 = 350°C$ and $P_1 = 20$ bar

Enthalpy of steam $= h_1 = hg_1 + Cps\ (T_{sup} - T_s)$

$h_1 = 2797.2 + 2.25(350 - 212.42)$

$h_1 = 3106.75$ kJ

(where $hg_1 = 2797.2$ kJ/kg, from steam table at 20 bar and 350°C)

In Boiler B_2, $P_2 = 20$ bar

$h_2 = hf_2 + x_2\, hfg_2$

$\therefore\ h_2 = 908.6 + x_1(1888.6)$

At 20 bar: $hf_2 = 908.6$ kJ/kg, $hfg_2 = 1888.6$ kJ/kg

In the main mass of steam is equal to 2 kg (1 kg from B_1 and 1 kg from B_2)

$\therefore\ h_3 = 2[hg_3 + Cps\ (T_{sup} - T_s)]$

$= 2[2797.2 + 2.25(250 - 21242)]$

$= 5763.51$ kJ/kg

At 20 bar and $T_3 = 250°C$, $hg_3 = 2797.2$ kJ/kg

By making energy balance

$$h_1 + h_2 = h_3$$

$$3106.75 + 908.6 + x_2(1888.6) = 5763.51$$

$$\boxed{x_2 = 0.925}$$

Ans.

Problem 35: *Steam expands isentropically from 70 bar and 500°C to 200 kpa in a steam turbine, calculate the dryness fraction and enthalpy of steam after expansion. If the steam flows steadily through the turbine with a mass flow rate of 150 kg/min, determine the turbine work. Neglect the heat losses, potential energy and kinetic energy effects. Draw T–S diagram and show the above process on it. (VTU, Feb. 2002)*

Solution: At state 1, $P_1 = 70$ bar and $T_1 = 500°C$, steam is in super heated condition from super heated steam table at 70 bar and 500°C.

$S_1 = 6.7993$ kJ/kgk

$$h_1 = 3410.6 \text{ kJ/kg}$$

At state 2, $P_2 = 200 \text{ kpa} = 2 \text{ bar}$

from steam table

$$Sf_2 = 1.5301 \text{ kJ/kgk}$$
$$hf_2 = 504.7 \text{ kJ/kg}$$
$$Sfg_2 = 5.5967 \text{ kJ/kgK}$$
$$hfg_2 = 2201.6 \text{ kJ/kg}$$

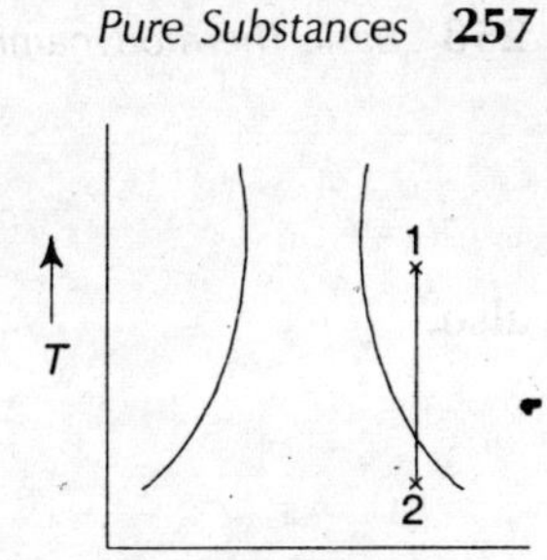

Fig. 6.37

During isentropic expansion 1 – 2, entropy of the steam remains constant

$$\therefore \quad S_1 = S_2$$
$$\therefore \quad S_1 = S_2 = Sf_2 + x_2 (Sfg_2)$$
$$6.7993 = 1.5301 + x_2 (5.5967)$$
$$\boxed{x_2 = 0.785} \qquad \textit{Ans.}$$

The SFEE can be written as

$$Q = \dot{m}\left[(h_2 - h_1) + \frac{V_2^2 - V_1^2}{2} + (z_2 - z_1)\, g\right] + W$$

$Q = 0$ [Isentropic expansion]

$$\Delta KE = 0$$
$$\Delta PE = 0$$

$\therefore$ also, $h_2 = hf_2 + x_2\, hfg_2$

$$= 504.7 + 0.785\,(2201.6)$$
$$h_2 = 2235 \text{ kJ/kg}$$

$$\therefore \quad \dot{m}\,[h_2 - h_1] + W = 0$$
$$W = m\,(h_1 - h_2)$$
$$= 2.5\,(3410.5 - 2235)$$
$$\boxed{W = 2939 \text{ kJ}} \qquad \textit{Ans.}$$

$$\dot{m} = 150 \text{ kg/min} = \frac{150}{60} = 2.5 \text{ kg/sec}$$

Problem 36: *A rigid tank of 2 m³ volume is filled with dry saturated steam at 0.2 MPa. Owing to poor insulation of the tank, the pressure of steam is found to be 0.1 MPa after some time. Determine the final state of steam and the amount of energy transferred as heat to the surroundings.* *(VTU Aug. 2002)*

Solution: $V = 2 \text{ m}^3$

$$P_1 = 0.2 \text{ MPa} = 200 \text{ kpa}$$
$$x_1 = 1 \text{ (dry saturated steam)}$$

from steam table

$$hg_1 = 2706.7 \text{ kJ/kg}$$
$$Vg_1 = 0.8857 \text{ m}^3\text{/kg}$$

$$\therefore \quad \text{mass of the steam} = \frac{V}{Vg_1} = \frac{2}{0.8857} = 2.258 \text{ kg}$$

At $P_2 = 0.1$ MPa;

$$Vg_2 = 1.694 \text{ m}^3\text{/kg}$$
$$hf_2 = 417.46 \text{ kJ/kg}$$
$$hfg_2 = 2258 \text{ kJ/kg}$$

Since volume is constant

$$V_1 = V_2$$
$$Vg_1 = x_2\, Vg_2$$

$$\therefore \quad x_2 = \frac{Vg_1}{Vg_2} = \frac{0.8857}{1.694} = \boxed{0.5228}$$ *Ans.*

also,

$$u_1 = hg_1 - p_1\, vg_1$$
$$= 2706.7 - 200 \times 0.8857$$
$$= 2529.56 \text{ kJ/kg}$$
$$u_2 = (hf_2 + x_2\, hfg_2) - P_2\, x_2\, Vg_2$$
$$= [417.46 + 0.5228\,(2258)] - 100 \times 0.5228 \times 1.694$$
$$u_2 = 1509.38 \text{ kJ/kg}$$

Heat transfered = change in internal energy

$$= u_2 - u_1 = 1509.38 - 2529.56$$
$$= -1020.179 \text{ kJ/kg}$$
$$= -1020.179 \times 2.258 \qquad \Big|\; m = 2.258 \text{ kg}$$
$$= \boxed{-2303.56 \text{ kJ}}$$ *Ans.*

Problem 37: *A rigid container is filled with steam at 600 kpa and 200°C. At what temperature the steam begins to condense when it is cooled? Determine the corresponding pressure.*

(VTU, Feb 2004)

Solution: As the container is rigid, during the execution of a process volume remains constant. From super heated steam table

$V_1 = 0.3520$ m^3/kg. [At 600 kpa and 200°C]

$V_1 = V_2$ when the steam starts condensing

But steam would just be saturated when it starts condensing from steam table

At $V_g = 0.3520$ the corresponding value of pressure is 535.3 kpa

and also the temperature is 154.4°C *Ans.*

Problem 38: *A vessel having a capacity of 0.85 m^3 contains steam at 12 bar and 0.92 dry. Steam is blown off until the pressure drops to 6 bar. The value is then closed and the vessel is cooled until the pressure is 4 bar. Assuming that the specific enthalpy of the steam to remain constant during the blowing off period, determine:*

(i) Quantity of steam blown off
(ii) Quantity of steam in the vessel after cooling
(iii) Quantity of steam in the vessel after cooling

(VTU, Old Scheme, Feb 2004)

Solution: At $P_1 = 12$ bar

$$h_1 = hf_1 + x_1\, hfg_1$$
$$\therefore \quad h_1 = 798.4 + 0.92\,(1984.3)$$
$$h_1 = 2623.25 \text{ kJ/kg}$$

At $P_2 = 6$ bar, $h_2 = hf_2 + x_2\, hfg_2$

$$\therefore \quad 2623.25 = 679.42 + x_2\,(2085)$$

At 12 bar
$hf_1 = 798.4$ kJ/kg
$hfg_1 = 1984.3$ kJ/kg
$x_1 = 0.92$ (given)

At 6 bar
$hf_2 = 679.42$ kJ/kg
$hfg_2 = 2085$ kJ/kg
$h_1 = h_2$ (given)

$$\boxed{x_2 = 0.9326}$$ *Ans.*

also

$$V = x_1\, Vg_1\, m_1$$
$$0.85 = 0.92 \times 0.1632 \times m_1$$
$$\therefore \quad \boxed{m_1 = 5.66 \text{ kg}}$$

$Vg_1 = 0.1632$ m^3/kg
(from steam table at 12 bar)

$$V = x_2\ Vg_2\ m_2$$
$$0.85 = 0.9326 \times 0.3153 \times m_2 \qquad Vg_2 = 0.3153 \text{ m}^3\text{/kg at 6 bar}$$
$$m_2 = 2.889 \text{ kg}$$

∴ Mass of steam blown off

$$= \Delta m = m_1 - m_2$$
$$= 5.66 - 2.889$$
$$\boxed{\Delta m = 2.771 \text{ kg}}$$

Ans.

also,

$$V = x_3\ Vg_3\ m_3$$
$$0.85 = x_3 \times 0.4622 \times 2.771 \qquad Vg_3 = 0.4622 \text{ m}^3\text{/kg at 4 bar.}$$
$$\boxed{x_3 = 0.6639}$$

Ans.

6.20 REVIEW QUESTIONS

1. What is a pure substance? Can air be treated as a pure substance?
2. What is meant by phase of a substance?
3. What are saturation states?
4. Explain the following
   (i) Critical state (ii) Triple point (iii) Normal boiling point
5. Draw the phase equilibrium diagram on *P-V* coordinates for a substance which shrinks in volume on melting and then for a substance which expands in volume on melting.
6. Draw the phase equilibrium diagram for a pure substance on *P-T* coor-dinate.
7. State and explain two property rule.
8. With temperature-enthalpy diagram explain formation of steam at constant pressure.
9. Define quality or dryness fraction of steam.
10. Explain the terms wet steam, dry steam, superheatead steam and degree of superheat.
11. Explain any one method of finding dryness fraction of steam.
12. What are the methods of determining quality of steam? Explain one of them.
13. With a neat sketch, explain separating and throttling calorimeter?

CHAPTER 7

Ideal Gases and Mixtures of Ideal Gases

7.1 INTRODUCTION

In the previous chapters, we have discussed first and second law of thermo- dynamics concerned with systems in general and more particularly to pure substances (steam). A large number of thermodynamic problems involve mixtures of different pure substances. There exists a relationship between various properties, say pressure, specific volume and temperature of a substance. This may be expressed as $P = f(V, T)$.

In the solid phase, the specific volume of a substance is quite insensitive to pressure changes and is not very sensitive to changes in temperature. When the substance exists in liquid phase, changes in pressure results in very small changes in specific volumes at very high temperatures. The change in specific volume of a liquid is relatively small with a large change in temperature. Thus, $f(V, T)$ is not readily available for most condensed phases.

In gaseous phase, variations in both pressure and temperature produces significant changes in the specific volume. The function $f(V, T)$ depends on the nature of the gas and its temperature and pressure. The equation $P = f(V, T)$ can be written as

$$Pv = ZRT, \text{ where } Z = \text{compressibility factor}$$

'Z' is a function of pressure and temperature and is a dimensionless ratio. R is called gas constant. The dimensions of gas constant depend on the units of pressure temperature and specific volume.

Most of the Engineering problems are concerned with total volumes rather than with specific volumes. Therefore, we can write

$$PV = mRT, \text{ where } V = \text{total volume}, m = \text{mass}$$

Ideal Gas

An Ideal or perfect gas is one which is having no intermolecular forces of attraction or repulsion. During a thermodynamic process, an Ideal gas does not change its phase and follows the gas laws at all ranges of pressure and temperature. No vapour or real gas behaves exactly like an Ideal gas. But under some conditions all gases closely obey the Ideal gas law. The real gases will tend to follow Ideal gas laws at low pressure or at high temperature or at both. The range of this low pressure is different for different gases.

In general, the gases which are difficult to liquify such as oxygen, nitrogen, air, carbon monoxide, hydrogen, and helium may be treated as Ideal gases.

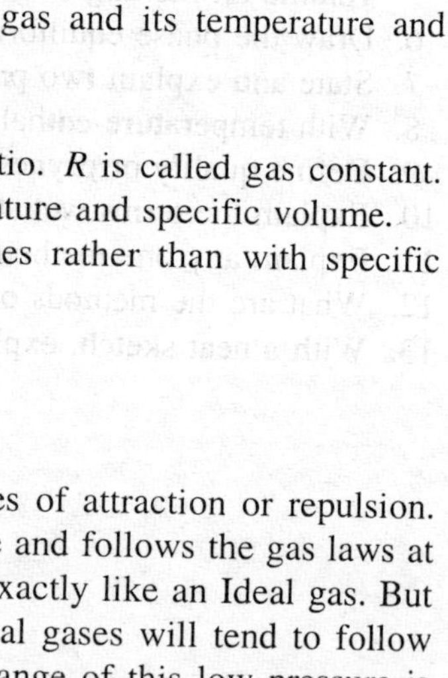

Fig. 7.1

7.2 GAS LAWS

7.2.1 Boyle's Law

It states that, during a process, for a constant gas temperature, the volume of the gas varies inversely with the absolute pressure. Mathematically, we can write

$$V \propto \frac{1}{P}, \text{ for } T = \text{constant}$$

$$\therefore \quad PV = C$$

for different state points, it can be written as

$$P_1V_1 = P_2V_2 = P_3V_3 = \ldots P_nV_n = C$$

This law is valid only at very low pressure or at moderately high temperature.

7.2.2 Charle's Law

It may be stated in two forms:

(a) During a change of state, the volume of the gas is directly proportional to the temperature when the pressure of the gas is maintained constant. Mathematically, we can write,

$$V \propto T, \text{ when } P = \text{constant}$$

$$\therefore \quad \frac{V}{T} = \text{constant}$$

For different state points, we can write

$$\frac{V_1}{T_1} = \frac{V_2}{T_2} = \ldots \frac{V_n}{T_n} = C$$

It was found that, at constant pressure, in a perfect gas, the change in volume corresponding to a unit degree temperature change is given by $B = 11273$ per °C of its volume at 0°C.

Let V_0 = Volume of gas at 0°C.

V_t = Volume of gas at t°C.

∴ At constant pressure

$$V_t = V_0 + V_0 \cdot \frac{t}{273} = V_0 \left[1 + \frac{t}{273}\right]$$

$$= V_0 \left[\frac{273 + t}{273}\right]$$

At $t = -273$°C, the volume of gas becomes zero. This is taken to be the point of absolute zero temperature. All gases liquify before this temperature.

Further, the above equation can be written as

$$\frac{V_t}{V_0} = \frac{T_t}{T_0} \text{ at constant pressure}$$

or

$$\frac{V_t}{T_t} = \frac{V_0}{T_0}$$

(b) Charle's law may be also stated as "During a change of state, the pressure of the gas is directly proportional to the absolute temperature, if the volume of the gas is maintained constant". Mathematically, we can write

$$P \propto T, \text{ at } V = \text{constant}$$

$$\frac{P}{T} = C$$

for different state points,

$$\frac{P_1}{T_1} = \frac{P_2}{T_2} = \ldots \frac{P_n}{T_n} = C.$$

7.3 CHARACTERISTIC GAS EQUATION [EQUATION OF STATE OF AN IDEAL GAS]

This equation gives the relation between pressure, volume and temperature of the gas. The equation can be obtained by using Boyle's law and Charle's law to a perfect gas.

Consider a *P-V* diagram which represents change of state of a gas as shown in Fig. 7.2.

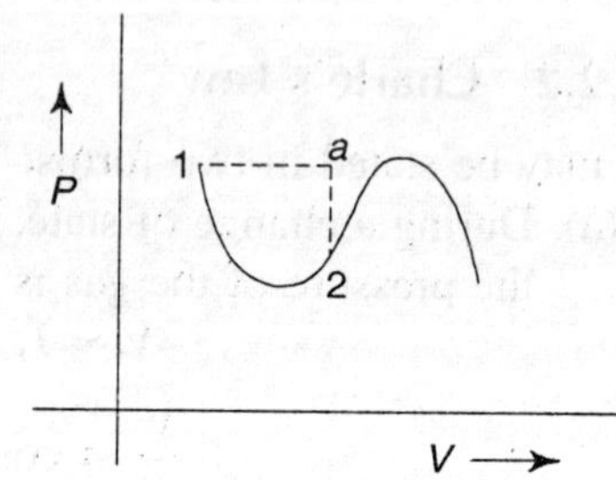

Fig. 7.2

Consider two state points 1 and 2. The change of state from 1-2 can take place by following the path 1-*a* and *a*-2. As the properties are independent of the path followed during a process, the change in properties from 1 to 2 will be same for both 1-2 or 1-*a*-2 . Consider path 1-*a*-2

Applying the charle's law for the path 1-*a*.

$$\frac{V_1}{T_1} = \frac{V_a}{T_a}$$

$$T_a = T_1\left(\frac{V_a}{V_1}\right)$$

$$T_a = T_1\left(\frac{V_2}{V_1}\right) \qquad [V_a = V_2] \qquad .(1)$$

Applying the Boyle's law for the path *a*-2

$$\frac{P_a}{T_a} = \frac{P_2}{T_2}$$

$$T_a = \left(\frac{P_a}{P_2}\right)T_2$$

$$T_a = T_2\left(\frac{P_1}{P_2}\right) \qquad [P_a = P_1] \qquad (2)$$

From equations (1) and (2), we can write

$$T_1\left(\frac{V_2}{v_1}\right) = T_2\left(\frac{P_1}{P_2}\right)$$

$$\therefore \qquad \frac{P_2 v_2}{T_2} = \frac{P_1 v_1}{T_1}$$

In general for '*n*' state points, we can write

$$\frac{P_1 v_1}{T_1} = \frac{P_2 v_2}{T_2} = \ldots \frac{P_n v_n}{T_n} = \text{constant } R$$

$$\therefore \qquad \frac{Pv}{T} = R$$

Let m = mass of the gas

$$\therefore \qquad V = mv$$

$$\therefore \qquad v = \frac{V}{m}$$

$$\therefore \qquad \frac{P \cdot \frac{V}{m}}{T} = R$$

$$PV = mRT$$

Where V = volume of mass 'm',

R = gas constant.

The above equation is known as characteristic gas equation of state for a perfect gas.

The gas which follows the law $PV = mRT$ for all pressure and temperature ranges is known as an Ideal or perfect gas.

The constant 'R' has a particular value for each gas and its units depends on units of pressure, volume, mass and temperature.

The values of gas constant for different gases is given in Table 7.1.

Table 7.1 Values of '*R*' for different gases

Gas	*Molecular mass*	*Gas constant J/Kg K*	*Gas const × molecular mass J/Kg mole K*
Air	28.97	287	
Hydrogen	2	4124	8314 approx.
Nitrogen	28	297	for all gases
Oxygen	32	260	
Helium	4	2077	
Carbon dioxide	44	189	

7.4 AVAGADRO'S LAW AND UNIVERSAL GAS CONSTANT

Different gases posses different values of gas constant. For uniformity, another unit of mass called kg mole is used, which has a mass equal to its molecular weight. For example, one kg mole of oxygen has a mass of 32 kg and one kg mole of nitrogen has a mass of 28 kg.

In 1811, Avagadro proved that equal volumes of all gases at the same conditions of pressure and temperature have same number of molecules." He further proved that at standard conditions or at NTP (t = 0°C or 273.14 k and P = 101325 N/m^2), the volume of one kilo mole of all gases is equal to 22.413 m^3. This 22.413 m^3 of volume is known as mole. It is a bigger unit of volume.

Let M_1 and M_2—molecular masses of two different ideal gases having same volume, pressure and temperature.

If $\qquad n$ = number of molecules.

Then according to Avagadro's law, both gases have same 'n' number of molecules. The masses of gases are

$$m_1 \propto nM_1 = KnM_1$$

$$m_2 \propto nM_2 = KnM_2$$

Let ρ_1 and ρ_2 – densities of gases, then

$$m_1 = \rho_1 V \text{ and } m_2 = \rho_2 V$$

where $V_1 = V_2 = V$ [V = volume of each gas]

$$\therefore \quad \frac{\rho_1 V}{\rho_2 V} = \frac{KnM_1}{KnM_2};$$

$$\frac{\rho_1}{\rho_2} = \frac{M_1}{M_2}$$

The gas densities are directly proportional to their molecular masses at the same temperature and pressure.

We know that, $\rho = \frac{1}{V}$.

$$\therefore \quad \frac{V_2}{V_1} = \frac{M_1}{M_2};$$

i.e the specific volume of each gas is inversely proportional to the molecular masses at the same temperature and pressure.

$$\therefore \quad V_1 M_1 = V_2 M_2 = \ldots V_n M_n = \text{constant}$$

At same pressure and temperature, the product of specific volume and molecular mass will remain constant for all ideal gases. Also we know that,

$$\text{Molar volume} = V_0 M = 22.4135 \text{ m}^3\text{/K mole}$$

$$\therefore \text{ Specific volume,} \quad V_0 = \frac{22.4135}{M} \text{ m}^3\text{/kg}$$

$$\text{density,} \quad \rho_0 = \frac{M}{22.4135} \text{ kg/m}^3$$

For molar volume of a gas, the characteristic gas equation can be written as

$$PV_{mol} = mRT$$

$$= R_{mol} T$$

where $R_{mol} = MR$ (molecular mass × gas constant)

The value of R_{mol} is same for all gases, it is called molar gas constant or universal gas constant. At NTP, we can write

$$R_{mol} = \frac{P \cdot V_{mol}}{T}$$

$$= \frac{10135 \times 22.4135}{273.15}$$

$$= 8314.31 \text{ J/kg moleK}$$

$$\therefore \quad P.V_{mol} = 8314.31 \times T$$

$$\approx 8314\ T.$$

This is the most general equation for all gases, because it relates three ideal gas laws (Boyle's, Charle's and Avagadro's) and also includes the universal gas constant which is independent of the nature of gas.

The characteristic gas constant can be represented as

$$R = \frac{R_{mol}}{\text{mol. wt}} \quad (R_{mol} = \bar{R})$$

For oxygen, $$R = \frac{8314}{32} = 262 \text{ J/kgK}$$

For air, $$R = \frac{8314}{28.96} = 287 \text{ J/ kgK}$$

There are 6.023×10^{23} molecules in a given mol of a substance. This is known as Avagadro's number (A).

7.5 INTERNAL ENERGY AND ENTHALPY AS FUNCTIONS OF TEMPERATURE

7.5.1 Internal Energy

For any substance, we can write

$$TdS = du + PdV \qquad [dQ = du + \delta W]$$

$$\therefore \quad dS = \frac{du}{T} + \frac{PdV}{T} \tag{1}$$

The internal energy '*u*' is a function of *T* and *V*.

i.e. $$u = f(T, V)$$

or $$du = \left(\frac{\partial u}{\partial T}\right)_v dT + \left(\frac{\partial u}{\partial V}\right)_T dV$$

Substitute the value of '*du*' in eqn (1), we get.

$$dS = \frac{1}{T}\left[\frac{\partial u}{\partial T}\right]_v dT + \frac{1}{T}\left(\frac{\partial u}{\partial V}\right)_T dV + \frac{P}{T} dV$$

$$dS = \frac{1}{T}\left[\frac{\partial u}{\partial T}\right]_v dT + \frac{1}{T}\left[\left(\frac{\partial u}{\partial V}\right)_T + P\right] dV \tag{2}$$

also, $$S = f(T, V)$$

$$dS = \left(\frac{\partial S}{\partial T}\right)_v dT + \left(\frac{\partial S}{\partial V}\right)_T dV \tag{3}$$

From Eqns. (2) and (3), we can write

$$\left(\frac{\partial S}{\partial T}\right)_v = \frac{1}{T}\left(\frac{\partial u}{\partial T}\right)_v \tag{4}$$

and $$\left(\frac{\partial S}{\partial V}\right)_T = \frac{1}{T}\left[\left(\frac{\partial u}{\partial V}\right)_T + P\right] \tag{5}$$

From (4), differentiate with respect to *V*, and *T* = const.

$$\frac{\partial^2 S}{\partial T \partial V} = \frac{1}{T}\frac{\partial^2 u}{\partial T \partial V} \tag{6}$$

From (5), Differentiate with respect to T and V = constant.

$$\frac{\partial^2 s}{\partial V \partial T} = \frac{1}{T}\frac{\partial^2 u}{\partial V \partial T} + \frac{1}{T}\left(\frac{\partial P}{\partial T}\right)_v - \frac{1}{T^2}\left(\frac{\partial u}{\partial V}\right)_T - \frac{P}{T^2} \tag{7}$$

From (6) and (7),

$$\frac{1}{T}\frac{\partial^2 u}{\partial T \partial V} = \frac{1}{T}\frac{\partial^2 u}{\partial V \partial T} + \frac{1}{T}\left(\frac{\partial P}{\partial T}\right)_v - \frac{1}{T^2}\left(\frac{\partial u}{\partial V}\right)_T - \frac{P}{T^2}$$

$$\therefore \quad \left(\frac{\partial u}{\partial V}\right)_T + P = T\left(\frac{\partial P}{\partial T}\right)_v$$

For Ideal gas,

$$PV = RT$$

$$V\left(\frac{\partial P}{\partial T}\right)_v = R$$

$$\left(\frac{\partial P}{\partial T}\right)_v = \frac{R}{V} = \frac{P}{T}$$

$$\therefore \quad \left(\frac{\partial u}{\partial V}\right)_T + P = T \cdot \left(\frac{P}{T}\right)$$

$$\therefore \quad \left(\frac{\partial u}{\partial V}\right)_T = 0$$

$$PV = RT$$

$$R = \frac{PV}{T}$$

$$R = \frac{P}{T}$$

Similarly, it can be shown that

$$\left(\frac{\partial u}{\partial P}\right)_T = 0.$$

i.e. $u = f(T)$, only for an Ideal gas.

This is known as Joule's law. This states that the internal energy of an Ideal gas depends on its absolute temperature only. The variation in the temperature of an Ideal gas results in change of internal energy which is same for all the values of P and V.

7.5.2 Enthalpy

For an Ideal gas, we can write

$$h = u + PV$$

$$h = u + RT \qquad [PV = RT]$$

As internal energy 'u' is a function of temperature only, Enthalpy is also a function of temperature.

$\therefore$ $h = f(T)$, only for an Ideal gas.

7.6 SPECIFIC HEAT

We know that, for an Ideal gas

$$h = u + RT$$

$$dh = du + RdT$$

$$\therefore \quad C_p dT = C_v dT + RdT$$

$$C_p = C_v + R$$

$$\therefore \quad \boxed{R = C_p - C_v}$$

We know that, from defination

$$C_P = \left(\frac{\partial u}{\partial T}\right)_P$$

$dh = C_p \cdot dT$ and

$$C_v = \left(\frac{\partial u}{\partial T}\right)_v$$

$$du = C_v dT$$

Also, the ratio of two specific heats can be written as

$$\frac{C_p}{C_v} = \gamma.$$

$\therefore$ The difference and ratio of two specific heats is a constant.

$$C_p = \gamma \times C_v$$

Also
$$R = C_p - C_v$$
$$= \gamma \cdot C_v - C_v$$
$$R = (\gamma - 1)\, C_v$$

$$\therefore \quad C_v = \frac{R}{\gamma - 1} \text{ kJ/kgK}$$

$$C_p = \gamma \cdot \left(\frac{R}{\gamma - 1}\right) \text{ kJ/kgK}$$

7.7 NON-FLOW PROCESSES USING IDEAL GAS

7.7.1 Constant Volume Process

Consider a constant volume process 1-2 as shown in figure.

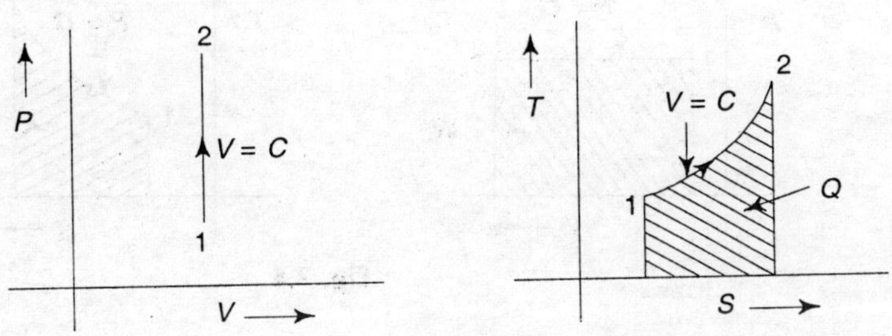

Fig. 7.3

(i) At 1, $P_1 V_1 = mRT_1$

At 2, $P_2 V_2 = mRT_2$

$\because \quad V_1 = V_2,$

$$\frac{P_1}{P_2} = \frac{T_1}{T_2}$$

or
$$\frac{P_1}{T_1} = \frac{P_2}{T_2} = \text{constant.}$$

From first law for a closed system

(ii) Heat,

$$Q = du + \int_1^2 PdV \qquad [V = \text{constant}]$$

$$Q = U_2 + U_1 = mC_v(T_2 - T_1)$$

As mechanical work is zero, the heat is equal to change in property.

(iii) Work,

$$\text{Work done} = \int_1^2 PdV = 0$$

(iv) The change in Internal energy is

$$\Delta u = mC_v\,(T_2 - T_1)$$

(v) The change in enthalpy

$$\Delta u = mC_p \cdot (T_2 - T_1)$$

(vi) The change in entropy

$$\int_1^2 dS = S_2 - S_1 = \int_1^2 \frac{dQ}{T} = \int_1^2 \frac{C_v dT}{T}$$

$$dS = C_v \log_e \left(\frac{T_2}{T_1}\right)$$

7.7.2 Constant Pressure Process

Consider a constant pressure process 1-2 as shown in Figure.

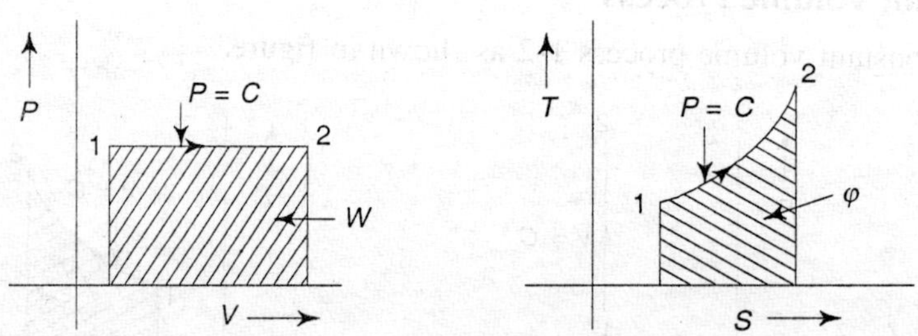

Fig. 7.4

(i) At, state 1, $P_1V_1 = mRT_1$

At state 2, $P_2V_2 = mRT_2$

As $P_1 = P_2$,

$$\frac{V_1}{V_2} = \frac{T_1}{T_2}$$

$$\frac{V_1}{T_1} = \frac{V_2}{T_2} = \ldots \frac{V_n}{T_n} = \text{constant}$$

(ii) **Heat**

from Ist law of TD

$$\delta Q = dV + PdV$$

$$\therefore \quad Q = (U_2 - U_1) + \int_1^2 PdV$$

But, $$H = U + PV$$

$$dH = dU + PdV + Vdp \qquad \text{(at constant pressure } VdP = 0)$$

$$\therefore \quad dH = dU + PdV$$

$$\therefore \quad dU = dH - PdV$$

$$\therefore \quad \delta Q = dU + PdV$$

$$\delta Q = dH - Pd\cancel{V} + Pd\cancel{V}$$

$$\boxed{\delta Q = dH} \text{ or } \boxed{Q = \Delta H}$$

(iii) Work

$$W = \int_1^2 PdV = P(V_2 - V_1)$$

$$W = P_2V_2 - P_1V_1$$

(iv) Internal energy

$$\Delta U = U_2 - U_1 = mC_v\,(T_2 - T_1)$$

(v) Enthalpy

$$\Delta H = H_2 - H_1 = mC_p\,(T_2 - T_1)$$

(vi) Entropy

$$\int_1^2 dS = S_2 - S_1 = \int_1^2 \frac{dQ}{T} = C_p \int_1^2 \frac{dT}{T}$$

$$dS = C_p \log_e\left(\frac{T_2}{T_1}\right)$$

7.7.3 Constant Temperature Process

Consider a change of state during which temperature is constant.

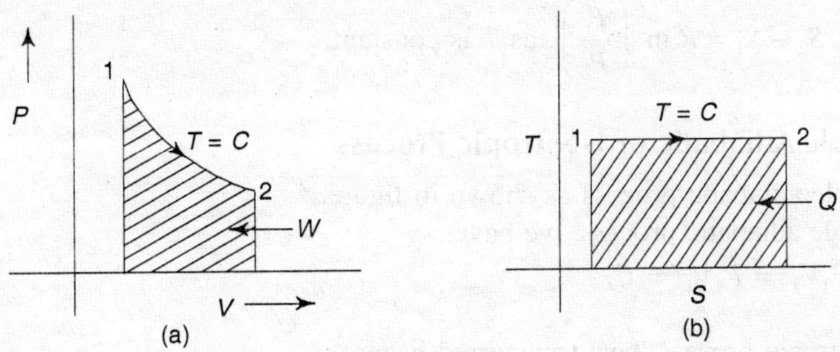

Fig. 7.5

(i) **At state 1,** $P_1V_1 = mRT_1$

At state 2, $P_2V_2 = mRT_2$

$\therefore$ $P_1V_1 = P_2V_2 = \text{constant}$

(ii) **Heat**

From first law for a closed system

$$Q = W + \Delta u$$

As $\Delta u = 0;$

$$Q = W = P_1V_1 \log_e \frac{V_2}{V_1}$$

$$= P_1V_1 \log_e \frac{P_1}{P_2} = mRT_1 \log_e \frac{P_1}{P_2}$$

All the heat added is converted into work during constant temperature expansion process.

(iii) **Work**

$$W = \int_1^2 PdV \qquad\qquad PV = C$$

$$= C \int_1^2 \frac{dV}{V} \qquad\qquad P = \frac{C}{V}$$

$$= C \ln\left(\frac{V_2}{V_1}\right) = P_1V_1 \ln\left(\frac{V_2}{V_1}\right) = RT_1 \ln\left(\frac{P_1}{P_2}\right)$$

(iv) **Internal energy**

The change in internal energy during an isothermal process is zero.

i.e $\Delta U = C_v(dT) = 0$, as $dT = 0$

(v) **Enthalpy**

The change in enthalpy is

$$\Delta H = C_p\,(dT) = 0, \text{ as } dT = 0.$$

(vi) **Entropy**

The entropy change is given by

$$\Delta S = S_2 - S_1 = \int_1^2 \frac{dQ}{T} = \frac{RT \ln(P_1/P_2)}{T}$$

$$S_2 - S_1 = R \ln\left(\frac{P_1}{P_2}\right), \text{ as } T \text{ is constant.}$$

7.7.4 Reversible Adiabatic or Isentropic Process

Consider reversible adiabatic process as shown in figure.

(i) For reversible adiabatic process, we have

$$P_1V_1^{\gamma} = P_1V_1^{\gamma} = C.$$

(ii) **Heat**

During isentropic process, heat transferred is zero.

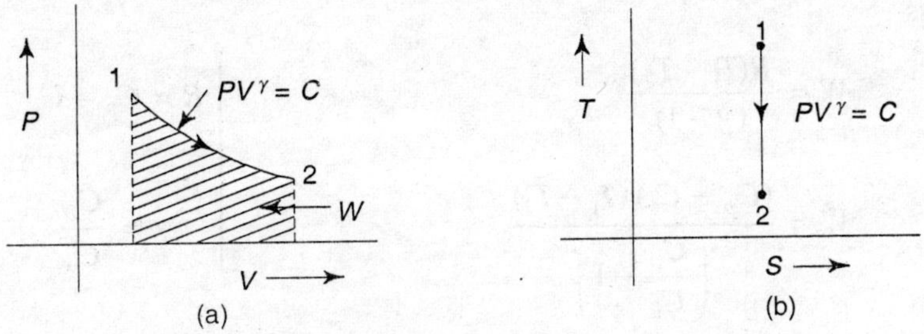

Fig. 7.6

(iii) **Work**

The work done during the process 1-2 is given by

$$W_{1\text{-}2} = \int_1^2 PdV \qquad PV^{\gamma} = C$$

$$= C\int_1^2 \frac{dV}{V^{\gamma}} \qquad P = C.V^{-\gamma}$$

$$= C\left[\frac{V^{-\gamma+1}}{-\gamma+1}\right]_1^2$$

$$= \frac{C}{1-\gamma}\left[V_2^{1-\gamma} - V_1^{1-\gamma}\right]$$

$$= \left(\frac{1}{1-\gamma}\right)\left[CV_2^{1-\gamma} - C\cdot V_1^{1-\gamma}\right]$$

$$(P_1V_1^{\gamma} = P_2V_2^{\gamma} = C)$$

$$\therefore \quad \left[\frac{1}{1-\gamma}\right]\left[P_2V_2^{\gamma}\cdot V_2^{1-\gamma} - P_1V_1^{\gamma}\cdot V_1^{1-\gamma}\right]$$

$$= \left[\frac{1}{1-\gamma}\right]\left[P_2 V_2^{\gamma} V_2^{1}\cdot V_2^{-\gamma} - P_1 V_1^{\gamma} V_1^{\gamma}\cdot V_1^{-\gamma}\right]$$

$$= \left[\frac{1}{1-\gamma}\right][P_2V_2 - P_1V_1]$$

$$W = \left(\frac{P_1V_1 - P_2V_2}{\gamma - 1}\right)$$

$$W = \frac{mR\,(T_1 - T_2)}{\gamma - 1}$$

For unit mass

$$W = \frac{R(T_1 - T_2)}{(\gamma - 1)} \qquad R = C_p - C_v$$

$$\therefore \quad W = \frac{(C_p - C_v)(T_1 - T_2)}{\left(\frac{C_p}{C_v} - 1\right)} \qquad \gamma = \frac{C_p}{C_v}$$

$$W = \frac{(C_p - C_v)(T_1 - T_2)}{\frac{C_p - C_v}{C_v}} = \frac{\cancel{(C_p - C_v)}(T_1 - T_2)}{\cancel{(C_p - C_v)}} \cdot C_v$$

$$W = C_v (T_1 - T_2)$$

i.e. the work done is equal to change in internal energy.

(iv) **Internal energy**

The change in internal energy

$$\Delta U = mC_v (T_2 - T_1)$$

(v) **Enthalpy**

The change in enthalpy is given by

$$\Delta H = mC_p (T_2 - T_1)$$

(vi) **Entropy**

The change in entropy during isentropic process is zero

i.e. $\Delta S = S_2 - S_1 = 0$

7.7.5 Polytropic Process

Consider a non-flow polytropic process as shown in figure.

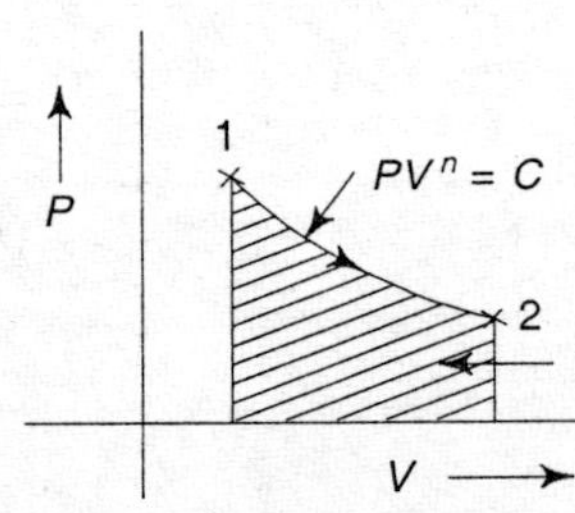

Fig. 7.7

(i) We have $P_1V_1^n = P_2V_2^n = C$

$$P_1 = \frac{RT_1}{V_1}$$

and
$$P_2 = \frac{RT_2}{V_2}$$

$$\therefore \quad \left(\frac{P_1}{P_2}\right) = \left(\frac{T_1}{T_2}\right)\left(\frac{V_2}{V_1}\right) \qquad (1)$$

also $P_1V_1^n = P_2V_2^n$

$$\therefore \quad \frac{P_1}{P_2} = \left(\frac{V_2}{V_1}\right)^n \qquad (2)$$

from (1) and (2), we can write

$$\left(\frac{V_2}{V_1}\right)^n = \left(\frac{T_1}{T_2}\right)\left(\frac{V_2}{V_1}\right)$$

$$\therefore \qquad \left(\frac{V_2}{V_1}\right)^{n-1} = \frac{T_1}{T_2}$$

$$\boxed{\left(\frac{V_2}{V_1}\right) = \left(\frac{T_1}{T_2}\right)^{1/n-1}} \qquad (3)$$

Substitute (3) in Eqn. (2), we get

$$\boxed{\frac{P_1}{P_2} = \left(\frac{T_1}{T_2}\right)^{n/n-1}}$$

The value of 'n' may be greater or less than γ.

(ii) **Heat**

We know that, $Q = W + C_v\, dT$

$$= \frac{R(T_1 - T_2)}{(n-1)} + C_v\,(T_2 - T_1)$$

$$Q = \frac{(C_p - C_v)(T_1 - T_2)}{(n-1)} + C_v\,(T_2 - T_1)$$

Divide and multiply by C_v

$$Q = C_v\,\frac{(\gamma - 1)(T_2 - T_1)}{(1-n)} + C_v\,(T_2 - T_1)$$

$$C_v\,(T_2 - T_1) = \left[\frac{\gamma - 1}{1-n} + 1\right] \qquad \left|\; \frac{\gamma-1}{1-n} + 1 = \frac{\gamma - 1 + 1 - n}{1-n} = \frac{\gamma - n}{1-n}\right.$$

$$\boxed{Q = C_v\,(T_2 - T_1)\left[\frac{\gamma - n}{1-n}\right]}$$

Or

$$Q = C_n\,(T_2 - T_1)$$

where $\qquad C_n = C_v\left[\frac{\gamma - n}{1-n}\right]$

which is known as polytropic specific heat

Also $\qquad Q = \left[\frac{\gamma - n}{\gamma - 1}\right] \times \text{polytropic work done}$

(iii) **Work**

The work done during a polytropic process is given by

$$W = \int_1^2 PdV \qquad \left| \; PV^n = C \right.$$

$$= C\int_1^2 \frac{dV}{V^n} \qquad \left| \; P = \frac{C}{V^n} \right.$$

$$= \frac{C \cdot V_2^{1-n} - CV_1^{1-n}}{(1-n)}$$

$$W = \frac{P_2V_2 - P_1V_1}{(1-n)} = \frac{P_1V_1 - P_2V_2}{(n-1)}$$

$$W = \frac{mR\,(T_1 - T_2)}{(n-1)}$$

(iii) Internal energy

The change in internal energy is

$$\Delta U = C_v\,(T_2 - T_1)$$

(iv) **Enthalpy**

The change in enthalpy is

$$\Delta H = C_p\,(T_2 - T_1)$$

(vi) **Entropy**

The change in entropy for a polytropic process can be written as

$$dS = S_2 - S_1 = C_p \ln\left(\frac{T_2}{T_1}\right) - R \ln\left(\frac{P_2}{P_1}\right)$$

7.8 STEADY FLOW (OPEN SYSTEM) PROCESSES

For a steady flow process in which both energy and mass transfer takes place.

We can write $h = u + PV$

In the differential form

$$dh = du + PdV + VdP.$$

But
$$dQ = du + PdV$$

$\therefore$
$$dh = dQ + VdP$$

$\therefore$
$$dQ - dh = -VdP.$$

and by ignoring kinetic and potential energy changes, we can write

$$dW = -VdP$$

or
$$W = -\int_1^2 VdP$$

Based on above equations, different processes are analysed in an open system.

7.8.1 Constant Volume Process

During this process, both specific volume and density are equal at inlet and exit

Example

Pumps

The work done is given by

$$W = -\int_1^2 VdP = -V\int_1^2 dP$$

$$W = V(P_1 - P_2)$$

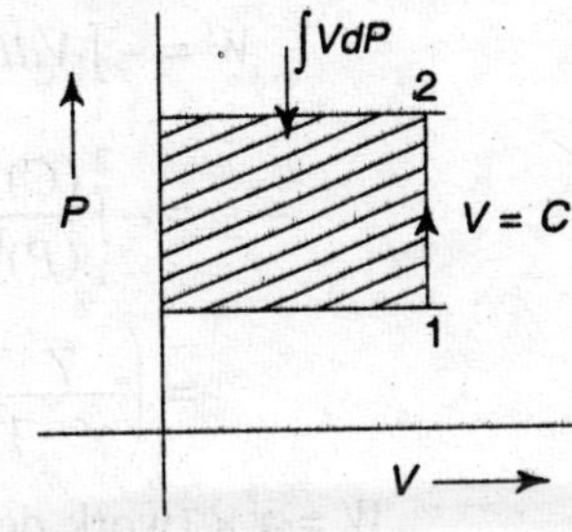

Fig. 7.8

7.8.2 Constant Pressure Process

During a constant pressure process as observed in heat exchangers, the work done is zero.

i.e. $$W = -\int_1^2 VdP = 0, \quad \text{as } dP = 0.$$

and heat transfer

$$dQ = dh - VdP$$

$$dQ = dh$$

$$\boxed{Q = H_2 - H_1}$$

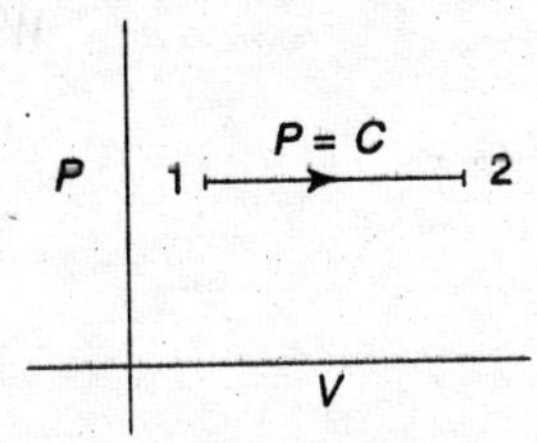

Fig 7.9

7.8.3 Constant Temperature Process

We know that, $PV = C$

$$\therefore \quad V = \frac{C}{P}$$

$$W = \int_1^2 VdP$$

$$= -C\int_1^2 \frac{dP}{P}$$

as $$C = PV$$

$$W = -C[\log_e P_2 - \log_e P_1]$$

$$W = -P_1V_1 \log_e\left(\frac{P_1}{P_2}\right)$$

$$= P_1V_1 \log_e\left(\frac{V_2}{V_1}\right) \quad \text{and}$$

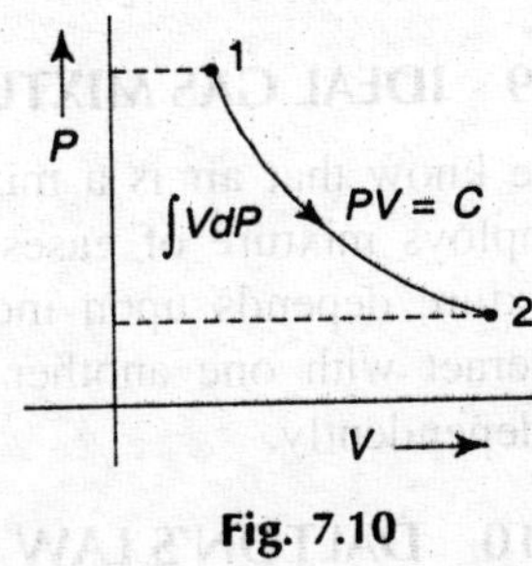

Fig. 7.10

Heat transfer

We know that, $dQ = dh - VdP$

as enthalpy is a function of temperature

$$dh = 0$$

$$\therefore \quad dQ = -VdP = \text{work done}$$

7.8.4 Isentropic Process

During isentropic process, there is no heat transfer and no friction losses.

$$W = -\int_1^2 VdP. \qquad PV^\gamma = C$$

$$= -\int_1^2 \frac{(C)^{1/\gamma}}{(P)^{1/\gamma}}\, dP \qquad V^\gamma = \frac{C}{P}$$

$$= \left(\frac{\gamma}{\gamma - 1}\right)(P_1V_1 - P_2V_2) \qquad V = \left(\frac{C}{P}\right)^{1/\gamma}$$

$W = \gamma \times$ (work done in a closed system process)

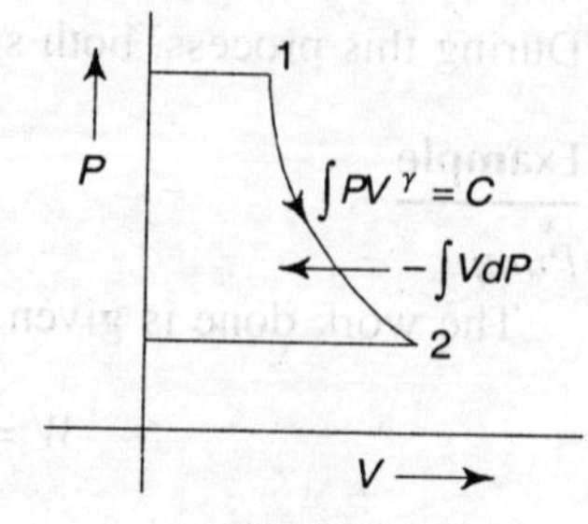

Fig. 7.11

7.8.5 Polytropic Process

$$W = -\int_1^2 VdP$$

$$= -\int_1^2 \frac{(C)^{1/n}}{(P)^{1/n}}\, dP$$

$$= \left(\frac{n}{n - 1}\right)(P_1V_1 - P_2V_2)$$

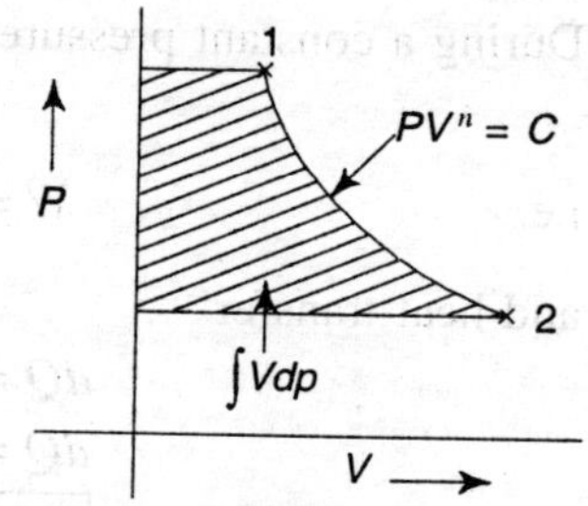

Fig 7.12

∴ Work done during flow process = $n \times WD$ during non-flow process.

7.9 IDEAL GAS MIXTURE

We know that air is a mixture of oxygen, nitrogen and other gases. Also most of the heat engines employs mixture of gases such as air, combustion products. The thermodynamic behaviour of gas mixture depends upon individual properties of its constituent gases. The constituent gases do not interact with one another chemically and are assumed to be ideal. The constituent gases behave independently.

7.10 DALTON'S LAW OF PARTIAL PRESSURE

It states that "The pressure of a gaseous mixture is the sum of the pressures which each constituent (component) would exert, if it alone occupied the volume of the mixture at the temperature of the mixture."

The pressure which a component would exert in these circumstances is known as the partial pressure of the constituent (component).

The Ideal gas mixture obeys this Dalton's law.

$$P = P_1 + P_2 + \ldots P_n = \sum_i^n P_i$$

where P = Pressure of the mixture

P_1, P_2, P_n = partial pressures

The assumption about temperature in Dalton's law can be written as

$$t = t_1 = t_2 = \ldots = t_n$$

Since by Dalton's law each component is thought of as occupying the whole volume V, we can write

$$V = V_1 = V_2 = \ldots V_n$$

In terms of specific volume

$$MV = m_1 V_1 = m_2 V_2 = \ldots = m_n V_n$$

where m's are masses.

Since the mass of the mixture is equal to the sum of the masses of the components, i.e

$$m = m_1 + m_2 + \ldots + m_n$$

from the above equations, the relation between specific volumes can be written as

$$\frac{1}{V} = \frac{1}{V_1} + \frac{1}{V_2} + \ldots + \frac{1}{V_n}$$

For density, we can write

$$\rho = \rho_1 + \rho_2 + \ldots + \rho_n$$

i.e., the density of the mixture is equal to the sum of the densities of individual constituents.

7.11 MASS, VOLUME AND MOLAR FRACTIONS

7.11(a) Mass Fraction

The mass fraction of individual constituent of a mixture is defined as the ratio of mass of a given constituent to the mass of the whole mixture.

Let m = total mass of the mixture

Then, we can write

$$m = m_1 + m_2 + \ldots + m_n$$

After dividing the above equation by m, we get

$$1 = \frac{m_1}{m} + \frac{m_2}{m} + \ldots + \frac{m_n}{m}$$

$$\therefore \quad 1 = g_1 + g_2 + \ldots + g_n$$

where g_1, g_2, g_n represents mass fractions of individual components and the sum of these mass fractions is equal to unity. If the gas mixture is analysed on the basis of mass, it is called a gravimetric analysis.

7.11(b) Volume Fraction

It is the ratio of partial volume (reduced volume) of individual constituent gas to the total volume of the mixture.

$$Vf_1 = \frac{V_1}{V};\ Vf_2 = \frac{V_2}{V};\ Vf_n = \frac{V_n}{V}$$

where Vf_1, Vf_2, Vf_n = Volume fractions

V_1, V_2, V_n = Partial volumes of individual constituent gas

V = Total volume of the gas mixture.

Consider a gas mixture with pressure P and temperature T, then according to characteristic gas equation,

$$PV_1 = m_1 R_1 T;\ PV_2 = m_2 R_2 T_1;\ PV_n = m_n R_n T$$

$$\therefore \quad P(V_1 + V_2 + \ldots V_n) = (m_1R_1 + m_2R_2 + \ldots + m_nR_n)\,T$$
$$= P_1V + P_2V + \ldots + P_nV$$
$$= (P_1 + P_2 + \ldots + P_n)V$$

where $P_1 + P_2 + \ldots P_n$ = Sum of partial pressures

According to Dalton's law of partial pressure

$$P_1 + P_2 + \ldots + P_n = P$$
$$\therefore \quad P(V_1 + V_2 + \ldots + V_n) = P.V$$
$$\therefore \quad V_1 + V_2 + \ldots V_n = V$$

i.e. the total volume of the mixture is equal to the sum of partial volumes of the constituent gases, when the partial volumes are obtained at pressure and temperature of the mixture. This is the Amagat's law of partial volume.

If we divide the total volume of the mixture in the above equation, we get

$$\frac{V_1}{V} + \frac{V_2}{V} \ldots + \frac{V_n}{V} = \frac{V}{V}$$
$$Vf_1 + Vf_2 + \ldots + Vf_n = 1.$$

i.e the sum of volume fractions is equal to unity. If the gas mixture is analysed on the basis of volume, it is called volumetric analysis.

7.11 (c) Mole Fraction

It is the ratio of number of moles of each constituent gas to the total number of moles of the gas mixture.

$$\therefore \quad \mu_1 = \frac{n_1}{n};\ \mu_2 = \frac{n_2}{n};\ \mu_n = \frac{n_n}{n}$$

where $\mu_1, \mu_2, \mu_n \ldots$ mole fractions of each constituent gas

$n_1, n_2, n_n \ldots$ number of moles of each constituent gas

n = total number of moles of the gas mixture

We know that,

$$n = n_1 + n_2 + \ldots + n_n$$

after dividing by n, we get

$$1 = \frac{n_1}{n} + \frac{n_2}{n} + \ldots \frac{n_n}{n}$$
$$\therefore \quad \mu_1 + \mu_2 + \ldots \mu_n = 1$$

It means the sum of mole fractions is unity. If the gas mixture is analysed on the basis of mole, it is called molar analysis.

Consider a gas mixture at pressure 'P' and temperature 'T'. Now, by applying characteristic gas equation, we can write

$$PV = n \cdot R_{\text{mol}}\ T \qquad (1)$$

For a component of a gas mixture, say 'y' component, we can write,

Similarly $$PV_y = n_y\, R_{\text{mol}}\ T \qquad (2)$$

By comparing equations (1) and (2), we can write

$$\frac{V_y}{V} = \frac{n_y}{n}$$

i.e. $Vf_y = \mu_y$

It means, for an Ideal gas mixture, the volume fraction and mole fraction are same.

7.12 SPECIFIC HEAT OF A GAS MIXTURE

Let $m_1, m_2, m_3 \ldots m_n$ = masses of constituent gases of a mixture.

$c_1, c_2, c_3 \ldots c_n$ = specific heats of constituent gases of a mixture.

Then the amount of heat absorbed by individual gas per unit degree temperature raise

$$q_1 = m_1c_1;\ q_2 = m_2c_2;\ q_n = m_nc_n$$

$$\therefore \quad \sum_i^n q = Q = m_1c_1 + m_2c_2 + \ldots m_nc_n$$

Let m = Total mass of mixture

c_m = Specific heat of gas mixture

Then, $Q = m \cdot c_m$

By comparing the above two equations, we can write

$$mc_m = m_1c_1 + m_2c_2 + \ldots m_nc_n$$

$$\therefore \quad c_m = \frac{m_1}{m}c_1 + \frac{m_2}{m}c_2 + \ldots \frac{m_n}{m}c_n$$

$$= g_1c_1 + g_2c_2 + \ldots g_nc_n$$

$$\boxed{c_m = \sum_i^n c_i g_i}$$

7.13 INTERNAL ENERGY, ENTHALPY AND ENTROPY OF A GAS MIXTURE

The internal energy, enthalpy and entropy of a gaseous mixture are respectively equal to the sums of internal energies, the enthalpies and the entropies of each individual constituent of the mixture, if each of them alone occupied the volume of the mixture at the temperature of the mixture.

The internal energy of the mixture can be written as

$$U = U_1 + U_2 + \ldots + U_n$$

Or on the mass basis,

$$mu = m_1u_1 + m_2u_2 + \ldots + m_nu_n$$

$$u = \frac{m_1u_1 + m_2u_2 + \ldots + m_nu_n}{m_1 + m_2 + \ldots + m_n}$$

where u = average specific internal energy of the mixture,

Similarly, the total enthalpy of a gas mixture can be written as

$$H = H_1 + H_2 + \ldots + H_n$$

or on the mass basis

$$m_h = m_1h_1 + m_2h_2 + \ldots + m_nh_n$$

$$h = \frac{m_1h_1 + m_2h_2 + \ldots + m_nh_n}{m_1 + m_2 + \ldots + m_n}$$

where, h = average specific enthalpy of the mixture

The total entropy of a gas mixture can be written as

$$S = S_1 + S_2 + \ldots + S_n$$

or

$$mS = m_1S_1 + m_2S_2 + \ldots + m_nS_n$$

$$S = \frac{m_1S_1 + m_2S_2 + \ldots + m_nS_n}{m_1 + m_2 + \ldots + m_n}$$

Also from the definitions of specific heats, it can be written as

$$C_{v\text{ mean}} = \frac{m_1C_{v1} + m_2C_{v2} + \ldots + m_nC_{vn}}{m_1 + m_2 + \ldots + m_n}$$

and

$$C_{p\text{ mean}} = \frac{m_1C_{p1} + m_2C_{p2} + \ldots + m_nC_{pn}}{m_1 + m_2 + \ldots + m_n}$$

7.14 PROCESSES OF GASEOUS MIXTURE

7.14 .1 Adiabatic Mixing of Gases

Consider an adiabatic container in which a thin diaphragm separates gases 1 and 2. If the diaphragm is punctured, then adiabatic mixing of gases take place and each of them occupies the total volume. This process is equivalent to a free expansion of each gas and is irreversible. As the container is perfectly insulated, the process is adiabatic and entropy of the system increases.

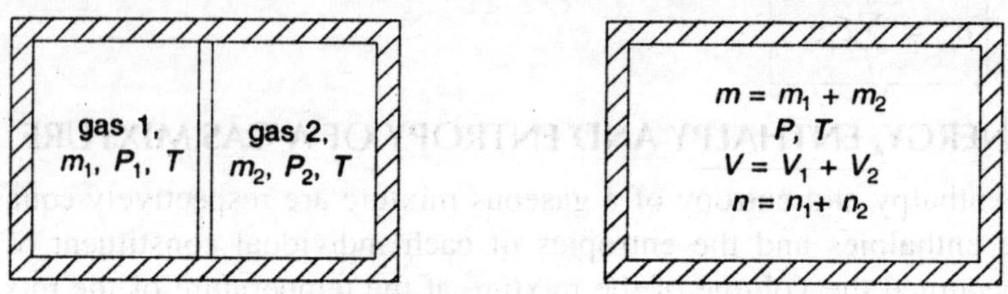

Fig. 7.13

As shown in *T-S* diagram, let '1' represent state '1' of gas 1 at pressure P and temperature T, before mixing. The state 1′ is the condition after mixing at temperature T and partial pressure $P_1 (P_1 < P)$. The increase in entropy of gas 1 is given by,

$$S_1' - S_1 = m_1R_1 \log_e \left(\frac{P}{P_1}\right)$$

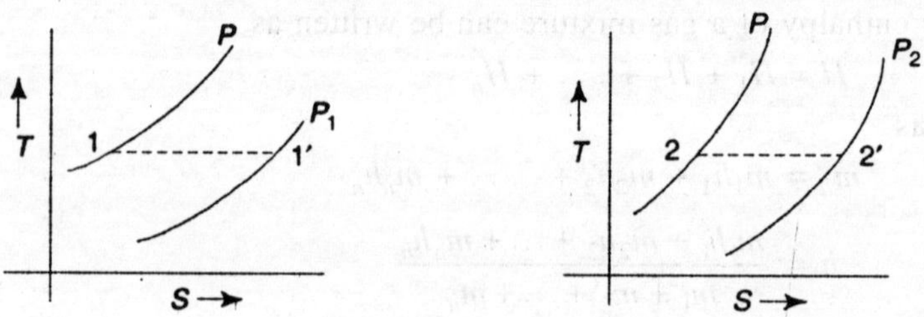

Fig. 7.14

Similarly for gas 2,

$$S_2' - S_2 = m_2 R_2 \log_e \left(\frac{P}{P_2}\right)$$

By the Gibb's-Dalton law, the entropy of $(m_1 + m_2)$ kg of mixture is given by

$$S_m = S_1' + S_2'$$

The entropy of $(m_1 + m_2)$ kg of constituents before mixing process is given by

$$S_c = S_1 + S_2$$

∴ The increase in entropy due to adiabatic mixing is given by

$$\Delta S = S_m - S_c = (S_1' + S_2') - (S_1 + S_2)$$
$$= (S_1' - S_1) + (S_2' - S_2)$$

$$S_m - S_c = m_1 R_1 \ln \left(\frac{P}{P_1}\right) + m_2 R_2 \ln \left(\frac{P}{P_2}\right)$$

$$= n_1 M_1 \frac{R_{mol}}{\mu_1} \ln \left(\frac{P}{P_1}\right) + n_2 M_2 \frac{R_{mol}}{\mu_2} \ln \left(\frac{P}{P_2}\right)$$

$$= R_{mol} \left[n_1 \log_e \left(\frac{P}{P_1}\right) + n_2 \log_e \left(\frac{P}{P_2}\right)\right]$$

But $$\frac{P_1}{P} = \mu_1$$

∴ $$S_m - S_c = R_{mol} \left[n_1 \log_e \frac{1}{\mu_1} + n_2 \log_e \frac{1}{\mu_2}\right]$$

It means, the increase in entropy of a mixture depends only on the number of moles of the component gases and it increases only when different gases mix.

There will be further increase of entropy, when the temperature of the constituent gases are initially different.

7.14.2 Mixtures of Perfect Gases at Different Initial Pressures and Temperatures

Consider two insulated containers having perfect gases '1' and '2', which are at different pressure and temperature. When the valve opens, the gases diffuse into one another and the resulting mixture attains some intermediate pressure and temperature.

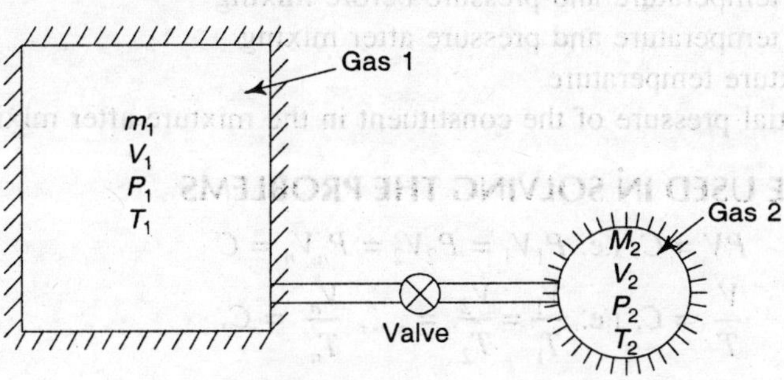

Fig. 7.15

We know that, the specific internal energy and enthalpy are the functions of temperature only.

$$mu = m_1u_1 + m_2u_2$$

$$\therefore \quad u = \frac{m_1}{m}u_1 + \frac{m_2}{m}u_2$$

$$u = g_1u_1 + g_2u_2$$

$$= \Sigma\, g_nu_n$$

$$= \Sigma\, g_nC_{Vn}\, T_n$$

We know that, for the mixture, the gas equation can be written as

$$PV = mRT$$

For constituent, we can write

$$P_1V = m_1R_1T_1$$

$$P_2V = m_2R_2T_2$$

$$\therefore \quad (P_1 + P_2)V = m_1R_1T_1 + m_2R_2T_2$$

But

$$P = P_1 + P_2$$

$$\therefore \quad PV = mRT = m_1R_1T_1 + m_2R_2T_2$$

$$\therefore \quad T = \frac{m_1R_1T_1 + m_2R_2T_2}{mR}$$

$$T = \frac{\Sigma m_nR_nT_n}{mR}$$

But

$$R = \Sigma g_n\, R_n$$

$$\therefore \quad T = \frac{\Sigma m_nR_nT_n}{m\Sigma g_nR_n}$$

$$\boxed{T = \frac{\Sigma m_nR_nT_n}{\Sigma m_nR_n}}$$

Also the entropy change of individual gas component is given by

$$\Delta S = m\left[\int_{T_1}^{T_2} C_p\frac{dT}{T} - R\log_e\left(\frac{P_2}{P_1}\right)\right]$$

where T_1, P_1 – temperature and pressure before mixing
T_2, P_2 – temperature and pressure after mixing
[T_2 – mixture temperature
P_2 – partial pressure of the constituent in the mixture after mixing]

7.15 FORMULAE USED IN SOLVING THE PROBLEMS

1. Boyle's law $\quad PV = C$, i.e. $P_1V_1 = P_2V_2 = P_nV_n = C$
2. Charle's law $\quad \frac{V}{T} = C$, i.e. $\frac{V_1}{T_1} = \frac{V_2}{T_2} = \ldots \frac{V_n}{T_n} = C.$

$$\frac{P}{T} = C$$

$$\frac{P_1}{T_1} = \frac{P_2}{T_2} = \ldots \frac{P_n}{T_n} = C$$

3. The characteristic gas equation is

$$PV = mRT$$

4. $$\frac{\rho_1}{\rho_2} = \frac{M_1}{M_2} \text{ i.e. } V_1M_1 = V_2M_2 = \ldots V_nM_n = C$$

5. $$PV_{mol} = MRT$$

$$= R_{mol}\, T, \quad R_{mol} = MR$$

6. $$\left(\frac{\partial u}{\partial u}\right)_T = 0 \text{ or } \left(\frac{\partial u}{\partial P}\right)_T = 0.$$

$\therefore$ $$u = f(T) \text{ and } h = f(T)$$

7. $$R = C_p - C_v$$

8. $$C_p/C_v = \gamma \text{ or } C_p = \gamma\left(\frac{R}{\gamma - 1}\right)$$

9. For constant volume process

$$\frac{P_1}{T_1} = \frac{P_2}{T_2} = \text{constant}$$

$$Q = U_2 - U_1 =- mC_v\,(T_2 - T_1)$$

$$WD = 0$$

$$\Delta u = mC_v\,(T_2 - T_1)$$

$$\Delta h = mC_p\,(T_2 - T_1)$$

$$dS = C_v \ln\,(T_2/T_1)$$

10. For constant pressure process

$$\frac{V_1}{T_1} = \frac{V_2}{T_2} = \ldots \frac{V_n}{T_n} = C$$

$$Q = DH$$

$$W = P(V_2 - V_1)$$

$$\Delta U = U_2 - U_1 = mC_p\,(T_2 - T_1)$$

$$\Delta H = H_2 - H_1$$

$$= mC_p\,(T_2 - T_1)$$

$$ds = C_p \ln\,(T_2/T_1)$$

11. For constant temperature process

$$P_1V_1 = P_2V_2 = C$$

$$Q = W = P_1V_1 \ln\,(V_2/V_1) = RT_1 \ln\,(P_1/P_2)$$

$$\Delta U = C_v(dT) = 0$$

$$\Delta H = C_p(dT) = 0$$

$$\Delta S = S_2 - S_1 = R \ln\,(P_1/P_2)$$

12. For Isentropic process

$$P_1V_1^{\gamma} = P_2V_2^{\gamma} = C$$

$$Q = 0$$

$$WD = \frac{P_1V_1 - P_1V_2}{\gamma - 1} = \frac{mR(T_1 - T_2)}{\gamma - 1} = C_v\,(T_1 - T_2)$$

$$\Delta U = mC_v\,(T_2 - T_1)$$

$$\Delta H = mC_p\,(T_2 - T_1)$$

$$\Delta S = S_2 - S_1 = 0$$

13. For polytropic process

$$P_1V_1^n = P_2V_2^n = C$$

$$Q = C_v\,(T_2 - T_1)\left[\frac{\gamma - n}{1 - n}\right]$$

or

$$Cn(T_2 - T_1) \text{ where } Cn = C_v\left(\frac{\gamma - n}{1 - n}\right)$$

or

$$Q = \left(\frac{\gamma - n}{\gamma - 1}\right) \times \text{Polytropic W.D.}$$

$$WD = \frac{P_1V_1 - P_2V_2}{n - 1} = \frac{mR(T_1 - T_2)}{n - 1}$$

$$\Delta U = C_v\,(T_2 - T_1)$$

$$\Delta H = C_p\,(T_2 - T_1)$$

$$dS = S_2 - S_1 = C_p \ln (T_2/T_1) - R \ln (P_2/P_1)$$

14. Steady flow (open system) processes

$W = V(P_1 - P_2)$: constant volume process

$Q = H_2 - H_1$: constant pressure process

$$\left.\begin{aligned} W &= -\,P_1V_1 \ln (P_1/P_2) \\ &= P_1V_1 \ln (V_2/V_1) \\ \text{and}\quad dQ &= VdP = WD \end{aligned}\right\}\ \text{constant temperature process}$$

$$W = \left(\frac{\gamma}{\gamma - 1}\right)(P_1/V_1 - P_2V_2) \text{ — Isentropic process}$$

$$W = \left(\frac{n}{n - 1}\right)(P_1V_1 - P_2V_2) \text{ — Polytropic process}$$

15. Dalton's law of partial pressure

$$P = P_1 + P_2 + P_3 + \ldots P_n$$

$$V = V_1 = V_2 = \ldots V_n$$

$$t = t_1 = t_2 = \ldots t_n$$

i.e. $m_1V_1 = m_2V_2 = \ldots m_nV_n$

$$m = m_1 + m_2 + \ldots m_n$$

16. $\frac{m_1}{m} + \frac{m_2}{m} + \ldots \frac{m_n}{m} = 1$

$g_1 + g_2 + \ldots g_n = 1$

17. $Vf_1 = \frac{V_1}{V}, Vf_2 = \frac{V_2}{V}, Vf_3 = \frac{V_3}{V}, Vf_n = \frac{V_n}{V}$

$\therefore \quad Vf_1 + Vf_2 + \ldots Vf_n = 1.$

18. $\mu_1 = \frac{n_1}{n}, \mu_2 = \frac{n_2}{n}, \mu_n = \frac{n_n}{n}$

$\therefore \mu_1 + \mu_2 + \ldots \mu_n = 1$

19. $C_m = \sum_i^n c_i g_i$

20. $U = U_1 + U_2 + \ldots U_n$

$$u = \frac{m_1 u_1 + m_2 u_2 + \ldots m_n u_n}{m_1 + m_2 + m_3 + \ldots m_n}$$

$$h = \frac{m_1 h_1 + m_2 h_2 + \ldots m_n h_n}{m_1 + m_2 + m_3 + \ldots m_n}$$

$$S = \frac{m_1 S_1 + m_2 S_2 + \ldots m_n S_n}{m_1 + m_2 + \ldots m_n}$$

$$C_{v\ \text{mean}} = \frac{m_1 C_{v1} + m_2 C_{v2} + \ldots m_n C_{vn}}{m_1 + m_2 + \ldots m_n}$$

$$C_{p\ \text{mean}} = \frac{m_1 C_{p1} + m_2 C_{p2} + \ldots m_n C_{pn}}{m_1 + m_2 + \ldots m_n}$$

21. $T = \frac{\Sigma m_n R_n T_n}{\Sigma m_n R_n}$

22. $\Delta S = m \left[\int_{T_1}^{T_2} C_p \cdot \frac{dT}{T} - R \ln \left(\frac{P_2}{P_1} \right) \right]$

7.16 NUMERICAL EXAMPLES

Problem 1: *Determine the molal specific volume for an Ideal gas at 1 atmosphere and 0°C.*

Solution: The Ideal gas equation can be written as

$$P\bar{V} = \bar{R}T$$

where $\bar{R} = 8.3144$ kJ/kg mol k.

$T = 0 + 273 = 273$ k

$P = 1 \text{ atm} = 1.01325 \times 10^5 \text{ N/m}^2$

$$\bar{V} = \frac{8314.4 \times 273}{1.01325 \times 10^5}$$

$$\boxed{\bar{V} = 22.40 \text{ m}^3/\text{kg mol}}$$ *Ans.*

Problem 2: *The specific volume of gas which is at 20°C and 1 bar is 0.38 m³/kg. During certain process, the pressure decreases to 0.5 bar and temperature of the gas remains same. Determine density of the gas at the end of process.*

Solution: As temperature is constant, according to Boyle's law

$$\frac{P_1}{V_1} = \frac{P_2}{V_2}$$

$$\therefore \quad V_2 = \frac{P_1 V_1}{P_2} = \frac{1 \times 0.38}{0.5} = 0.76 \text{ m}^3/\text{kg}$$

∴ Density of the gas at the end of process

$$\rho_2 = \frac{1}{V_2} = \frac{1}{0.76}$$

$$\boxed{\rho_2 = 1.315 \text{ kg/m}^3}$$ *Ans.*

Problem 3: *During a certain process, the temperature of Ideal gas increases from 25°C to 130°C, while the pressure is constant. The initial volume of the gas is 0.65 m³ and mass is 0.75 kg. Determine density of gas after the process, assuming that the pressure remains constant during the process.*

Solution: As pressure remains constant, according to Charle's law, we can write

$$\frac{V_1}{T_1} = \frac{V_2}{T_2}$$

$$V_2 = V_1 \frac{T_2}{T_1}$$

$$V_2 = 0.866 \times \frac{(130 + 273)}{(25 + 273)}$$

$$V_2 = 1.171 \text{ m}^3/\text{kg}$$

$$V_1 = \frac{V}{m} = \frac{0.65}{0.75}$$

$$V_1 = 0.866 \text{ m}^3/\text{kg}$$

∴ Density at the end of process

$$\rho_2 = \frac{1}{V_2} = \frac{1}{1.171} = 0.85 \text{ kg/m}^3$$ *Ans.*

Problem 4: *A quantity of gas whose original volume and temperature are 0.2 m³ and 300°C. It is cooled at constant pressure until its volume becomes 0.1 m³. Determine the final temperature of the gas?*

Solution: According to Charle's law

$$= \frac{V_2}{T_2}$$

$$T_2 = T_1 \frac{V_2}{V_1} = 573 \times \frac{0.1}{0.2} = 286.5 \text{ k}$$

$$\boxed{T_2 = 13.5 \text{ °C}}$$ **Ans.**

Problem 5: *A quantity of gas has a pressure of 350 kN/m² when its volume is 0.03 m³ and its temperature is 35°C. If the value of R = 0.29 kJ/kgK, determine mass of the gas. If the gas pressure is increased to 1.05 MN/m² while the volume remains constant, determine new temperature of the gas?*

Solution:

(i) The characteristic equation is

$$PV = mRT$$

$$m = \frac{PV}{RT} = \frac{350 \times 1000 \times 0.03}{0.29 \times 1000 \times (35 + 273)}$$

$$\boxed{m = 0.1175 \text{ kg}}$$ **Ans.**

(ii) We can write

$$\frac{P_1V_1}{T_1} = \frac{P_2V_2}{T_2} \qquad [\because V_1 = V_2]$$

$$\frac{P_1}{T_1} = \frac{P_2}{T_2}$$

$$\therefore \qquad T_2 = T_1 \cdot \frac{P_2}{P_1} = 308 \left(\frac{1.05 \times 10^6}{350 \times 10^3}\right)$$

$$T_2 = 924 \text{ k}$$

$$\boxed{T_2 = 651\text{°C}}$$ **Ans.**

Problem 6: *1.5 kg of gas, occupying 0.7 m³, has original temperature of 20°C. It was then heated at constant volume until its temperature rises to 130°C. Determine heat transferred to the gas and its final pressure.*
Take C_v = 0.72 kJ/kgK and R = 0.29 kJ/kgK.

Solution:

(i) Heat transferred at constant volume

$$= Q = mC_v (T_2 - T_1)$$

$$= 1.5 \times 0.72 \,(130 - 20)$$

$$\boxed{Q = 118.8 \text{ kJ}}$$ **Ans.**

(ii) From the Ideal gas equation, we can write

$$P_1V_1 = mRT_1$$

$$P_1 = \frac{mRT_1}{P_1} = \frac{1.5 \times 0.29 \times (20 + 273)}{0.7}$$

$$P_1 = 182.07 \text{ kN/m}^2$$

Since heat addition takes place at constant volume,

$$\frac{P_1 \cancel{V_1}}{T_1} = \frac{P_2 \cancel{V_2}}{T_2}$$

$$\frac{P_1}{T_1} = \frac{P_2}{T_2}$$

$$\therefore \qquad P_2 = T_2 \frac{P_1}{T_1} = (130 + 273)\left(\frac{182.07}{20 + 273}\right)$$

$$\boxed{P_2 = 250.42 \text{ kN/m}^2} \qquad \textit{Ans.}$$

Problem 7: *0.1 m³ of gas is at a pressure of 260 kN/m² and 180°C. During a constant pressure process, its temperature decreases to 20°C. Determine heat transferred from the gas and work done on the gas during the process.*
Take R = 0.29 kJ/kgK, C_p = 1.005 kJ/kgK.

Solution: From the Ideal gas equation we can write

$$P_1 V_1 = mRT_1$$

$$m = \frac{P_1 V_1}{RT_1} = \frac{260 \times 1000 \times 0.1}{0.29 \times 1000 \times 453}$$

$$\boxed{m = 0.1979 \text{ kg}}$$

$$\therefore \qquad \text{Heat transferred} = Q = mC_p (T_2 - T_1)$$

$$= 0.1979 \times 1.005\ (20 - 180)$$

$$\boxed{Q = -31.82 \text{ kJ}} \qquad \textit{Ans.}$$

Negative sign indicates that the heat has been transferred from the gas.

We can write

$$\frac{\cancel{P_1} V_1}{T_1} = \frac{\cancel{P_2} V_2}{T_2}$$

$$\frac{V_1}{T_1} = \frac{V_2}{T_2}$$

$$V_2 = T_2 \left(\frac{V_1}{T_1}\right) = (20 + 273)\left(\frac{0.1}{180 + 273}\right)$$

$$V_2 = 0.064 \text{ m}^3$$

$$\therefore \qquad \text{Work done} = P(V_2 - V_1)$$

$$= 260 \times (0.064 - 0.1) = -9.36 \text{ kJ.}$$

Problem 8: *The initial pressure and temperature of a gas are 2.5 bar and 20°C respectively. If the temperature of the gas rises to 170°C due to compression, which follows the law $PV^{1.4} = C$, determine the pressure of the gas at the end of compression.*

Solution: Since the compression process follows the law $PV^{1.4} = C$, the relation between pressure and temperature can be expressed as

$$\frac{T_1}{T_2} = \left(\frac{P_1}{P_2}\right)^{n-1/n}$$

$$\therefore \quad \frac{P_1}{P_2} = \left(\frac{T_1}{T_2}\right)^{n/n-1}$$

$$P_2 = P_1\left(\frac{T_2}{T_1}\right)^{n/n-1} = 2.5 \times \left(\frac{170+273}{20+273}\right)^{\frac{1.4}{1.4-1}}$$

$$= 10.62 \text{ bar}$$

$$\boxed{P_2 = 10.62 \text{ bar}}$$ **Ans.**

Problem 9: *0.75 kg of gas at 14 bar and 260°C is expanded to four times the original volume by following the equation $PV^{1.35}$ = °C. Determine (i) The original and final volume of the gas, (ii) The final pressure and temperature of the gas. Take R = 0.287 kJ/kgK.*

Solution:

(i) The Ideal gas equation can be written as

$$P_1V_1 = mRT_1$$

$$V_1 = \frac{mRT_1}{P_1} = \frac{0.75 \times 0.287 \times 10^3 \times (260+273)}{14 \times 10^5}$$

$$\boxed{V_1 = 0.0891 \text{ m}^3}$$ **Ans.**

After expansion of the gas $V_2 = 4V_1$.

$$\therefore \quad V_2 = 4 \times 0.0819$$

$$\boxed{V_2 = 0.3276 \text{ m}^3}$$ **Ans.**

The initial and final volumes of the gas are

$$V_1 = 0.0819 \text{ m}^3 \quad \text{and} \quad V_2 = 0.3276 \text{ m}^3.$$

(ii) ∴ we know that, for a polytropic process,

$$P_1V_1^n = P_2V_2^n = C$$

$$\therefore \quad \frac{P_2}{P_1} = \left(\frac{V_1}{V_2}\right)^n$$

$$P_2 = 14\left(\frac{0.0819}{0.3276}\right)^{1.35}$$

$$\boxed{P_2 = 2.15 \text{ bar}}$$ **Ans.**

Also, we can write

$$\frac{P_1V_1}{T_1} = \frac{P_2V_2}{T_2}$$

$$\therefore \quad T_2 = \left(\frac{P_2V_2}{P_1V_1}\right)T_1 = \left(\frac{2.15\times10^5\times0.3276}{14\times10^5\times0.0819}\right)(260+273)$$

$$T_2 = 327.41 \text{ k} = 54.41°\text{C}$$ **Ans.**

Problem 10: *0.25 kg of air at a pressure of 1.5 bar and volume 0.12 m³, is compressed to 10 bar according to the law $PV^{1.35} = C$. Determine:*

(i) Change in internal energy of the air

(ii) The workdone on or by the air

(iii) The heat received or rejected by the air

Take C_p = 1.005 kJ/kgK. C_v = 0.718 kJ/kgK

Solution:

(i)
$$R = C_p - C_v$$
$$= 1.005 - 0.718 = 0.287 \text{ kJ/kgK}$$

$$\therefore \quad P_1V_1 = mRT_1$$

$$T_1 = \frac{P_1V_1}{mR} = \frac{1.5\times10^5\times0.12}{0.25\times0.287\times10^3}$$

$$\boxed{T_1 = 250.87 \text{ k}}$$

Also we know that,

$$\frac{T_1}{T_2} = \left(\frac{P_1}{P_2}\right)^{\frac{n-1}{n}}$$

$$T_2 = T_1\left(\frac{P_2}{P_1}\right)^{\frac{n-1}{n}} = 250.87\left(\frac{10}{1.5}\right)^{\frac{1.35-1}{1.35}}$$

$$\boxed{T_2 = 410.25 \text{ k}}$$

$\therefore$ The change in internal energy is

$$\Delta U = U_2 - U_1 = mC_v(T_2 - T_1)$$
$$= 0.25\times0.718\,(410.25 - 250.87)$$

$$\boxed{\Delta U = 28.608 \text{ kJ}}$$ **Ans.**

The positive sign indicates that there is a gain of internal energy to the air.

(ii)
$$\text{Work done} = W = \frac{mR(T_1 - T_2)}{n-1}$$

$$= \frac{0.25\times0.287\,(250.87 - 410.25)}{1.35 - 1}$$

$$\boxed{WD = -32.67 \text{ kJ}}$$ **Ans.**

The negative sign indicates that the work is done on the air.

(iii)
$$Q = \Delta U + W$$
$$Q = 28.608 - 32.67$$

$$\boxed{Q = -4.062 \text{ kJ}}$$ **Ans.**

The negative sign indicates that the heat is rejected by the air.

Problem 11: *A gas at a pressure and volume of 8 bar and 0.015 m³ respectively undergoes adiabatic expansion, during which the pressure decreases to 2 bar. Determine,*

(i) Volume and work done by the gas after expansion

(ii) Change in internal energy

Take $C_p = 1.046$ *kJ/kgK and*

$C_v = 0.752$ *kJ/kgK*

Solution: For adiabatic expansion, the index of process can be calculated as

$$\gamma = \frac{C_p}{C_v} = \frac{1.046}{0.752} = 1.39$$

(i) Also, $P_1V_1^{\gamma} = P_2V_2^{\gamma} = C$

$$\frac{P_1}{P_2} = \left(\frac{V_2}{V_1}\right)^{\gamma}$$

$$\therefore \quad \left(\frac{V_2}{V_1}\right) = \left(\frac{P_1}{P_2}\right)^{1/\gamma}$$

$$V_2 = V_1\left(\frac{P_1}{P_2}\right)^{1/\gamma}$$

$$V_2 = 0.015\left(\frac{8}{2}\right)^{1/1.39}$$

$$\boxed{V_2 = 0.0406 \text{ m}^3} \qquad \textit{Ans.}$$

$$WD = \frac{P_1V_1 - P_2V_2}{\gamma - 1} = \frac{8\times10^5 \times 0.015 - 2\times10^5(0.0406)}{1.39-1}$$

$$= 9948.71 \text{ J}$$

$$\boxed{WD = 9.948 \text{ kJ}} \qquad \textit{Ans.}$$

(ii) For an adiabatic process

$$W = -\,\Delta U$$

$$\therefore \quad \Delta U = -W$$

$$\boxed{\Delta U = -\,9.948 \text{ kJ}}$$

Problem 12: *A certain quantity of gas occupies a volume of 0.35 m³ at a pressure of 1.5 bar and a temperature of 25°C. The gas is compressed isothermally to a pressure of 5 bar and then expanded adiabatically to its initial volume. Determine:*

(i) Mass of the gas

(ii) The heat transferred during compression

(iii) Change in internal energy during expansion

Assume $\gamma = 1.4$, $C_p = 1.0$ *kJ/kgK*

Solution:

(i) $\therefore$ We know that $R = C_p\left(1 - \dfrac{1}{\gamma}\right)$

$$R = 1.0\left(1 - \frac{1}{1.4}\right)$$

$$R = 0.286 \text{ kJ/kgK}$$

$$P_1V_1 = mRT_1$$

$$m = \frac{P_1V_1}{RT_1}$$

$$= \frac{1.5 \times 10^5 \times 0.35}{0.286 \times 10^3 \times (25 + 273)}$$

$$= 0.615 \text{ kg}$$

$$C_p - C_v = R$$

$$\frac{C_p}{C_v} = \gamma$$

$$\frac{C_p}{\gamma} = C_v$$

$$c\left(C_p - \frac{C_p}{\gamma}\right) = R$$

$$C_p\left(1 - \frac{1}{\gamma}\right) = R$$

$$\frac{N}{m^2} \cdot m^3$$

$$\frac{J}{kg\,k} \times k$$

(ii) During Isothermal compression i.e., Process 1–2

$$P_1V_1 = P_2V_2 = C$$

$$V_2 = \frac{P_1}{P_2}V_1 = 0.35\left(\frac{1.5}{5}\right)$$

$$\boxed{V_2 = 0.105 \text{ m}^3}$$

From 1st law of thermodynamics

$$Q = \Delta U + W$$

$\Delta U = 0$ for isothermal process

$$\therefore \quad Q = W = P_1V_1 \ln\frac{P_1}{P_2} = 1.5 \times 10^5 \times 0.35 \ln\left(\frac{1.5}{5}\right)$$

$$= -63208.57 \text{ J}$$

$$\boxed{Q = -63.208 \text{ kJ}}$$

Ans.

The negative sign indicates that the heat is rejected from the gas.

(iii) For adiabatic expansion, i.e. process 2–3

$$P_2V_2^{\gamma} = P_3V_3^{\gamma} = C$$

$$P_3 = P_2\left(\frac{V_2}{V_3}\right)^{\gamma} \qquad (\because V_3 = V_1)$$

$$= 5\left(\frac{0.105}{0.35}\right)^{1.4}$$

$$\boxed{P_3 = 0.926 \text{ bar}}$$

From I[st] law, we can write

$$Q = \Delta U + W$$

For adiabatic expansion $Q = 0$

$$\therefore \quad W = -\Delta U$$

$$\therefore \quad \Delta U = -W = -\left[\frac{P_2V_2 - P_3V_3}{\gamma - 1}\right]$$

$$= -\left[\frac{5 \times 10^5 \times 0.105 - 0.926 \times 10^5 \times 0.35}{1.4 - 1}\right] = -\,50225 \text{ J}$$

$$\boxed{\Delta U = -\,50.225 \text{ kJ}}$$

The negative sign indicates that, there is loss of internal energy

Problem 13: *Air at a pressure of 15 bar and a temperature of 240°C expands according to $PV^{1.3} = C$ to a pressure of 1.5 bar Show the process on P-V and T-S diagrams. Also determine the work-done, heat transfer and change in entropy of the system if it contains 0.85 kg of air.*

Solution: The *P-V* and *T-S* diagrams for the above process are as shown in figures.

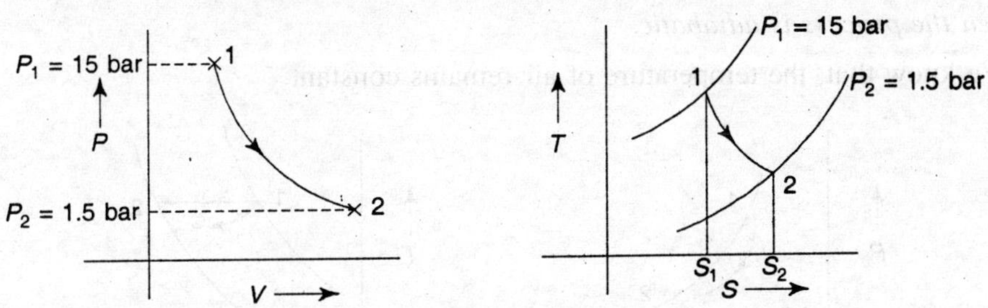

Fig. 7.16

For the process 1–2, we can write

$$\frac{T_2}{T_1} = \left(\frac{P_2}{P_1}\right)^{\frac{n-1}{n}}$$

$$T_2 = (240 + 273)\left(\frac{1.5}{15}\right)^{\frac{1.3-1}{1.3}}$$

$$\boxed{T_2 = 301.54 \text{ k}}$$

$\therefore$ Work done

$$W_{1-2} = \frac{P_1V_1 - P_2V_2}{(n-1)} = \frac{mR(T_1 - T_2)}{(n-1)}$$

$$= \frac{0.9 \times 0.287\,[(240 + 273) - (301.54)]}{(1.3 - 1)}$$

$$\boxed{W_{1-2} = 182.06 \text{ kJ}}$$ ***Ans.***

This is the work done by the system.

Also we know that, $Q = \Delta U + W$

$$= m\,C_v\,(T_2 - T_1) + W$$

$$= 0.9 \times 0.718 \times (301.54 - 513) + 182.06$$

$$\boxed{Q = 45.41 \text{ kJ}}$$ *Ans.*

The positive sign indicates that the heat is added to the system. The change in entropy is

$$S_2 - S_1 = m\,C_v \ln \frac{T_2}{T_1} + mR \ln \frac{V_2}{V_1}$$

$$= 0.9 \times 0.718 \ln \left(\frac{301.54}{513}\right) + 0.9 \times 0.287 \times \ln \left(\frac{15}{1.5}\right)^{1/1.3}$$

$$P_1V_1^n = P_2V_2^n = C$$

$$\left(\frac{V_2}{V_1}\right) = \left(\frac{P_1}{P_2}\right)^{1/n} = \left(\frac{15}{1.5}\right)^{1/1.3}$$

$$= -0.3433 + 0.4575$$

$$\boxed{S_2 - S_1 = 0.1142 \text{ kJ/K}}$$ *Ans.*

Problem 14: *Air in a vessel of 0.05 m³ and is at a pressure and temperature of 3 bar and 20°C respectively. If the air is allowed to expand in to an evacuated vessel of 0.05 m³, determine change in entropy, when the process is adiabatic.*

Solution: We know that, the temperature of air remains constant

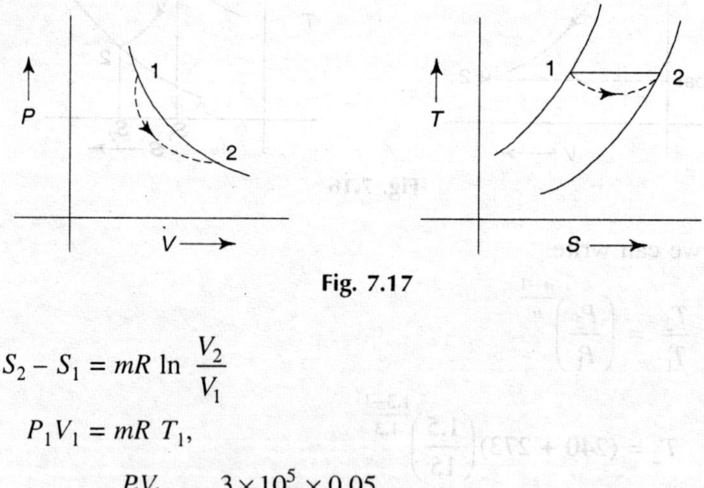

Fig. 7.17

$$\therefore \quad S_2 - S_1 = mR \ln \frac{V_2}{V_1}$$

We know that, $P_1V_1 = mR\,T_1$,

$$m = \frac{P_1V_1}{RT_1} = \frac{3 \times 10^5 \times 0.05}{0.287 \times 10^3 \times 293}$$

$$\boxed{m = 0.178 \text{ kg}}$$

$$\therefore \quad S_2 - S_1 = 0.178 \times 0.287 \times \ln \left(\frac{0.10}{0.05}\right)$$

$$\boxed{S_2 - S_1 = 0.0354 \text{ kJ/kgK}}$$ *Ans.*

This entropy increase is due to irreversibility (as there is no heat exchange).

Problem 15: *0.5 m^3 of H_2S is contained in a tank at 25°C. when 2.5 kg of gases are removed, the temperature in the tank decreases to 15°C and pressure becomes 8 bar. Determine the mass and pressure of the gas at the initial state.*

Solution: We know that, $R = \frac{\bar{R}}{M} = \frac{8.3144}{34.09} \frac{\text{kJ/kg mol K}}{\text{mol}}$

$$R = 0.243 \text{ kJ/kgK}$$

From table C-6/58, [for H_2S, molecular weight $M = 34.09$]

Mass of the gas after removing 2.5 kg of gases from the tank,

$$m = \frac{P_2V_2}{RT_2}$$

$$= \frac{8 \times 10^5 \times 0.5}{0.243 \times 10^5 \ (288)}$$

$$\boxed{m = 5.71 \text{ kg}}$$

∴ Original mass of the gas in the tank

$$= 5.71 + 2.5$$

$$m_1 = 8.21 \text{ kg}$$ **Ans.**

The Ideal gas equation is

$$PV = m\,RT$$

$$P = \frac{m_1RT_1}{V} = \frac{8.21 \times 0.243 \times 10^3 (25 + 273)}{0.5}$$

$$P = 1189037.88 \text{ N/m}^2$$

$$P = 1189.037 \text{ kN/m}^2$$

$$\boxed{P = 11.89 \text{ bar}}$$ **Ans.**

Problem 16: *A cylinder of capacity 25 m^3 contained oxygen. When it is put in to operation by reducing pressure from 1.5 bar to 0.35 bar and temperature from 25 to 5°C. Assuming Ideal gas behaviour, determine mass of O_2 removed from the cylinder Express this mass in terms of volume measured at 1.5 bar and 25°C.*

Solution: From table C-6/58

For O_2, $M = 32$

$$R = \frac{\bar{R}}{m} = \frac{8.3144}{32} = 0.2598 \text{ kJ/kgK}$$

The Ideal gas equation is

$$PV = m\,RT$$

Initial mass of O_2,

$$m_1 = \frac{P_1V_1}{RT_1} = \frac{1.5 \times 10^5 \times 25}{0.2598 \times 10^3 (25 + 273)}$$

$$\boxed{m_1 = 48.43 \text{ kg}}$$

Final mass of the O_2

$$m_2 = \frac{P_2V_2}{RT_2} = \frac{0.35 \times 10^5 \times 25}{0.2598 \times 10^3 \ (5 + 273)}$$

$$\boxed{m_2 = 12.11 \text{ kg}}$$

$\therefore$ The mass of O_2 removed is

$$m_1 - m_2 = 48.43 - 12.11$$

$$\boxed{m = 36.32 \text{ kg}}$$ *Ans.*

Volume of this mass at 1.5 bar and 25°C is

$$V = \frac{mRT}{P} = \frac{36.32 \times 0.2598 \times 10^3 \ (25 + 273)}{1.5 \times 10^5}$$

$$\boxed{V = 18.74 \text{ m}^3}$$ *Ans.*

Problem 17: *A container stores 5 m^3 of an inert gas at 100 bar and 290 k. Determine mass, density, specific volume and molar volume of the gas. Assume molecular mass of the gas = 28.*

Solution: We know that,

Characteristic gas constant $R = \frac{\bar{R}}{M}$

$\bar{R}$ = universal gas const. (R_{mol})

M = molecular mass

$$R = \frac{8.3144}{28} = 0.2969 \text{ kJ/kgK}$$

We know that, from the ideal gas equation

$$PV = mRT$$

$$m = \frac{PV}{RT} \qquad \frac{\frac{N}{m^2} \cdot m^3}{\frac{J}{kg\,K} \cdot K}$$

$$= \frac{100 \times 10^5 \times 5}{0.2969 \times 10^3 \times (290)}$$

$$\boxed{m = 580.71 \text{ kg}}$$ *Ans.*

Density of the gas

$$\rho = \frac{\text{mass}}{\text{unit volume}}$$

$$\rho = \frac{580.71}{5} = \boxed{116.14 \text{ kg/m}^3}$$ *Ans.*

$\therefore$ Specific volume

$$V = \frac{1}{\rho} = \frac{1}{116.14}$$

$$\boxed{V = 0.00861 \text{ m}^3\text{/kg}}$$

$$\text{Molar volume} = V_{mol} = \frac{MRT}{P} = \frac{28 \times 0.2969 \times 10^3 \times 290}{100 \times 10^5}$$

$$\boxed{V_{mol} = 0.241 \text{ m}^3}$$ *Ans.*

[Molar volume →The volume of 1 k mole of the gas i.e., volume corresponding to mass equal to molecular mass.]

Problem 18: *1.5 kg of nitrogen at 300 k and 20 bar is contained in a rigid vessel. If the oxygen is added to the rigid vessel so that total pressure becomes 35 bar for the same temperature, determine mass of the oxygen added.*

Solution: For N_2, number of moles $= \dfrac{\text{Mass of } N_2}{\text{Molar mass}}$

$$= \frac{1.5}{28} = 0.0535$$

Partial pressure of N_2 = 20 bar

By Dalton's law, we can write

Partial pressure of $O_2 = P$ – partial pressure of N_2

where P = Total pressure = 35 – 20 = 15 bar

Now
$$\frac{(P)_{O_2}}{(n)_{O_2}} = \frac{(P)_{N_2}}{(n)_{N_2}}$$

Number of moles of O_2

$$(n)_{O_2} = \frac{(P)_{O_2} \cdot (n)_{N_2}}{(P)_{N_2}} = \frac{15 \times 0.0535}{20} = 0.040125$$

∴ The mass of O_2 added;

$$(n)_{O_2} = \frac{\text{mass of } O_2}{\text{molar mass}}$$

∴ mass of $O_2 = 0.040125 \times 32 = 1.284$ kg. *Ans.*

Problem 19: *Determine the gas constant, density and partial pressures of the constituents of a gas mixture having 10 mass fractions of air, 1 mass fraction of lighting gas. Take ρ for lighting gas as 0.5 kg/m³ at 101325 N/m² and 273 K.*

Solution: For air, $R_1 = 0.287$ kJ/kgK

For lighting gas, $R_2 = \dfrac{PV}{T} = \dfrac{101325}{0.5 \times 273}$ $\quad \left| \; V = \dfrac{1}{\rho} \right.$

$R_2 = 742.30$ J/kgK *Ans.*

For the mixture, gas constant is

$$R_{mix} = g_1 R_1 + g_2 R_2 \qquad \left| \; g_1 = \frac{\text{Mass of constuent 1}}{\text{Total mass}} \right.$$

$$= \frac{10}{11}(287) + \frac{1}{11}(742.30)$$

$$\boxed{R_{mix} = 328.39 \text{ J/kgK}}$$

$$\rho_{mix} = \frac{P}{R_{mix}\,T} = \frac{101325}{328.39 \times 273} = 1.130 \text{ kg/m}^3 \qquad \textit{Ans.}$$

For air, partial pressure P_1

$$P_1 = Pg_1 \frac{R_1}{R_{mix}} = 101325 \times \frac{10}{11}\,\frac{287}{328.39}$$

$$P_1 = 80503.7 \text{ N/m}^2 \qquad \textit{Ans.}$$

For lighting gas, partial pressure P_2

$$P_2 = Pg_2 \frac{R_2}{R_{mix}} = 101325 \times \frac{1}{11} \times \frac{742.30}{328.39}$$

$$P_2 = 20821.6 \text{ N/m}^2 \qquad \textit{Ans.}$$

or Alternate method
By Dalton's law

$$P = P_1 + P_2$$

$$P_2 = P - P_1$$

$$= 101325 - 80503.7 = 20821.3 \text{ N/m}^2$$

Problem 20: *A vessel contains 11 kg of oxygen, 9 kg of nitrogen and 20 kg of carbon dioxide at 370 k temperature and 4 bar. Determine capacity of the vessel, partial pressure of each constituent and total pressure in the vessel when the temperature is increased to 430 k.*

Solution: We know that

$$R_{mix} = g_1 R_1 + g_2 R_2 + g_3 R_3 + \ldots$$

$$= \frac{11}{40} \times \frac{8314}{32} + \frac{9}{40} \times \frac{8314}{28} + \frac{20}{40}\left(\frac{8314}{44}\right)$$

$$= 71.448 + 66.808 + 44.477$$

$$= 232.733 \text{ J/kgK}$$

Now, for the gas mixture in the vessel the Ideal gas equation is

$$PV = mRT$$

$$V_{mix} = \frac{mRT}{P} = \frac{40 \times 232.733 \times 370}{4 \times 10^5} \qquad \left|\ \frac{\cancel{kg} \times \text{J}\ \cancel{K}\,\text{m}^2}{\cancel{kg}\ \cancel{K} \times \text{N}}\right.$$

$$\boxed{V_{mix} = 8.611 \text{ m}^3}$$

Now, for heating at constant volume of fixed mass of gas mixture,

$$\frac{P_1}{T_1} = \frac{P_2}{T_2}$$

$$\therefore \quad P_2 = P_1 \frac{T_2}{T_1} = 4 \times \frac{430}{370}$$

$$\boxed{P_2 = 4.64 \text{ bar}}$$ **Ans.**

Now the partial pressure of each constituent in the mixture is for oxygen,

Partial pressure $$P_i = P \times g_1 \times \frac{R_i}{R_{mix}}$$

$$= 4 \times \frac{11}{40} \times \frac{8314/32}{232.733} = 1.22 \text{ bar}$$ **Ans.**

For nitrogen

$$P_i = 4 \times \frac{9}{40} \times \frac{8314/28}{232.733} = 1.148 \text{ bar}$$ **Ans.**

For carbon dioxide

$$P_i = 4 \times \frac{20}{40} \times \frac{8314/44}{232.733} = 1.623 \text{ bar}$$ **Ans.**

Problem 21: *Two chambers A and B, both containing nitrogen are connected by a valve. When the valve is opened, the contents of A and B mix and achieve an equilibrium temperature of 28°C. Before mixing, the data available were as follows:*

Chamber A	*Chamber B*
P = 15 bar	*P = 8 bar*
t = 50°C	*t = 25°C*
Contents = 0.5 kg mol	*Contents = 2.5 kg mol*

Determine the final equilibrium pressure, the amount of heat transferred to the surroundings. If the chamber had been perfectly insulated, calculate the final temperature and pressure which would have been reached.

Take γ = 1.4.

Solution: For the gas in the chamber *A*,

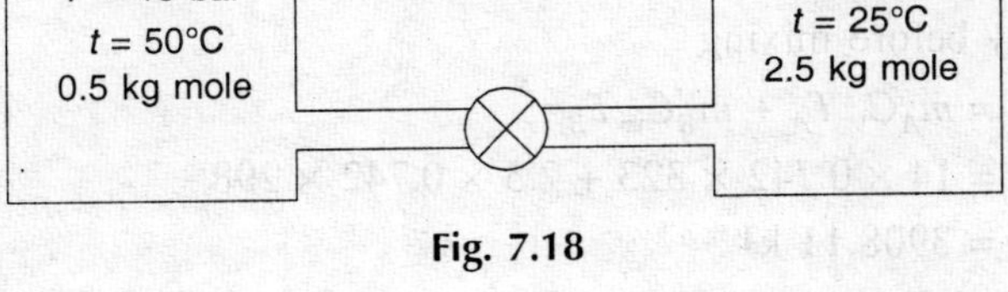

Fig. 7.18

$$P_A V_A = M_A R T_A$$

$$15 \times 10^5 \times V_A = 0.5 \times 8314 \times 323$$

$$\boxed{V_A = 0.895 \text{ m}^3}$$

$$\frac{\text{N m}}{\frac{\text{kg mol J K m}^2}{\text{kg mol K N}}}$$

∴ The mass of the gas

$$m_A = n_A M_A$$
$$= 0.5 \text{ (kg mol)} \times 28 \text{ (kg/kg mol)}$$
$$= 14 \text{ kg} \qquad \text{(for} \quad N_2{:}M = 28)$$

The characteristic gas constant

$$R = \frac{8314.4}{28} = 296.9 \text{ J/kgK}$$

For the chamber 'B',

$$P_B V_B = M_B R\, T_B$$
$$8 \times 10^5 \times V_B = 2.5 \times 296.9 \times (25 + 273)$$
$$\boxed{V_B = 0.276 \text{ m}^3}$$

∴ Total volume of A and B is

$$V = V_A + V_B$$
$$V = 0.895 + 0.276$$
$$\boxed{V = 1.171 \text{ m}^3}$$

∴ The total mass of gas

$$m = m_A + m_B = 14 + 2.5 = 16.5 \text{ kg}$$

final temperature after mixing

$$T = 28 + 273 = 301 \text{ k}$$

∴ After mixing, we can write

$$PV = mRT$$
$$P = \frac{mRT}{V} = \frac{16.5 \times 296.9 \times 301}{1.171}$$
$$P = 1259.22 \text{ kN/m}^2$$
$$\boxed{P = 12.5 \text{ bar}}$$

Now
$$C_v = \frac{R}{\gamma - 1} = \frac{0.2969}{1.4 - 1}$$
$$C_v = 0.742 \text{ kJ/kgK}$$

Since there is no work transfer, the amount of heat transfer

$$Q = \Delta U \qquad U_2 - U_1$$

At $T = \text{K}, \qquad u = \text{kJ/kg}$

U_1—initial internal energy before mixing

$$= m_A C_v\, T_A + m_B C_v\, T_B$$
$$= 14 \times 0.742 \times 323 + 2.5 \times 0.742 \times 298$$
$$= 3908.11 \text{ kJ}$$

after mixing, the final internal energy

$$U_2 = m\, C_v\, T$$
$$= 16.5 \times 0.742 \times 301 = 3685.143 \text{ kJ}$$

∴ $Q = 3685.143 - 3908.11$

$$\boxed{Q = -222.967 \text{ kJ}}$$

Ans.

When A and B are insulated,

$$U_1 = U_2$$

$$M_A C_v T_A + M_B C_v T_B = m C_v T$$

where, T = final temperature

$$T = \frac{m_A T_A + m_B T_B}{m}$$

$$= \frac{14 \times 323 + 2.5 \times 298}{16.5}$$

$$\boxed{T = 319.21 \text{ k}} = 46.21°\text{C}$$ **Ans.**

and the final pressure is

$$PV = mRT$$

$$P = \frac{mRT}{V} = \frac{16.5 \times 0.2969 \times 319.21}{1.171}$$

$$\boxed{= 1335.40 \text{ kN/m}^2}$$

$$P = 13.35 \text{ bar}$$ **Ans.**

Problem 22: *1 kg of air at 1 bar and 50°C is compressed adiabatically to a pressure of 5 bar. The air is then expanded at constant pressure to the initial volume. Represent these processes on P-V and T-S diagrams. Determine work transfer and heat transfer for the whole path.*

Solution: The process have been shown on *P-V* and *T-S* diagrams. From the Ideal gas equation

$$P_1V_1 = mRT_1$$

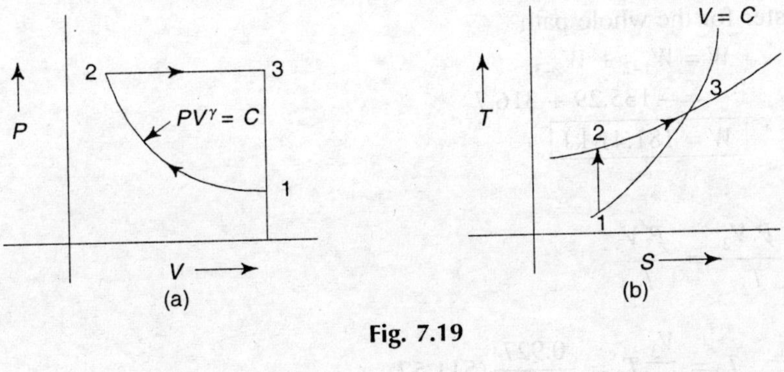

Fig. 7.19

$$V_1 = \frac{mRT_1}{P_1} = \frac{1 \times 0.287 \times 10^3 \times 323}{1 \times 10^5}$$

$$\left| \frac{\cancel{kg} \times \cancel{N} m \cancel{K}}{\cancel{kg}\ \cancel{K} \times \frac{\cancel{N}}{m^2}} \right.$$

$$\boxed{V_1 = 0.927 \text{ m}^3}$$

For the process (Isentropic) 1-2

$$\frac{T_2}{T_1} = \left(\frac{P_2}{P_1}\right)^{(\gamma-1)/\gamma}$$

$$T_2 = 323 \left(\frac{5}{1}\right)^{\frac{1.4-1}{1.4}}$$

$$\boxed{T_2 = 511.57 \text{ k}}$$

$$W_{1-2} = \frac{P_1V_1 - P_2V_2}{\gamma - 1} = \frac{mR(T_1 - T_2)}{\gamma - 1}$$

$$= \frac{1 \times 0.287\,(323 - 511.57)}{1.4 - 1}$$

$$\boxed{W_{1-2} = -\ 135.29 \text{ kJ}}$$

Also, for isentropic process

$$P_1V_1^{\gamma} = P_2V_2^{\gamma} = C$$

$$\left(\frac{V_2}{V_1}\right) = \left(\frac{P_1}{P_2}\right)^{1/\gamma}$$

$$V_2 = 0.927\left(\frac{1}{5}\right)^{1/1.4}$$

$$\boxed{V_2 = 0.2936 \text{ m}^3}$$

For the constant pressure process 2–3

$$W_{2-3} = \int PdV = P_2\,(V_3 - V_2)$$

But $\quad V_3 = V_1 = 0.927 \text{ m}^3$

$\therefore \quad W_{2-3} = 5 \times 10^5\,(0.927 - 0.2936) = 316700 \text{ J}$

$$\boxed{W_{2-3} = 316.7 \text{ kJ}}$$

$\therefore$ Total work transfer for the whole path

$$W = W_{1-2} + W_{2-3}$$

$$= -\ 135.29 + 316.7$$

$$\boxed{W = 181.41 \text{ kJ}}$$

Ans.

Also we can write

$$\frac{P_2V_2}{T_2} = \frac{P_3V_3}{T_3} \qquad [P_2 = P_3]$$

$$T_3 = \frac{V_3}{V_2}T_2 = \frac{0.927}{0.2936}(511.57)$$

$$\boxed{T_3 = 1615.2 \text{ k}}$$

$\therefore$ Total heat transfer for the whole path

$$Q = Q_{1-2} + Q_{2-3}$$

$Q_{1-2} = 0,$ (adiabatic process)

$$Q = Q_{2-3} = m\,C_p\,(T_3 - T_2)$$

$$Q = 1 \times 1.005\,(1615.2 - 511.57)$$

$$\boxed{Q = 1109.14 \text{ kJ}}$$

Ans.

Problem 23: *A mixture of Ideal gases consists of 2.5 kg of N_2 and 4.5 kg of carbon dioxide at a pressure of 4 bar and a temperature of 25°C. Determine:*

(i) *Mole fraction of each constituent,*
(ii) *Equivalent molecular weight of the mixture,*
(iii) *Equivalent gas constant of the mixture,*
(iv) *The partial pressure and partial volumes,*
(v) *The volume and density of the mixture,*
(vi) *The C_p and C_v of the mixture,*

If the mixture is heated at constant volume to 35°C, what will be the changes in internal energy, enthalpy and entropy of the mixture. Also determine changes in same properties when heating is done at constant pressure.

Take $\gamma = 1.285$ *for* CO_2
$\gamma = 1.4$ *for* N_2

Solution:

(i) The mole fraction can be calculated as

$$\mu_1 = \frac{n_1}{n} \text{ for I}^{\text{st}} \text{ constituent}$$

for N_2, $$n_{N_2} = \frac{2.5}{28} = 0.0892$$

n – total number of moles of the gas mixture

$$n = \frac{2.5}{28} + \frac{4.5}{44} = 0.1915$$

$$\mu_{N_2} = \frac{0.0892}{0.1915} = 0.4657$$

$$\boxed{\mu_{N_2} = 0.4657}$$ *Ans.*

$$\mu_{CO_2} = \frac{n_{CO_2}}{n} = \frac{0.10227}{0.1915} \qquad \left[n_{CO_2} = \frac{4.5}{44} = 0.10227\right]$$

$$\boxed{\mu_{CO_2} = 0.534}$$ *Ans.*

(ii) Equivalent molecular weight of the mixture

$$M = \mu_1 M_1 + \mu_2 M_2$$
$$= 0.4657\ (28) + 0.534\ (44)$$
$$\boxed{M = 36.535 \text{ kg/kg mol}}$$ *Ans.*

(iii) eq ivalent gas constant of the mixture,
Total mass of mixture

$$= m = m_{N_2} + m_{CO_2}$$
$$= 2.5 + 4.5$$
$$= 7 \text{ kg}$$

The characteristic gas constant of the mixture can be calculated as

$$R = \frac{m_{N_2} \times R_{N_2} + m_{CO_2} \times R_{CO_2}}{m}$$

$$= \frac{2.5\left(\frac{8.314}{28}\right) + 4.5\left(\frac{8.314}{44}\right)}{7} = \frac{0.7423 + 0.8502}{7}$$

$$\boxed{R = 0.2275 \text{ kJ/kg K}}$$ **Ans.**

(iv) $P_{N_2} = \mu_{N_2} \times P$

$= 0.4657 \times 4$

$$\boxed{P_{N_2} = 1.862 \text{ bar}}$$

$P_{CO_2} = \mu_{CO_2} \times P$

$= 0.534 \times 4$

$$\boxed{P_{CO_2} = 2.136 \text{ bar}}$$

∴ The Ideal gas equation can be written as

$PV = m\,RT$ for N_2

∴ $V_{N_2} = \left(\frac{m\,RT}{P}\right)_{N_2}$

$$= \frac{2.5 \times \frac{8.314 \times 10^3}{28} \times (25 + 273)}{4 \times 10^5} \qquad \left| \; \frac{\text{kg}\,\frac{\text{kJ}}{\text{kg k}} \times \text{k}}{\frac{\text{N}}{\text{m}^2}} \right.$$

$$\boxed{V_{N_2} = 0.553 \text{ m}^3}$$ **Ans.**

For CO_2,

$$V_{CO_2} = \left(\frac{m\,RT}{P}\right)_{CO_2} = \frac{4.5 \times \frac{8.314 \times 10^3}{44} \times (25 + 273)}{4 \times 10^5}$$

$$\boxed{V_{CO_2} = 0.\ 6334 \text{ m}^3}$$ **Ans.**

(v) Total volume of the mixture

$$V = \frac{m\,RT}{P}$$

$$= \frac{7 \times 0.22975 \times 10^3 \times 298}{4 \times 10^5}$$

$$\boxed{V = 1.188 \text{ m}^3}$$ **Ans.**

Density of the mixture

$$\rho = \frac{m}{V} = \frac{7}{1.188} = \boxed{5.892 \text{ kg/m}^3}$$ **Ans.**

(vi) ∴ C_p and C_v of the mixture

For N_2;

$$(C_p - C_v) = R$$

$$C_v = \frac{R}{\gamma - 1} = \frac{8.3144}{28\,(1.4 - 1)}$$

$$\boxed{C_v = 0.742 \text{ kJ/kg K}}$$

$$\frac{C_p}{C_v} = \gamma$$

$$C_p = 1.4 \times 0.742$$

$$\boxed{C_p = 1.0388 \text{ kJ/kg K}}$$

For Co_2;

$$\gamma = 1.285$$

$$C_v = \frac{R}{\gamma - 1} = \frac{8.3144}{44\,(1.285 - 1)}$$

$$\boxed{C_v = 0.663 \text{ kJ/kg K}}$$

$$C_p = 1.285 \times 0.663 = \boxed{0.8519 \text{ kJ/kg K}}$$

For the mixture;

$$C_p = \frac{(mC_p)_{N_2} + (mC_p)_{CO_2}}{(m)_{N_2} + (m)_{CO_2}} = \frac{(2.5 \times 1.0388) + (4.5 \times 0.8519)}{2.5 + 4.5}$$

$$\boxed{C_p = 0.918 \text{ kJ/kg K}}$$ *Ans.*

$$C_v = \frac{(mC_v)_{N_2} + (mC_v)_{CO_2}}{(m)_{N_2} + (m)_{CO_2}} = \frac{(2.5 \times 0.742) + (4.5 \times 0.663)}{2.5 + 4.5}$$

$$\boxed{C_v = 0.691 \text{ kJ/kg K}}$$ *Ans.*

For constant volume heating of the mixture.

Change in internal energy

$$U_2 - U_1 = mC_v\,(T_2 - T_1)$$

$$= 7 \times 0.691\,(35 - 25)$$

$$\boxed{U_2 - U_1 = 48.37 \text{ kJ}}$$ *Ans.*

Change in enthalpy;

$$H_2 - H_1 = m\,C_p\,(T_2 - T_1)$$

$$= 7 \times 0.918\,(35 - 25)$$

$$\boxed{H_2 - H_1 = 64.26 \text{ kJ}}$$ *Ans.*

Change in entropy;

$$S_2 - S_1 = m\,C_v \ln \frac{T_2}{T_1} + m\,R \ln \overbrace{\left(\frac{V_2}{V_1}\right)}^{0}$$

$$S_2 - S_1 = m\,C_v \ln \frac{T_2}{T_1} = 7 \times 0.691 \ln \left(\frac{308}{298}\right)$$

$$\boxed{S_2 - S_1 = 0.1596 \text{ kJ/kg K}}$$ **Ans.**

For constant pressure heating of the mixture, ΔU, ΔH values will remain same. The entropy change in this case will be

$$S_2 - S_1 = m\,C_p \ln \frac{T_2}{T_1} - mR \ln \frac{P_2}{P_1} \quad (\to 0)$$

$$= m\,C_p \ln \frac{T_2}{T_1} = 7 \times 0.918 \ln \left(\frac{308}{298}\right)$$

$$\boxed{S_2 - S_1 = 0.212 \text{ kJ/kgK}}$$ **Ans.**

Problem 24: *A mixture of gases has the following volumetric composition:*
CO_2 = 12%. O_2 = 4%, N_2 = 82%, CO = 2%. Calculate the
(i) gravimetric composition,
(ii) Molecular weight of the mixture
(iii) R of the mixture.

Solution:

(i) We know that, $x = \frac{n_i}{n} = \frac{V_i}{V}$

Constituent	Fraction by volume x_i	Molecular wt M_i	Product x_iM_i	Fraction by Mass. $= \frac{M_i\,x_i}{\Sigma M_i\,x_i}$	Gravimetric composition or % by mass $= \frac{M_i\,x_i}{\Sigma M_i\,x_i} \times 100$
N_2	0.82	28	22.96	$\frac{22.96}{30.08} = 0.763$	76.3%
CO_2	0.12	44	5.28	0.1755	17.55
O_2	0.04	32	1.28	0.0425	4.25
CO	0.02	28	0.56	0.0186	1.86
			$\Sigma M_ix_i = 30.08$		

(ii) Molecular weight of the mixture

$$M_{\text{mixture}} = \Sigma M_ix_i = 30.08 \text{ kJ/kg mol}$$ **Ans.**

$$[M_{\text{mixture}} = X_{N_2} \times \mu_{N_2} + X_{CO_2}\,\mu_{CO_2} + X_{O_2}\,\mu_{O_2} + X_{CO} \times \mu_{CO})$$

(iii) $$R_{\text{mixture}} = \frac{\bar{R}}{M_{mix}} = \frac{8314.4}{30.08}$$

$$\boxed{R_{\text{mix}} = 276.40 \text{ J/kgK}}$$ **Ans.**

Problem 25: *The gravimetric analysis of air is given below.*

Component	*Percentage*
N_2	75%
Argon	1.3
CO_2	0.05
O_2	23.65

What will be the volume analysis and partial pressure of each component for a total pressure of 1 bar.

Solution: Consider 1kg of mixture

Component	m_i	M	$n_i = \frac{m_i}{M}$	$\frac{n_i}{n} \times 100 = \frac{V_i}{V} \times 100$	Volume %
N_2	0.75	28	0.02678	$\frac{0.02678}{0.03450} \times 100$	77.62 %
Argon	0.013	40	3.25×10^{-4}	$\frac{3.25 \times 10^{-4}}{0.0345} \times 100$	0.942%
CO_2	0.0005	44	1.136×10^{-5}	$\frac{1.136 \times 10^{-5}}{0.0345} \times 100$	0.0329%
O_2	0.2365	32	7.39×10^{-3}	$\frac{7.39 \times 10^{-3}}{0.0345} \times 100$	21.42%
			$n = \Sigma n_i = 0.03450$		

∴ **Partial pressure of each component can be calculated as**

$$\text{(i)}\ P_{N2} = \frac{n_i}{n} \times P = \frac{0.02678}{0.0345} \times 1 = 0.7762 \text{ bar}$$

$$\text{(ii)}\ P_{argon} = \frac{n_i}{n} \times P = \frac{3.25 \times 10^{-4}}{0.0345} \times 1 = 0.00942 \text{ bar}$$

$$\text{(iii)}\ P_{CO2} = \frac{n_i}{n} \times P = \frac{1.136 \times 10^{-5}}{0.0345} \times 1 = 0.000329 \text{ bar}$$

$$\text{(iv)}\ P_{O2} = \frac{n_i}{n} \times P = \frac{7.39 \times 10^{-3}}{0.0345} \times 1 = 0.2142 \text{ bar}$$

Problem 26: *An engine cylinder has a diameter 22.5 cm and stroke 37.5 cms. Swept volume is 4 times the clearance volume. The pressure at the beginning of the expansion stroke is 1569.6 kpa. Find the work done during the expansion stroke, assuming the process as reversible adiabatic. Take γ = 1.4.* (VTU March 2001)

Solution: Swept volume $= V_s = \frac{\pi d^2}{4} l$

$$= \frac{\pi}{4}(0.225)^2 \times 0.375$$

$$V_s = 0.0149 \text{ m}^3$$

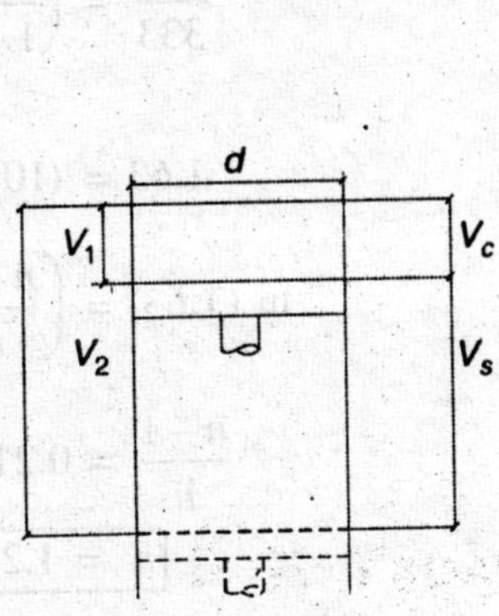

$$\text{Clearance volume} = V_c = \frac{V_s}{4} = \frac{0.0149}{4}$$

$$V_c = 0.003725 \text{ m}^3$$

$$V_1 = V_c = 0.003725 \text{ m}^3$$

$$V_2 = V_c + V_s = 0.003725 + 0.0149$$

$$= 0.018625 \text{ m}^3$$

At the beginning of expansion,

$$P_1 = 1569.6 \text{ kpa}$$

$$\frac{P_2}{P_1} = \left(\frac{V_1}{V_2}\right)^{\gamma}$$

$$\therefore \quad P_2 = 1569.6\left(\frac{0.003725}{0.018625}\right)^{1.4}$$

$$P_2 = 164.93 \text{ kpa}$$

$$\text{Work done} = \frac{P_1V_1 - P_2V_2}{\gamma - 1}$$

$$= \frac{(1569.6 \times 0.003725) - (164.93 \times 0.018625)}{1.4 - 1}$$

$$\boxed{W.D = 6.93 \text{ kJ}}$$

Ans.

Problem 27: *0.2 m³ of mixture of fuel and air at 1.2 bar and 60°C is compressed until its pressure becomes 12 bar and temperature becomes 270°C. Then it is ignited suddenly at constant volume and its pressure become twice the pressure at the end of compression. Find the maximum temperature reached and change in internal energy. Also find the heat transfer during the compression process for the mixture. Take C_p = 1.072 kJ/kgk. R = 0.2943 kJ/kg* *(VTU, Aug. 2001)*

Given:
$V_1 = 0.2 \ m^3$
$P_1 = 1.2$ bar
$T_1 = 60°C$
$P_2 = 12$ bar
$T_2 = 270°C$
$P_3 = 2 \times 12 = 24$ bar

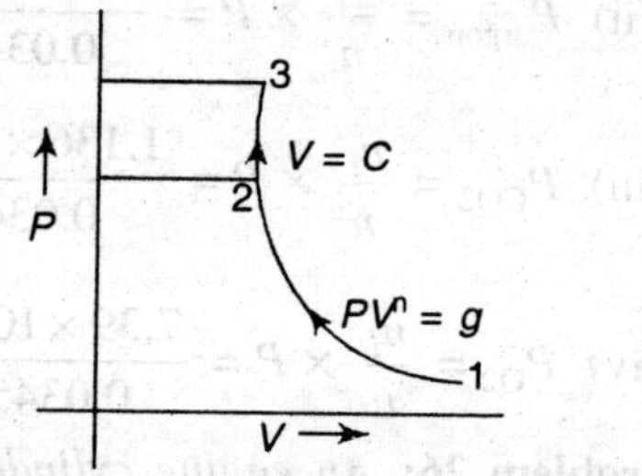

Solution: For the process 1-2 which is polytropic

$$\frac{T_2}{T_1} = \left(\frac{P_2}{P_1}\right)^{\frac{n-1}{n}}$$

$$\frac{543}{333} = \left(\frac{12}{1.2}\right)^{\frac{n-1}{n}}$$

$$1.63 = (10)^{\frac{n-1}{n}}$$

$$\ln(1.63) = \left(\frac{n-1}{n}\right)\ln(10)$$

$$\frac{n-1}{n} = 0.212$$

$$\boxed{n = 1.27}$$

where n is the index of compression

also, $$V_2 = V_1 \left(\frac{P_1}{P_2}\right)^{\frac{1}{n}} = 0.2 \left(\frac{1.2}{12}\right)^{\frac{1}{1.27}}$$

$$\boxed{V_2 = 0.0326 \text{ m}^3}$$

$$\text{WD} = \frac{P_1V_1 - P_2V_2}{n-1} = \frac{[(1.2 \times 0.2) - (12 \times 0.0326)]10^2}{1.27 - 1}$$

$$\boxed{\text{WD} = -56.33 \text{ kJ}}$$ **Ans.**

$$\delta Q = \left(\frac{V - n}{V - 1}\right) \times \text{polytropic workdone}$$

$$= \left(\frac{1.4 - 1.27}{1.4 - 1}\right) \times (-56.33)$$

$$\boxed{Q = -18.3 \text{ kJ}}$$ **Ans.**

Change in constant energy = $\delta Q = \delta W$

$$= -18.3 - (-56.33)$$

$$\boxed{\Delta U = 38.03 \text{ kJ}}$$ **Ans.**

for the constant volume process 2-3

$$\frac{P_2V_2}{T_2} = \frac{P_3V_3}{T_3}$$

$$\therefore \quad T_3 = \frac{24 \times 543}{12} = \boxed{1086 \text{ K}}$$ **Ans.**

$$P_1V_1 = mRT_1$$

$$\therefore \quad m = \frac{1.2 \times 10^2 \times 0.2}{0.287 \times 333} = 0.251 \text{ kg}$$

$$C_p - C_v = R$$

$$\therefore \quad C_v = 1.072 - 0.2943 = 0.7777$$

Change in internal energy = $mC_v\,(T_3 - T_2)$

$$= 0.251 \times 0.777\ (1086 - 543)$$

$$= 103.4 \text{ kJ}$$

$\therefore$ Total change in internal energy

$$= 38.03 + 103.4$$

$$= 141.48 \text{ kJ}$$ **Ans.**

Problem 28: *A cylinder contains 1 kg of certain fluid at an initial pressure of 20 bar. The fluid is allowed to expand reversibly behind a piston according to the law $PV^2 = 5$ until the volume is doubled. The fluid is then cooled reversibly at constant pressure until the piston regains its original position; heat is then supplied reversibly with the piston firmly locked in position until the pressure rises to the original value of 20 bar. Calculate net work done by the fluid, for an initial volume of 0.5 m^3.* *(VTU, Aug 2001)*

Solution: For polytropic process 1-2

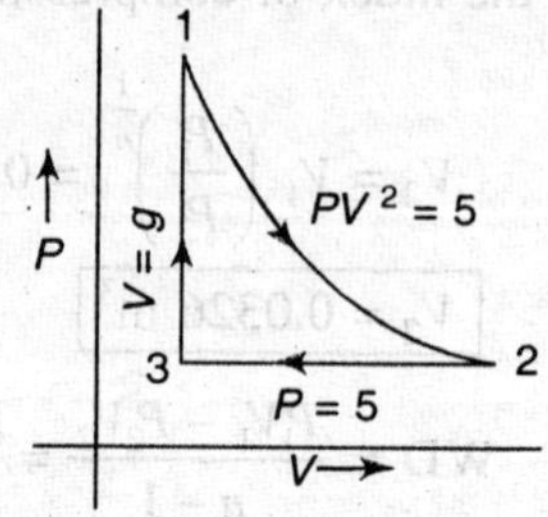

$$P_1V_1^2 = P_2V_2^2$$

$$P_2 = P_1\left(\frac{V_1}{V_2}\right)^2$$

$$= P_1\left(\frac{V_1}{2V_2}\right)^2 = 20\left(\frac{1}{2}\right)^2$$

$$P_2 = 5 \text{ bar}$$

$$\text{WD} = \frac{P_1V_1 - P_2V_2}{n-1} = \frac{(20\times 0.05\times 10^2)-(5\times 0.1\times 10^2)}{2-1}$$

$$\boxed{\text{WD} = 50 \text{ kJ}}$$

For the constant pressure process 2-3,

$$\text{WD} = \int_2^3 PdV = P_2(V_3 - V_2)$$

$$= 5\times 10^2 (0.1 - 0.05)$$

$$\boxed{\text{WD} = 25 \text{ kJ}}$$

For the constant volume process 3-1

$$\text{WD} = 0$$

∴ Net work done by the the fluid area under 1231

$$50 - 25 = 25 \text{ kJ}$$

Ans.

Problem 29: *Two kg of an ideal gas are contained in a rigid cylinder at a temperature of 32°C. If 21.1 kJ of heat are added to the gas. Calculate (i) final temperature (ii) change in enthalpy (iii) change in entropy.* *(VTU Feb. 2002)*

Solution: During a constant volume process

$$Q = \text{heat transfer} = \text{change in internal energy}$$

$$= mC_v(T_2 - T_1)$$

$$21.2 = 2\times 0.718\,(T_2 - 305)$$

$$\boxed{T_2 = 319.69 \text{ K}}$$

Ans.

$$\text{Enthalpy change} = \Delta H = m\,C_p(T_2 - T_1)$$

$$= 2\times 1.005\,(319.69 - 305)$$

$$\boxed{\Delta H = 29.5 \text{ kJ}}$$

Ans.

Entropy change

$$\Delta S = m\,C_v \ln\frac{T_2}{T_1} + mR\cancelto{0}{\ln\frac{V_2}{V_1}}$$

$$= m\,C_v \ln\frac{T_2}{T_1}$$

$$= 2\times 0.718\times \ln\left(\frac{319.69}{305}\right)$$

$$\boxed{\Delta S = 0.0675 \text{ kJ/k}}$$

Ans.

Problem 30: *In a turbine 75 kg/s of air expands polytropically from 4.25 bar and 1360 K to 1.01 bar. The index of expansion n = 1.45. Calculate the work and heat interactions.* *(VTU Feb. 2002)*

Solution:
$$\frac{T_2}{T_1} = \left(\frac{P_2}{P_1}\right)^{\frac{n-1}{n}}$$

$$T_2 = 1360\left(\frac{1.01}{4.25}\right)^{\frac{1.45-1}{1.45}}$$

$$T_2 = 775 \text{ K}$$

$$WD = \frac{P_1V_1 - P_2V_2}{n-1} = \frac{mR\,(T_1 - T_2)}{n-1}$$

$$= \frac{75 \times 0.287\,(1360 - 775)}{1.45 - 1}$$

$$\boxed{WD = 27982.5 \text{ kJ/s}}$$ *Ans.*

Positive sign indicates that wotk is done by the air

$$\text{Heat transfer} = \left(\frac{\gamma - n}{\gamma - 1}\right) \times \text{Polytropic work done}$$

$$Q = \left(\frac{1.4 - 1.45}{1.4 - 1}\right) \times 27982.5$$

$$\boxed{Q = -3497.8 \text{ kJ/s}}$$ *Ans.*

Negative sign indicates that the heat is rejected by the air.

Problem 31. *One kg of ideal gas is heated from 50°C to 150°C. Determine*

(i) Change in internal energy
(ii) Change in enthalpy
(iii) Change in flow energy
(iv) C_p and C_v, Take R = 280 kJ/kgK and γ = 1.32 for gas. *(VTU, Feb. 2003)*

Solution:
$$\gamma = \frac{C_p}{C_v} = 1.32$$

$$R = C_p - C_v = 0.28$$

$$1.32\,C_v - C_v = 0.28$$

$$C_v = 0.875 \text{ kJ/kgK}$$ *Ans.*

$$C_p = 1.32 \times 0.875 = 1.155 \text{ kJ/kgK}$$ *Ans.*

$$\text{Change in intenal energy} = mC_v\,(T_2 - T_1)$$

$$= 1 \times 0.875\,(150 - 50)$$

$$= \boxed{87.5 \text{ kJ/kg}}$$ *Ans.*

Change in enthelpy

$$= mC_p (T_2 - T_1)$$
$$= 1 \times 1.155 (150 - 50)$$
$$= \boxed{115.5 \text{ kJ/kg}}$$ *Ans.*

Change in flow energy $= P_2V_2 - P_1V_1$

$$= mR (T_2 - T_1)$$
$$= 1 \times 0.28 (150 - 50) = \boxed{28 \text{ kJ/kg}}$$ *Ans.*

also

$$\overline{C_p} - \overline{C_v} = \overline{R}$$

$$\overline{C_p} = 1.32 \overline{C_v}$$

$$\therefore \quad 1.32 \overline{C_v} - \overline{C_v} = 8.314$$

$$\overline{C_v} = 25.98 \text{ kJ/kg mol K}$$ *Ans.*

$$\overline{C_p} = 34.29 \text{ kJ/kg mol K}$$ *Ans.*

Problem: 32. *A gas mixture consists of 6 Kmol of H_2 and 4 Kmol of N_2. Determine the mass of each gas and the gas constant of the mixture.* *(VTU, Feb 2004)*

Solution: Total mol numbers = 6 + 4 = 10 K mols.

Mass of $H_2 = N_{H_2} \times \mu H_2 = 6 \times 2 = 12$ kg *Ans.*

Mass of $N_2 = N_{N_2} \times \mu_{N_2} = 4 \times 28 = 112$ kg *Ans.*

Molar Mass of the mixture

$$M_{mix} = \frac{M_m}{N_m} = \frac{12 + 112}{10} = 12.4 \text{ kg mol}$$ *Ans.*

Gas constant of the mixture

$$= \frac{R}{M_{mix}} = \frac{8.314}{12.4} = 0.6701 \text{ kJ/kg K}$$ *Ans.*

7.17 REVIEW QUESTIONS

1. What is an Ideal gas?
2. State and explain the following:
   (i) Charle's law (ii) Boyle's law
   (iii) Avogadro's law
3. What is an equation of state of an Ideal gas?
4. What is the characteristic gas constant and universal gas constant?
5. When is the relation $PV = RT$ expected to be applicable?
6. What are the conditions necessary for a substance to be defined as perfect gas?
7. Show that for an Ideal gas the internal energy depends only on its temperature.
8. Show that the enthalpy of an Ideal gas is a function of temperature only.
9. Show that for an Ideal gas $C_p - C_v = R$.

9. Show that for an Ideal gas $C_p - C_v = R$.
10. Derive an expression of work transfer for an Ideal gas in a reversible Isothermal process.
11. State and explain Dalton's law of partial pressure.
12. Explain the following terms:
    (i) Mole fraction, (ii) Volume fraction (iii) Mass fraction
13. Derive an expression for specific heat of a gas mixture.
14. What is partial pressure and partial or reduced volume of a gas mixture component?
15. Determine entropy change during adiabatic mixing of two gases.

CHAPTER 8

Real Gases

8.1 INTRODUCTION

The real or actual gas disobey the law of perfect gas or the equation of state for Ideal gas; $PV = mRT$. Under certain conditions (i.e. at elevated pressures and at low temperatures) the real gases show marked deviations from the Ideal gas laws. While deriving the Ideal gas equation $PV = mRT$, it was assumed that the intermolecular attractions are negligible, and that the molecular size is insignificant in comparision with total volume of the gas. (The volume occupied by the molecules themselves is negligibly small compared to the total volume of the gas). Any deviations from these assumptions cause the actual gas to deviate from the Ideal gas laws. This necessitates the suitable equation of state which can be used for relating the thermodynamic properties P, V and T. Vander wall, Berthelot, Beattie-Bridgeman and others suggested a number of mathematical equations of state for the analysis of real gas behaviour. For the given pressure and temperature, the compressibility factor may be used to modify the perfect gas equation and it can be written as $PV = Z\bar{R}T$; where Z is the compressibility factor which is a function of temperature and pressure. Therefore, $z = f(P, T)$. It is a correction factor and indicates deviation of the actual gas from the Ideal gas behaviour. The values are available in the form of charts for different substances.

8.2 EQUATIONS OF STATE FOR REAL GASES

A large number of equations of state for real gases have been proposed to over- come the limitations of Ideal gas characteristic equation. Most of these equations are simple and are based on either analytical considerations or empirical. Some equations produce acceptable results for some ranges of pressure and temperature. No equation gives fully satisfactory results for all ranges of pressure and temperature.

The following table shows the list of equations of state for actual gases.

Table 8.1

Name	Equation	Constants
1. Vander Walls	$\left(p+\frac{a}{\bar{V}^2}\right)(\bar{V}-b)=\bar{R}T$	a, b
2. Clausius	$P(\bar{V}-b)=\bar{R}T$	b
3. Callender	$V=\frac{\bar{R}T}{P}-\frac{aC_p}{n-1}\left(\frac{1}{T^n}\right)$	a, n
4. Dieterici	$P=\left(\frac{\bar{R}T}{V-b}\right)e^{-\frac{a}{\bar{R}TV}}$	a, b
5. Berthelot	$P=\left(\frac{\bar{R}T}{V-b}\right)-\left(\frac{a}{TV^2}\right)$	a, b

(Contd.)

Name	Equation	Constants
6. Beattie-Bridgeman	$P = \frac{\bar{R}T\left(1-\frac{C}{VT^3}\right)}{V^2}\left[V+B_0\left(1-\frac{b}{V}\right)\right] - \frac{A_0}{V^2}\left(1-\frac{a}{V}\right)$	a, A_0, b, B_0, C

8.3. VANDER WALL'S EQUATION OF STATE

In 1873, J.D. Vander Walls, proposed the following equation of state for real or actual gases.

$$\left(P+\frac{a}{V^2}\right)(V-b) = RT$$

The above equation represents in a qualitative way both liquid and vapour phase conditions. a and b are specific constants and the values of a and b depends upon the type of fluid considered.

V – Volume per unit mass

R – Gas constant

If we consider volume of one mole, the above equation can be written as

$$\left(P+\frac{a}{\bar{V}^2}\right)(\bar{V}-b) = \bar{R}\,T$$

where P – Pressure of gas, N/m^2

$\bar{V}$ – Molal volume of gas, m^3/kg mol

T – Temperature of gas, K.

$\bar{R}$ – Universal gas constant, 8314.4 Nm/kg mol K

a – Nm^4/ (kg mol)^2

b – m^3/kg mol

The Vander wall's equation is simple one. It gives results which are accurate when compared to the results obtained by using Ideal gas equation of state. The errors in predicting properties by use of this equation are too large over a wide range of pressure and temperature.

8.3.1 To Find Constants 'a' and 'b'

In the Vander Wall's equation, the values of 'a' and 'b' can be evaluated in terms of the properties at critical points.

The first and second derivatives of the pressure with respect to volume at critical point are zero.

i.e $\left(\frac{\partial P}{\partial V}\right)_C = 0$ and

$$\left(\frac{\partial^2 P}{\partial V^2}\right)_C = 0$$

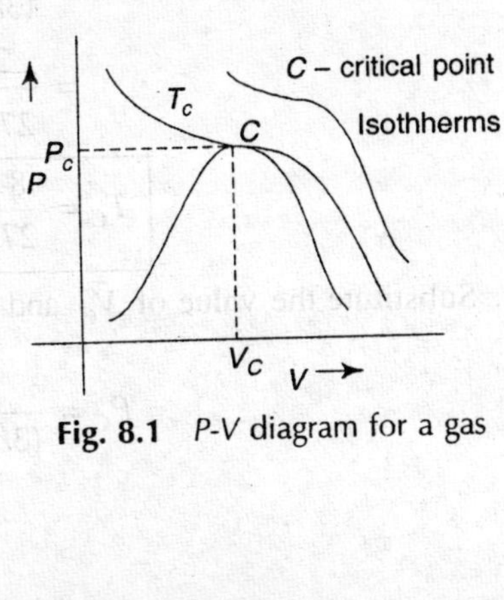

Fig. 8.1 P-V diagram for a gas

From Vander Wall's equation, we can write

$$P = \frac{\bar{R}T}{(\bar{V}-b)} - \frac{a}{\bar{V}^2} \qquad (1)$$

Differentiate equation (1) into V,

$$\left(\frac{\partial P}{\partial V}\right)_C = 0 = \frac{-\bar{R}T_C}{(\bar{V}_C - b)^2} + \frac{2a}{\bar{V}_C^3} \qquad (2)$$

$$\therefore \quad \bar{R}\, T_C = \frac{2a}{\bar{V}_C^3}\,(\bar{V}_C - b)^2 \qquad (3)$$

Differentiate Eqn. (2) with respect to V and equate it to zero, we get

$$\left(\frac{\partial^2 P}{\partial V^2}\right)_C = 0 = \frac{2\bar{R}T_C}{(\bar{V}_C - b)^3} - \frac{6a}{\bar{V}_C^4}$$

$$\therefore \quad \bar{R}\, T_C = \frac{3a}{\bar{V}_C^4}\,(\bar{V}_C - b)^3 \qquad (4)$$

From (3) and (4), we can write

$$\frac{2a}{\bar{V}_C^3}\,(\bar{V}_C - b)^2 = \frac{3a}{\bar{V}_C^4}\,(\bar{V}_C - b)^3$$

$$2 = \frac{3}{\bar{V}_C}\,(\bar{V}_C - b)$$

$$2 = \frac{3}{\bar{V}_C}\bar{V}_C - \frac{3}{\bar{V}_C}\cdot b$$

$$2 = 3 - \frac{3}{\bar{V}_C}\, b$$

$$\therefore \quad -1 = -\frac{3}{\bar{V}_C}\, b$$

$$\therefore \quad \boxed{\bar{V}_C = 3b}$$

Substitute the value of $\bar{V}_C$ in eqn. (3), we get

$$\bar{R}\, T_C = \frac{2a}{(3b)^3}\,(3b - b)^2$$

$$= \frac{2a}{27\, b^3}\,(4b^2) = \frac{8}{27}\cdot\frac{a}{b}$$

$$\therefore \quad \boxed{T_C = \frac{8}{27}\,\frac{a}{b}\cdot\frac{1}{\bar{R}}}$$

$\therefore$ Substitute the value of $\bar{V}_C$ and T_C in eqn. (1), we get

$$P_C = \frac{\bar{R}}{(3b - b)} \times \frac{8}{27}\,\frac{a}{b}\cdot\frac{1}{\bar{R}} - \frac{a}{(3b)^2}$$

$$= \frac{8}{2b} \cdot \frac{1}{27}\frac{a}{b} - \frac{a}{9b^2} = \frac{8}{27} \times \frac{a}{2b^2} - \frac{1}{9}\frac{a}{b^2} = \frac{4}{27}\left(\frac{a}{b^2}\right) - \frac{1}{9}\left(\frac{a}{b^2}\right)$$

$$\boxed{P_C = \frac{a}{27b^2}}$$

We know that, $\bar{V}_C = 3b$

$$\therefore \quad b = \left(\frac{\bar{V}_C}{3}\right)$$

$$\therefore \quad P_C = \frac{a}{\cancel{27}_3\left(\frac{\bar{V}_C^2}{\cancel{9}}\right)}$$

$$\therefore \quad \boxed{a = 3P_C \bar{V}_C^2}$$

$$= \frac{9}{8}\bar{R}\,T_C\,\bar{V}_C^2 = \frac{27}{64} \cdot \frac{\bar{R}^2 T_C^2}{P_C}$$

and
$$b = \frac{\bar{V}_C}{3} = \frac{\bar{R}T_C}{8P_C}$$

$$\therefore \quad \bar{R} = \frac{8}{3}\frac{P_C\bar{V}_C}{T_C}$$

$$\therefore \quad \boxed{\frac{P_C\bar{V}_C}{\bar{R}T_C} = \frac{3}{8}}$$

The values of constants 'a' and 'b' for different gases are as shown in the table below.

Table 8.2 Vander Waal's Constants

Gas	a kN m^4/(kg mol)^2	b m^3/kg mol	$Z = \frac{PV}{RT}$
Air	135.8	0.0365	0.284
O_2	138.0	0.0318	0.29
N_2	136.7	0.0386	0.291
H_2O	551.7	0.0304	0.23
CH_4	228.6	0.0427	0.29
CO	147.9	0.0393	0.293
CO_2	365.6	0.0373	0.242
NH_3	424.9	0.0373	0.242
H_2	24.8	0.0266	0.304
He	3.42	0.0235	0.30

The values of critical properties of several gases are shown in the table given below.

Table 8.3 Critical properties for different gases

Gas	T_C K	P_C MPa	$\bar{V}_C$ (m^3/kg mol) $\times 10^2$
Air	132.41	3.774	8.30
O_2	154.78	5.080	7.43
N_2	126.20	3.398	8.99
H_2O	647.27	22.105	5.68
CH_4	190.70	4.640	9.93
CO	132.91	3.496	9.30
CO_2	304.20	7.386	9.55
NH_3	405.40	11.277	7.24
H_2	33.24	1.297	6.49
He	5.19	0.229	5.81

8.3.2 Limitations of Vander Waal's Equations

1. The values 'a' and 'b' are assumed to be constant in this equation and the effect of temperature which causes changes in values of 'a' and 'b' is not considered. Therefore, when the temperature effect is significant on the values of 'a' and 'b', the results obtained by this equation is not accurate.
2. We know that

$$\frac{P_C\bar{V}_C}{RT_C} = \frac{3}{8} \text{ (Vander Wall's Eqn.)}$$

and

$$\frac{P_C\bar{V}_C}{RT_C} = 1 \text{ (for Ideal gas equation)}$$

But for most of the gases the value of $\frac{P_C T_C}{RT_C}$ varies from 0.2 to 0.3. Hence the results obtained by Vander waal's equation is not accurate in the critical region.

8.4 COMPRESSIBILITY FACTOR

Many investigators have proposed a number of equations to represent behaviour of real gases. These equations are complicated and gives accurate results only in certain ranges of pressure and temperatures and constants of equations have to be determined experimentally. A universal equation of state, eliminates all these problems and is represented by

$$PV = ZRT,$$

where Z = compressibility factor

The compressibility factor is one which shows the extent of deviation of the gas from a perfect gas. For a perfect gas Z = 1, for all ranges of pressure and temperatures, for real gases Z > 1 or Z < 1, depending on pressure and temperature of the gas.

The reduced or Non-dimensional pressure can be written as the ratio of actual pressure of the gas to the critical pressure of the same gas

i.e
$$Pr = \frac{P}{P_C}$$

Similarly, for, temperature and volume, we can write

$$Tr = \frac{T}{T_C} \text{ and}$$

$$Vr = \frac{V}{V_C} = \frac{V}{\frac{RT_C}{P_C}} \qquad \left| \; V_C = \frac{RT_C}{P_C} \right.$$

$$Vr = \frac{P_C V}{RT_C}$$

The value of 'Z' for any substance depends on only two properties, usually temperature and pressure.

$$\therefore \qquad Z = f(Tr, Pr)$$

We know that, $\qquad PV = ZRT$

$$Z = \frac{PV}{RT}$$

But $\qquad P = PrP_C$

$$T = TrT_C \text{ and}$$

$$V = VrV_C$$

$$Z = \frac{(PrP_C)(VrV_C)}{R\,(TrT_C)}$$

$$Z = \frac{P_C V_C}{RT_C} \cdot \frac{PrVr}{Tr} = Z_C \cdot \frac{PrVr}{Tr}$$

where Z_C = compressibility factor at critical point and Pr, Vr and Tr are non dimensional.

8.5 COMPRESSIBILITY CHART

Compressibility chart is one which is obtained by plotting Z on Y axis and 'Pr' on X-axis. Different curves are obtained for different values of 'Tr'.

It is not possible to represent the behaviour of actual gases at low pressure and high temperature. The compressibility chart, overcomes this difficulty and the properties of each real gas is represented by means of a compressibility factor, denoted by 'Z'. For a perfect gas the value of 'Z' is unity and for real gases it may be lesser or more than unity, depending on the temperature and pressure of the gas.

8.6 CORRESPONDING STATES

Each gas needs a different compressibility chart. A single generalised chart can be obtained by describing the gases in reduced properties viz. reduced tempera-ture, pressure and volume by dividing the critical temperature, pressure and volume of the gas respectively and the values are expressed in dimensionless forms.

i.e Reduced temperature

$$Tr = \frac{T}{T_C}$$

Reduced pressure

$$Pr = \frac{P}{P_C}$$

Reduced volume

$$Vr = \frac{V}{V_C}$$

where Tr, Pr and Vr are known as the reduced temperature, pressure and volume respectively, and T_C, P_C and V_C are the critical temperature, pressure and volume respectively, T, P and V—actual temperature, pressure and volume of gas considered.

When substances exist at the same reduced conditions, they are said to be in corresponding states. This concept has been used to obtain *P-V-T* relationship through use of compressibility factor. The concept of corresponding states offers many possibilities of predicting unknown properties of substances based on properties of known substances.

8.7 LAW OF CORRESPONDING STATES

It states that, "If two or more substances have the same reduced pressure 'Pr', and reduced temperature 'Tr', then they will have the same reduced volume 'Vr'.

Mathematically, it can be written as

$$Z = f(Tr, Pr)$$

The law of corresponding states indicates that Vr is a universal function of Pr and Tr. The compressibility factor is also a universal function of Pr and Tr for all gases that have the same critical compressibilities.

8.8 GENERALISED COMPRESSIBILITY CHART

A generalised compressibility chart can be obtained by plotting z on y axis and Pr on X-axis. The figure shows a generalised compressibility chart, for different values of Tr and it can be used for all gases at any condition provided its properties (P_C, V_C and T_C) at critical point are known. If the compressibility factor of a gas is less than unity, then it is said to be more compressible than a perfect

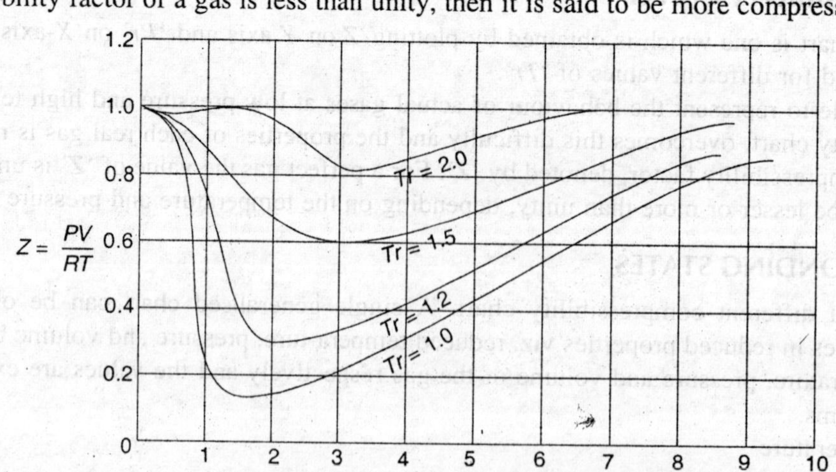

Fig. 8.2 Generalised Compressibility Chart

gas. Also, it is shown that, for all temperatures, the compressibility factor tends to unity as Pr approaches zero. At the critical point, the compressibility factor can be written as

$$Zc = \frac{P_C V_C}{R T_C} = \frac{P_C \bar{V}_C}{\bar{R} T_C} = 0.275 \text{ for all gases}$$

where Zc = critical compressibility factor.

8.9 FORMULAE USED IN SOLVING PROBLEMS

1. Vander Waal's equation of state

$$\left(P + \frac{a}{V^2}\right)(V - b) = RT$$

$$\left(P + \frac{a}{\bar{V}^2}\right)(\bar{V} - b) = \bar{R}\,T \text{ for molar volume}$$

2. $\bar{R} = \frac{8}{3}\,\frac{P_C \bar{V}_C}{T_C}$

3. $\frac{P_C \bar{V}_C}{\bar{R} T_C} = \frac{3}{8}$

4. $Tr = \frac{T}{T_C}$

$$Pr = \frac{P}{P_C}$$

$$Vr = \frac{V}{V_C}$$

5. $PV = ZRT$

6. $Z = Z_C \frac{PrVr}{Tr}$

8.10 NUMERICAL EXAMPLES

Problem 1: *The specific volume of CO_2 at 100°C is 1 m^3/kg.*

(a) Use Vander Waal's equation and determine pressure exerted by CO_2.

(b) Compare the result by treating CO_2 as a perfect gas.

Solution:

(a) $\bar{V} = 1 \times 44 \quad [\bar{V} = 44 \text{ for } CO_2]$

$= 44\ m^3/kg\ mol$

$T = 100 + 273 = 373$ K

From table C-8/59, data hand book

$a = 365.6\ kNm^4/(kg\ mol)^2$

$b = 0.0428\ m^3/kg\ mol$

and $\bar{R} = 8314.4$ Nm/kg mol K

We know that, the Vander Wall's equation is

$$\left(P + \frac{a}{\bar{V}^2}\right)(\bar{V} - b) = \bar{R}T$$

$$\therefore \qquad P = \frac{\bar{R}T}{(\bar{V} - b)} - \frac{a}{\bar{V}^2} = \left(\frac{8.3144 \times 373}{44 - 0.0428}\right) - \frac{365.6}{(44)^2}$$

$$P = 70.552 - 0.18887$$

$$\boxed{P = 70.36 \text{ kPa}}$$

Ans.

(b) Using Ideal gas equation

$$P\bar{V} = \bar{R}\,T \qquad R: \frac{\bar{R}}{M}$$

$$P = \frac{8.3144 \times 373}{44}$$

$$\boxed{P = 70.48 \text{ kPa}}$$

Ans.

Problem 2: *Determine the pressure exerted by CO_2 in a container of 1.5 m^3 capacity when it contains 5 kg at 27°C by using*

(a) Ideal gas equation

(b) Vander Waal's equation

(VTU, Aug. 2000)

Solution:

(a) We know that, the Ideal gas equation can be written as

$$PV = mRT$$

$$P = \frac{mRT}{V} = \frac{m\bar{R}T}{VM} \qquad (\text{for } CO_2\ M = 44)$$

$$P = \frac{5 \times 8.3144 \times 300}{1.5 \times 44}$$

$$\boxed{P = 188.963 \text{ kN/m}^2}$$

Ans.

(b) By using Vander Waal's equation

$$P = \left(\frac{\bar{R}T}{\bar{V} - b}\right) - \left(\frac{a}{\bar{V}^2}\right)$$

From table C-8/59, the Vander Waal's constant are obtained as

$$a = 365.6 \text{ kN m}^4/(\text{kg mol})^2$$

$$b = 0.0428 \text{ m}^3/\text{kg mol}$$

$$\bar{V} = \frac{M \cdot V}{m}$$

$$= \frac{44 \times 1.5}{5} = 13.2 \text{ m}^3/\text{kg mol}$$

$$\therefore \quad P = \left(\frac{8.3144 \times 300}{13.2 - 0.0428}\right) - \left(\frac{365.6}{(13.2)^2}\right)$$

$$\boxed{P = 187.480 \text{ kN/m}^2} \qquad \textit{Ans.}$$

Problem 3: *Determine the specific volume of CO_2 at 200°C and at 60 bar by using*
(i) Perfect gas equation
(ii) Compressibility chart

Solution:

(i) The perfect gas equation can be written as

$$PV = mRT$$

$$V = \frac{mRT}{P} = \frac{m \times \bar{R} \times T}{MP} = \frac{1 \times 8.3144 \times (200 + 273)}{44 \times \dfrac{60 \times 10^5}{1000}}$$

$$\boxed{V = 0.01489 \text{ m}^3\text{/kg}}$$

(ii) By using compressibility chart from table C-6/58, the critical parameters of 'CO_2' can be written as

$$T_{Cr} = 31°\text{C}$$
$$P_{Cr} = 73.85 \text{ bar}$$
$$\rho_{Cr} = 460.0 \text{ kg/m}^3$$

$$\therefore \quad P_r = \frac{P}{P_{Cr}} = \frac{60}{73.85}$$

$$\boxed{P_r = 0.812}$$

$$T_r = \frac{T}{T_{Cr}} = \frac{(200 + 273)}{(31 + 273)}$$

$$\boxed{T_r = 1.555}$$

From generalized compressibility chart (chart–7), corresponding to $Pr = 0.812$ and $T_r = 1.555$, the value of $Z = 0.96$.

$$Z = \frac{PV}{RT}$$

$$\therefore \quad V = \frac{ZRT}{P}$$

$$V = \frac{0.96 \times \dfrac{8.3144}{44} (200 + 273)}{\dfrac{60 \times 10^5}{1000}}$$

$$\boxed{V = 0.0173 \text{ m}^3\text{/kg}}$$

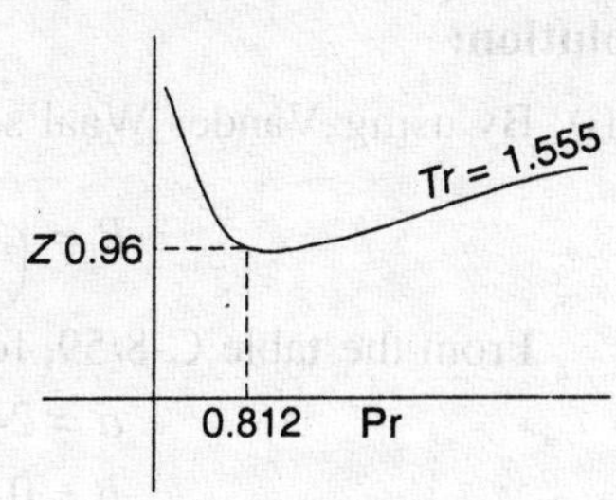

Fig. 8.3

Problem 4: *Determine the critical pressure of oxygen using perfect gas equation and Vander Waal's equation.*

Take $\bar{V}_C = 0.0745\ m^3/kg\ mol$, *and*

$T_C = -118.10°C$

Solution:

(i) From perfect gas equation, at critical point, we can write

$$P_C \bar{V}_C = m\bar{R}T_C$$

$$P_C = \frac{m\bar{R}T_C}{\bar{V}_C} = \frac{1 \times 8.3144 \times (-118.10 + 273)}{0.0745}$$

$$= 17287.25 \text{ kN/m}^2$$

$$\boxed{P_C = 172.87 \text{ bar}}$$ *Ans.*

(ii) The Vander Waal's equation in terms of critical parameters can be written as

$$\bar{R} = \frac{8}{3} \times \frac{P_C \bar{V}_C}{T_C}$$

$$\therefore \quad P_C = \frac{3}{8} \frac{\bar{R}T_C}{\bar{V}_C} = \frac{3}{8}\left[\frac{8.3144 \times (-118.1 + 273)}{0.0745}\right]$$

$$P_C = 6482.72 \text{ kN/m}^2$$

$$\boxed{P_C = 64.82 \text{ bar}}$$

From table C-6/58, P_C for O_2 is 50.77 bar.

It means, the Vander Waal's equation gives more accurate value compared to the Ideal gas equation.

Problem 5: *Determine the specific volume of Hydrogen gas when its pressure is 60 bar and temperature is – 173°C*

(i) By using Vander Waal's equation

(ii) By using compressibility chart.

Solution:

(i) By using Vander Waal's equation,

$$P = \left(\frac{\bar{R}T}{\bar{V} - b}\right) - \left(\frac{a}{\bar{V}^2}\right)$$

From the table C-8/59, for H_2,

$$a = 24.8 \text{ kN m}^4/(\text{kg mol})^2$$

$$b = 0.0266 \text{ m}^3/\text{kg mol}$$

$$\frac{60 \times 10^5}{1000} = \left[\frac{8.3144 \times (-173 + 273)}{\bar{V} - 0.0266}\right] - \left[\frac{24.8}{\bar{V}^2}\right]$$

$$60 \times 100 = \left[\frac{831.44}{\bar{V} - 0.0266}\right] - \left[\frac{24.8}{\bar{V}^2}\right]$$

By trial and error method, the value of $\bar{V}$ can be obtained as $\bar{V}$ = 0.145

$$\bar{V}_S = \frac{\bar{V}}{M} = \frac{0.145}{2} = 0.0725 \text{ m}^3/\text{kg}.$$

Problem 6: *For Nitrogen, determine the compressibility factor at 10 bar and – 85°C and at 5 bar and 35°C.*

Solution: From table C-6/58, the critical parameters for Nitrogen are

$$T_{Cr} = -147.1°\text{C} = 125.9 \text{ K}$$

$$P_{Cr} = 33.94 \text{ bar}$$

$$\rho_{Cr} = 310.96 \text{ kg/m}^3$$

At 10 bar and – 85°C, reduced properties are

$$Pr = \frac{P}{P_{Cr}} = \frac{10}{33.94} = \boxed{0.2946}$$

$$Tr = \frac{T}{T_{Cr}} = \frac{85 + 273}{125.9}$$

$$\boxed{Tr = 1.49}$$

By referring generalized compressibility chart (Chart-7), we get

$$Z = 0.98$$

At 5 bar and 35°C, the reduced properties are

$$Pr = \frac{5}{33.94} = 0.147$$

$$Tr = \frac{+35 + 273}{125.9} = 2.44$$

From the chart-7, corresponding to Pr = 0.147 and Tr = 2.44,

$$Z = 0.99$$ *Ans.*

Problem 7: *One kg mol of O_2 undergoes a reversible non-flow isothermal compression and the volume of O_2 decreases from 0.14 m³/kg to 0.05 m³/kg and the initial temperature is 50°C. If the gas compression obeys Vander Wall's equations, find work done during isothermal compression and the pressure at the end of compression.*

Solution: For O_2, the Vander Waal's constant are (from table C-8/59)

$$a = 138.0 \text{ kN m}^4/(\text{kg mol})^2 = 138.0 \times 10^3 \text{ Nm}^4/(\text{kg mol})^2$$

$$b = 0.0318 \text{ m}^3/\text{kg mol}$$

The work done/kg mol is

$$WD = \int_1^2 P d\bar{V}$$

$$= \int_1^2 \left[\left[\frac{\bar{R}T}{\bar{V} - b}\right] - \left[\frac{a}{\bar{V}^2}\right]\right] d\bar{V}$$

$$= \bar{R}T\int_1^2 \frac{1}{\bar{V}-b}d\bar{V} - a\int_1^2 \frac{1}{\bar{V}^2}d\bar{V}$$

$$= \bar{R}T\left[\log_e(\bar{V}-b)\right]_{\bar{V}_1}^{\bar{V}_2} + a\left[\frac{1}{\bar{V}}\right]_{\bar{V}_1}^{\bar{V}_2}$$

$$= \bar{R}T\log_e\left(\frac{\bar{V}_2-b}{\bar{V}_1-b}\right) + a\left[\frac{1}{\bar{V}_2} - \frac{1}{\bar{V}_1}\right]$$

here, $T = 50 + 273 = 323$ K

$\bar{R} = 8314.4$ Nm/kg mol K

$\bar{V}_1 = 0.14 \times 32 = 4.48$ m^3/kg mol

$\bar{V}_2 = 0.05 \times 32 = 1.6$ m^3/kg mol

$$= 8314.4\ (323)\ \log_e\left(\frac{1.6-0.0318}{4.48-0.0318}\right) + 138.0 \times 10^3\left[\frac{1}{1.6} - \frac{1}{4.48}\right]$$

$$= -\ 2799877.94 + 55446.428$$

$$\boxed{WD = -\ 27.44 \times 10^5 \text{ Nm/kg mol}}$$ *Ans.*

At $\bar{V}_2 = 1.6$ m^3/kg mol and 323 K, i.e. at the end of Isothermal compression,

$$P_2 = \left(\frac{\bar{R}T}{\bar{V}_2-b}\right) - \left(\frac{a}{\bar{V}_2^2}\right)$$

$$P_2 = \left(\frac{8314.4\times 323}{1.6-0.0318}\right) - \left(\frac{138.0\times 10^3}{(1.6)^2}\right)$$

$$P_2 = 1712505.548 - 53906.25$$

$$P_2 = 1658599.29 \text{ N/m}^2$$

$$P_2 = 16.585 \times 10^5 \text{ N/m}^2$$

$$\boxed{P_2 = 16.585 \text{ bar}}$$ *Ans.*

Problem 8: *A vessel contains nitrogen at 150 bar and 200 K. The volume of the vessel is 32 m^3. Determine mass of the nitrogen by using (i) The Ideal gas equation, (ii) The compressibility chart.*

Solution:

(i) The Ideal gas equation can be written as

$$PV = mRT \qquad R = \frac{\bar{R}}{M} = \frac{8.3144}{28} \text{ kJ/kg molK}$$

$$m = \frac{PV}{RT} \qquad R = 0.296 \text{ kJ/kgK}$$

$$= \frac{150 \times 10^5 \times 32}{1000 \times 0.296 \times 200}$$

$$\boxed{m = 8108.10 \text{ kg}}$$ **Ans.**

(ii) From compressibility chart, we can write from data hand book, table C-6/58, for N_2,

$$T_{Cr} = -147.1°C = -147.1 + 273 = 125.9 \text{ K}$$

$$P_{Cr} = 33.94 \text{ bar}$$

$$\rho_{Cr} = 310.96 \text{ kg/m}^3$$

∴ The reduced properties are

$$Pr = \frac{150}{33.94} = 4.41$$

$$Tr = \frac{200}{125.9} = 1.588$$

From compressibility chart, chart-7

$$Z = 0.86$$

$$\therefore \quad v = \frac{ZRT}{P} = \frac{0.86\,(0.296) \times 200}{\frac{150 \times 10^5}{1000}}$$

$$\boxed{V = 3.39 \times 10^{-3} \text{ m}^3/\text{kg}}$$

$$m = \frac{V}{v} = \frac{32}{3.39 \times 10^{-3}}$$

$$\boxed{m = 9439.52 \text{ kg}}$$ **Ans.**

Problem 9: *Determine density of steam at 15,000 MPa and 773 K by using*

(i) Perfect gas equation

(ii) Compressibility chart.

Solution:

(i) The perfect gas equation gives

$$PV = RT$$

$$P = \rho RT$$

$$\rho = \frac{P}{RT}$$

$$\rho = \frac{15{,}000 \times 1000}{461.91 \times 773}$$

$$\boxed{\rho = 42.01 \text{ kg/m}^3}$$

$$R = \frac{\bar{R}}{M} = \frac{8314.4}{18} \text{ J/kg mol K}$$

$$R = 461.91 \text{ J/kg K}$$

$$\frac{Pa = \text{N/m}^2}{\frac{\text{J}}{\text{kg}\,\cancel{\text{K}}} \times \cancel{\text{K}}}$$

$$\frac{\cancel{\text{N}}/\text{kg}}{\text{m}^2 \cdot \cancel{\text{N}}\text{m}}$$

(ii) Using compressibility chart,
from table, C-6/58, data hand book, the critical parameters for steam are

$$T_{Cr} = 374.3°C = 374.3 + 273 = 647.3 \text{ K}$$
$$P_{Cr} = 221.36 \text{ bar}$$
$$\rho_{Cr} = 345.00 \text{ kg/m}^3$$

The reduced properties are

$$Pr = \frac{P}{P_{Cr}} = \frac{150}{221.36} = 0.6776$$

$$Tr = \frac{T}{T_{Cr}} = \frac{773}{647.3} = 1.194$$

From chart-7, corresponding to values of Pr and Tr the value of Z is found to be $Z = 0.86$.

$$\therefore \quad PV = ZRT$$

$$V = \frac{ZRT}{P} = \frac{0.86 \times 461.91 \times 773}{150 \times 10^5}$$

$$\boxed{V = 0.0204 \text{ m}^3/\text{kg}}$$

$$\therefore \quad \boxed{\rho = \frac{1}{V} = 48.84 \text{ m}^3/\text{kg}}$$

Problem 10: *A spherical ballon of 1m diameter contains H_2 at 25°C and 1 bar. Determine mass of the H_2 in the ballon, by using real gas equation.*

Solution: The critical parameters for H_2, are

$$T_{Cr} = -239.9 \text{ °C}$$
$$P_{Cr} = 12.97 \text{ bar}$$
$$\rho_{Cr} = 31.0 \text{ kg/m}^3$$

from data hand book

The reduced properties are

$$Pr = \frac{P}{P_{Cr}} = \frac{1}{12.97} = 0.07710$$

$$Tr = \frac{25 + 273}{-239.9 + 273} = 9.00$$

From Chart-7, compressibility factor $Z = 1$. i.e. for higher values of P_{Cr} and lower values of T_{Cr}, the gas behaves like an Ideal gas at normal pressure and temperature conditions.

$$\therefore \quad PV = ZmRT$$

$$m = \frac{PV}{ZRT}$$

$$= \frac{1 \times 10^5 \times 0.5235}{1 \times 4157.2 \times 298}$$

$$\boxed{m = 0.0422 \text{ kg}}$$

$$V = \frac{4}{3}\pi r^3$$

$$= \frac{4}{3}\pi (0.5)^3$$

$$V = 0.5235 \text{ m}^3$$

$$R = \frac{R}{M} = \frac{8314.4}{2}$$

$$R = 4157.2 \text{ J/kgK}$$

Problem 11: *Determine the pressure in a vessel having a volume of 15 litres and containing 3.4 kg Nitrogen at 400°C by using (i) Ideal gas equation and (ii) Vander waal's equation. Also calculate the compressibility factor by using the answer obtained from the Vander waal's equation.*

(VTU, Feb. 2003)

Solution: For N_2; $R = \frac{\overline{R}}{M} = \frac{8.314}{28} = 0.2969$ kJ/kgK

$$PV = MRT$$

$$P = \frac{3.4 \times 0.2969 \times 673}{0.015} = 45295.46 \text{ kpa} \qquad \textit{Ans.}$$

The constants '*a*' and '*b*' are

$$a = 136.4 \frac{\text{KNM}^4}{(\text{kg mol})^2} \text{ from table}$$

$$b = 0.0386 \text{ m}^3/\text{kg mol}$$

Molar specific volume $\overline{V} = \frac{V}{n}$

$$n = \text{number of moles} = \frac{m}{M} = \frac{3.4}{28} = 0.1214$$

$$\therefore \quad \overline{V} = \frac{0.015}{0.1214} = 0.1235 \text{ m}^3/\text{kg mol}$$

The Vander waal's equation is

$$\left(P + \frac{a}{\overline{V}^2}\right)(\overline{V} - b) = RT$$

$$P = \frac{\overline{R}T}{\overline{V} - b} - \frac{a}{\overline{V}^2}$$

$$= \frac{8.314 \times 673}{0.1235 - 0.0386} - \frac{136.4}{(0.1235)^2}$$

$$= 56961.91 \text{ kpa} \qquad \textit{Ans.}$$

$$PV = ZRT$$

$$V = \frac{V}{m} = \frac{0.015}{3.4} = 0.0044 \text{ m}^3/\text{kg}$$

$$\therefore \quad z = \frac{56961.91 \times 0.0044}{0.2969 - 673}$$

$$\boxed{z = 1.25} \qquad \textit{Ans.}$$

8.11 REVIEW QUESTIONS

1. Differentiate between a perfect gas and a real gas.
2. Differentiate between universal gas constant and gas constant.
3. Write a short note on: (i) Compressibility factor (ii) Compressibility chart.
4. How does a real gas deviate from a perfect gas?
5. What is a generalised compressibility chart?
6. Write down Vander Waal's equation.
7. Derive Vander Waal's equation in terms of reduced parameters.
8. Prove that $\dfrac{P_C\bar{V}_C}{\bar{R}T_C} = 3/8$
9. What are the limitations of Vander Waal's equation?
10. Explain the law of corresponding states.